Linear and Non-Linear System Theory

Linear and Non-Linear System Theory

T. Thyagarajan and D. Kalpana

CRC Press
Taylor & Francis Group
Boca Raton London New York

CRC Press is an imprint of the
Taylor & Francis Group, an **informa** business

First edition published 2021
by CRC Press
6000 Broken Sound Parkway NW, Suite 300, Boca Raton, FL 33487-2742

and by CRC Press
2 Park Square, Milton Park, Abingdon, Oxon, OX14 4RN

© 2021 Taylor & Francis Group, LLC

CRC Press is an imprint of Taylor & Francis Group, LLC

ISBN: 978-0-367-34014-8 (hbk)
ISBN: 978-0-429-32345-4 (ebk)

Typeset in Palatino
by codeMantra

To my adorable grandson, R. Aarav

T. Thyagarajan

To my beloved husband, Mr. C. Kamal, ME

D. Kalpana

Contents

Preface

This textbook is intended for undergraduate and postgraduate students of electrical, chemical, electronics and instrumentation, and instrumentation and control branches of engineering disciplines pursuing advanced control system and linear and non-linear system theory as a subject. It will also be useful to research scholars pursuing research in similar disciplines. This book focuses on the basics of state space approach, analysis of non-linear systems, optimal control and optimal estimation.

Chapter 1 explains the limitations of transfer function approach and introduces the state model for electrical, mechanical and other physical systems. As a recap, general description about determinants and matrices are dealt with suitable numerical. Chapter 2 focuses on the advantages of state space approach and skills needed to represent the time-invariant systems in state space form which is essential for analysing the system. Chapter 3 aims at the design of state feedback controller and state observers which will help to analyse, whether the system is stabilizable, controllable, observable and detectable. Chapter 4 presents the different types of nonlinearity existing in the system and method to analyse the non-linear system using phase plane analysis and linearizing the non-linear systems using describing function approach. Chapter 5 provides the concepts of stability and introduces technique to assess the stability of certain class of non-linear system using describing function, Lyapunov stability, Popov's stability and circle criterion. Chapter 6 describes bifurcation theory, types of bifurcation, introduction to chaos: Lorentz equations, stability analysis of the Lorentz equations and chaos in chemical systems. Chapter 7 deals with comparison of optimal control with classical control theory, basic definitions, performance measures, design of linear quadratic regulator (LQR), linear quadratic tracking (LQT) and linear quadratic Gaussian (LQG). Chapter 8 discusses the review of statistical tools, state estimation in linear time-invariant systems using Kalman filter, continuous-time and discrete-time Kalman filter and introduction to Extended Kalman Filter (EKF).

Errors might have crept in, despite utmost care to avoid them. We shall be grateful to the readers if any errors are brought to our attention, along with suggestions for improvements to the book. We solicit your ever-granted support for the success of this professional endeavour.

Acknowledgements

At the outset, we express our gratitude to Almighty for his blessings. We are indebted to our Vice Chancellor, Prof. M. K Surappa, and to our Registrar, Prof. L. Karunamoorthy, of Anna University for their ever-granted patronage and encouragement. Our special thanks to our colleagues from the Department of Instrumentation Engineering for their valuable suggestions. We are also indebted to the teaching fraternity and student community for their support.

Our family members (wife, Mrs. Meera Thyagarjan, and daughter, Mrs. T. Priyanka, of the first author and mother, Mrs. D. Sarasu, son, K. Rahul Ramesh, and daughter, K. Vakshika, of the coauthor), who missed our quality time during holidays, weekends, vacations, and late evenings, while we were preparing the manuscript, deserve our wholehearted approbation.

Dr. Gagandeep Singh and his team at Taylor & Francis deserve our encomiums for their cooperation in bringing out this book in record time.

T. Thyagarajan
D. Kalpana

Authors

T. Thyagarajan is a Professor in the Department of Instrumentation Engineering at MIT Campus, Anna University. He earned a PhD in intelligent control at Anna University. With a National Science Council Fellowship, he pursued postdoctoral research at NTU, Taiwan. He is an erudite professor and a socially conscious researcher and has been engaged in teaching, research and administration for 33 years. He has published over 105 research papers, authored/coauthored 3 textbooks, guided 14 PhD scholars, coordinated R&D funding worth Rs 15.5 crores through various funding agencies. He made technical visits to the United States, Europe, Southeast Asia and the Middle East. He received 13 awards, including DTE award for guiding the best BE project, Sisir Kumar Award for publishing the best research paper, 2 best paper awards in IEEE International Conferences, Best Teacher award and Best Researcher award. He is a Senior Member of IEEE. His current teaching and research interests include process modelling and control, auto-tuning, applied soft computing and healthcare instrumentation.

D. Kalpana is an Assistant Professor of instrumentation engineering at MIT Campus, Anna University. She earned a BE (EIE), an ME (instrumentation engineering), and a PhD at Anna University. She received a Young Scientist Fellowship from Tamilnadu State Council for Science and Technology and pursued postdoctoral research work at NIT, Warangal. She has been engaged in teaching and research for 14 years. She has received the Dr A.P.J. Abdul Kalam Award for her contribution to teaching. She has published 25 papers in peer-reviewed journals and conference proceedings. She is a member of ISA. Her teaching and research interests include linear and non-linear system theory, auto-tuning, modelling and control of industrial processes and applied soft computing.

1

Introduction

Determinants

Determinants find their usefulness in solving linear systems. The mathematical expressions of linear systems can be simplified by representing them in the form of determinants.

A determinant contains elements that are arranged along horizontal and vertical lines.

Rows: The elements of a determinant that are arranged along horizontal lines form the 'rows'.

Columns: The elements of a determinant that are arranged along vertical lines form the 'columns'.

Second-Order Determinant

It consists of four elements, namely, a_{11}, a_{12}, a_{21} and a_{22} which are arranged in two rows and two columns and is expressed as follows:

$$\begin{vmatrix} a_{11} & a_{12} \\ a_{21} & a_{22} \end{vmatrix} = a_{11}a_{22} - a_{21}a_{12}$$

The above determinant consists of two rows and two columns.

Third-Order Determinant

It consists of nine elements, namely, a_{11}, a_{12}, a_{13}, a_{21}, a_{22}, a_{23}, a_{31}, a_{32} and a_{33} which are arranged in three rows and three columns and is expressed as follows:

$$\begin{vmatrix} a_{11} & a_{12} & a_{13} \\ a_{21} & a_{22} & a_{23} \\ a_{31} & a_{32} & a_{33} \end{vmatrix} = a_{11}(a_{22}a_{33} - a_{32}a_{23}) - a_{12}(a_{21}a_{33} - a_{31}a_{23}) + a_{13}(a_{21}a_{33} - a_{31}a_{22})$$

The above determinant consists of three rows and three columns.

Minor

The minor M_{ij} of an element a_{ij} of a determinant is the determinant obtained by deleting the ith row and jth column which intersect the element a_{ij}.

For example, if $\Delta = \begin{vmatrix} a_{11} & a_{12} & a_{13} \\ a_{21} & a_{22} & a_{23} \\ a_{31} & a_{32} & a_{33} \end{vmatrix}$

Then, the minor of a_{22} is given by $M_{22} = \begin{vmatrix} a_{11} & a_{13} \\ a_{31} & a_{33} \end{vmatrix} = \left[(a_{11} \times a_{33}) - (a_{13} \times a_{31}) \right].$

Similarly, the minor of a_{33} is given by $M_{33} = \begin{vmatrix} a_{11} & a_{12} \\ a_{21} & a_{22} \end{vmatrix} = \left[(a_{11} \times a_{22}) - (a_{12} \times a_{21}) \right]$

NOTE

Thus, it is clear that the minor of an element of a determinant of order $n(n \geq 2)$ will be a determinant of order $(n-1)$.

Illustration 1

Find the minor of element 8 in the determinant:

$$\Delta = \begin{vmatrix} 1 & 4 & 7 \\ 2 & 5 & 8 \\ 3 & 6 & 9 \end{vmatrix}$$

Solution

Since element 8 lies in the second row and the third column, its minor M_{23} is given by $M_{23} = \begin{vmatrix} 1 & 4 \\ 3 & 6 \end{vmatrix} = (1 \times 6) - (4 \times 3) = -6$

Cofactor

The cofactor (A_{ij}) of an element a_{ij} is defined by $A_{ij} = (-1)^{i+j} M_{ij}$. Here, M_{ij} is the minor of a_{ij}.

For example, if $\Delta = \begin{vmatrix} a_{11} & a_{12} & a_{13} \\ a_{21} & a_{22} & a_{23} \\ a_{31} & a_{32} & a_{33} \end{vmatrix}$,

Then, the cofactor of a_{22} is given by

$$A_{22} = (-1)^{2+2} \begin{vmatrix} a_{11} & a_{13} \\ a_{31} & a_{33} \end{vmatrix} = -\begin{vmatrix} a_{11} & a_{13} \\ a_{31} & a_{33} \end{vmatrix} = -[(a_{11} \times a_{33}) - (a_{13} \times a_{31})]$$

NOTE

Thus, it is clear that the cofactor of any element in a determinant is obtained by deleting the row and column which intersect that element with the proper sign.

Illustration 2

Find the cofactor of element 8 in the determinant:

$$\Delta = \begin{vmatrix} 1 & 4 & 7 \\ 2 & 5 & 8 \\ 3 & 6 & 9 \end{vmatrix}$$

Solution

Since element 8 lies in the second row and the third column, its cofactor is given by $A_{23} = (-1)^{2+3} M_{23}$

From Illustration 1, M_{23} is given as -6.

$$A_{23} = (-1) \times (-6) = 6$$

NOTE

If elements of a row (or column) are multiplied with cofactors of any other row (or column), then their sum is zero.

Illustration 3

Find minors and cofactors of the elements of the given determinant and verify that $a_{11} A_{31} + a_{12} A_{32} + a_{13} A_{33} = 0$

$$\Delta = \begin{vmatrix} 1 & 4 & 7 \\ 2 & 5 & 8 \\ 3 & 6 & 9 \end{vmatrix}$$

Solution

$$M_{11} = \begin{vmatrix} 5 & 8 \\ 6 & 9 \end{vmatrix} = 45 - 48 = -3; \quad A_{11} = (-1)^{1+1}(-3) = -3$$

$$M_{12} = \begin{vmatrix} 2 & 8 \\ 3 & 9 \end{vmatrix} = 18 - 24 = -6; \quad A_{12} = (-1)^{1+2}(-6) = 6$$

$$M_{13} = \begin{vmatrix} 2 & 5 \\ 3 & 6 \end{vmatrix} = 12 - 15 = -3; \quad A_{13} = (-1)^{1+3}(-3) = -3$$

$$M_{21} = \begin{vmatrix} 4 & 7 \\ 6 & 9 \end{vmatrix} = 36 - 42 = -6; \quad A_{21} = (-1)^{2+1}(-6) = 6$$

$$M_{22} = \begin{vmatrix} 1 & 7 \\ 3 & 9 \end{vmatrix} = 9 - 21 = -12; \quad A_{22} = (-1)^{2+2}(-12) = -12$$

$$M_{23} = \begin{vmatrix} 1 & 4 \\ 3 & 6 \end{vmatrix} = 6 - 12 = -6; \quad A_{23} = (-1)^{2+3}(-6) = 6$$

$$M_{31} = \begin{vmatrix} 4 & 7 \\ 5 & 8 \end{vmatrix} = 32 - 35 = -3; \quad A_{31} = (-1)^{3+1}(-3) = -3$$

$$M_{32} = \begin{vmatrix} 1 & 7 \\ 2 & 8 \end{vmatrix} = 8 - 14 = -6; \quad A_{32} = (-1)^{3+2}(-6) = 6$$

$$M_{33} = \begin{vmatrix} 1 & 4 \\ 2 & 5 \end{vmatrix} = 5 - 8 = -3; \quad A_{33} = (-1)^{3+3}(-3) = -3$$

Now, $a_{11} = 1$; $a_{12} = 4$; $a_{13} = 7$; $A_{31} = -3$; $A_{32} = 6$; $A_{33} = -3$.

$\therefore a_{11}A_{31} + a_{12}A_{32} + a_{13}A_{33} = 1(-3) + 4(6) + 7(-3) = -3 + 24 - 21 = 0$.

Properties of Determinants

The properties of determinant will help in simplifying its evaluation. The properties of determinant are true for determinants of any order.

The evaluation of determinants can be performed in a simplified manner by using certain properties of determinants, which are illustrated below, for a third-order determinant.

Property 1

The value of the determinant remains the same by changing its rows into columns and columns into rows.

Proof

$$\text{Let } \Delta_1 = \begin{vmatrix} a_{11} & a_{12} & a_{13} \\ b_{21} & b_{22} & b_{23} \\ c_{31} & c_{32} & c_{33} \end{vmatrix}$$

Expanding along the first row (R_1), we get

$$\Delta_1 = a_{11} \begin{vmatrix} b_{22} & b_{23} \\ c_{32} & c_{33} \end{vmatrix} - a_{12} \begin{vmatrix} b_{21} & b_{23} \\ c_{31} & c_{33} \end{vmatrix} + a_{13} \begin{vmatrix} b_{21} & b_{22} \\ c_{31} & c_{32} \end{vmatrix}$$

$$= a_{11}(b_{22}c_{33} - b_{23}c_{32}) - a_{12}(b_{21}c_{33} - b_{23}c_{31}) + a_{13}(b_{21}c_{32} - b_{22}c_{31})$$

By interchanging the rows and columns of Δ_1, we get the determinant

$$\Delta_2 = \begin{vmatrix} a_{11} & b_{21} & c_{31} \\ a_{12} & b_{22} & c_{32} \\ a_{13} & b_{23} & c_{33} \end{vmatrix}$$

Expanding along the first column (C_1), we get

$$\Delta_2 = a_{11} \begin{vmatrix} b_{22} & c_{32} \\ b_{23} & c_{33} \end{vmatrix} - a_{12} \begin{vmatrix} b_{21} & c_{31} \\ b_{23} & c_{33} \end{vmatrix} + a_{13} \begin{vmatrix} b_{21} & c_{31} \\ b_{22} & c_{32} \end{vmatrix}$$

$$= a_{11}(b_{22}c_{33} - b_{23}c_{32}) - a_{12}(b_{21}c_{33} - b_{23}c_{31}) + a_{13}(b_{21}c_{32} - b_{22}c_{31})$$

$$\Delta_1 = \Delta_2$$

Remark: From Property 1, thus, it is clear that if A is a square matrix, then $\det(A) = \det(A')$ where $A' =$ transpose of A.

Illustration 4

Given $\Delta_1 = \begin{vmatrix} 1 & 4 & 7 \\ 2 & 5 & 8 \\ 3 & 6 & 9 \end{vmatrix}$, prove that the value of the determinant remains the same by changing its rows into columns and columns into rows.

Solution

By expanding the determinant along the first row, we get

$$\Delta_1 = 1\begin{vmatrix} 5 & 8 \\ 6 & 9 \end{vmatrix} - 4\begin{vmatrix} 2 & 8 \\ 3 & 9 \end{vmatrix} + 7\begin{vmatrix} 2 & 5 \\ 3 & 6 \end{vmatrix}$$

$$= 1(45-48) - 4(18-24) + 7(12-15)$$

$$= 1(-3) - 4(-6) + 7(-3)$$

$$= -3 + 24 - 21 = 0$$

By interchanging the elements of rows and columns, we get

$$\Delta_2 = \begin{vmatrix} 1 & 2 & 3 \\ 4 & 5 & 6 \\ 7 & 8 & 9 \end{vmatrix}$$

By expanding the determinant along the first column, we get

$$\Delta_2 = 1\begin{vmatrix} 5 & 6 \\ 8 & 9 \end{vmatrix} - 4\begin{vmatrix} 2 & 3 \\ 8 & 9 \end{vmatrix} + 7\begin{vmatrix} 2 & 3 \\ 5 & 6 \end{vmatrix}$$

$$= 1(45-48) - 4(18-24) + 7(12-15)$$

$$= 1(-3) - 4(-6) + 7(-3)$$

$$= -3 + 24 - 21 = 0$$

$$\therefore \Delta_1 = \Delta_2$$

Hence, the property is proved.

Property 2

If any two rows or columns of a determinant are interchanged, the numerical value of the determinants remains the same. However, there is a change in the sign of the determinant.

Proof

$$\text{Let } \Delta_1 = \begin{vmatrix} a_{11} & a_{12} & a_{13} \\ b_{21} & b_{22} & b_{23} \\ c_{31} & c_{32} & c_{33} \end{vmatrix}$$

By expanding the elements along the first row, we get

$$\Delta_1 = a_{11}(b_{22}c_{33} - b_{23}c_{32}) - a_{12}(b_{21}c_{33} - b_{23}c_{31}) + a_{13}(b_{21}c_{32} - b_{22}c_{31})$$

By interchanging the elements of the first and third rows, the new determinant obtained is given by

$$\Delta_2 = \begin{vmatrix} c_{31} & c_{32} & c_{33} \\ b_{21} & b_{22} & b_{23} \\ a_{11} & a_{12} & a_{13} \end{vmatrix}$$

By expanding the elements along the third row, we get

$$\Delta_2 = a_{11}(c_{32}b_{23} - b_{22}c_{33}) - a_{12}(c_{31}b_{23} - c_{33}b_{21}) + a_{13}(b_{22}c_{31} - b_{21}c_{32})$$

$$\Delta_2 = -[a_{11}(c_{32}b_{23} - b_{22}c_{33}) - a_{12}(c_{31}b_{23} - c_{33}b_{21}) + a_{13}(b_{22}c_{31} - b_{21}c_{32})]$$

$$\therefore \Delta_2 = -\Delta_1$$

Similarly, the results can be verified by interchanging the element of any two columns.

Illustration 5

$$\text{Given } \Delta_1 = \begin{vmatrix} 1 & -4 & 7 \\ 2 & 5 & 8 \\ 3 & 6 & -9 \end{vmatrix}, \text{ verify Property 2 for the given determinant.}$$

Solution

By expanding the determinant along the first row, we get

$$\Delta_1 = 1 \begin{vmatrix} 5 & 8 \\ 6 & -9 \end{vmatrix} - (-4) \begin{vmatrix} 2 & 8 \\ 3 & -9 \end{vmatrix} + 7 \begin{vmatrix} 2 & 5 \\ 3 & 6 \end{vmatrix}$$

$$= 1(-45 - 48) + 4(-18 - 24) + 7(12 - 15)$$

$$= -93 - 168 - 21$$

$$= -282$$

By interchanging the elements of the first and third rows, the new determinant obtained is given by

$$\Delta_2 = \begin{vmatrix} 3 & 6 & -9 \\ 2 & 5 & 8 \\ 1 & -4 & 7 \end{vmatrix}$$

By expanding elements along the third row, we get

$$\Delta_2 = 1\begin{vmatrix} 6 & -9 \\ 5 & 8 \end{vmatrix} - (-4)\begin{vmatrix} 3 & -9 \\ 2 & 8 \end{vmatrix} + 7\begin{vmatrix} 3 & 6 \\ 2 & 5 \end{vmatrix}$$

$$= 1(48 - (-45)) + 4(24 - (-18)) + 7(15 - 12)$$

$$= 1(48 + 45) + 4(24 + 18)) + 7(3)$$

$$= 93 + 168 + 21$$

$$= 282$$

$$\therefore \Delta_2 = -\Delta_1$$

Hence, it is verified.

Property 3

If the elements of any two rows or columns are identical, then the numerical value of the determinant will be zero.

Illustration 6

Given $\Delta = \begin{vmatrix} 4 & 5 & 7 \\ 2 & 5 & 7 \\ 4 & 5 & 7 \end{vmatrix}$, verify Property 3 for the given determinant.

Solution

Here, the elements of the first and third rows are identical.
   By expanding the determinant along the first row, we get

$$\Delta = 4\begin{vmatrix} 5 & 7 \\ 5 & 7 \end{vmatrix} - 5\begin{vmatrix} 2 & 7 \\ 4 & 7 \end{vmatrix} + 7\begin{vmatrix} 2 & 5 \\ 4 & 5 \end{vmatrix}$$

$$= 4(35 - 35) - 5(14 - 28) + 7(10 - 20)$$

$$= 4(0) - 5(-14) + 7(-10) = 0 + 70 - 70 = 0$$

$$\Delta = 0$$

Hence, it is verified.

Property 4

If each element of a row or a column is multiplied by a factor 'p', then the value of the determinant will get multiplied by the same factor p.

Proof

$$\text{Let } \Delta_1 = \begin{vmatrix} a_{11} & a_{12} & a_{13} \\ b_{21} & b_{22} & b_{23} \\ c_{31} & c_{32} & c_{33} \end{vmatrix}$$

Expanding the determinant along the second column, we get

$$\Delta_1 = a_{12}(b_{21}c_{33} - b_{23}c_{31}) - b_{22}(a_{11}c_{33} - a_{13}c_{31}) + c_{32}(a_{11}b_{23} - a_{13}b_{21})$$

Let Δ_2 be the determinant obtained by multiplying the elements of the second column by p.
Then,

$$\Delta_2 = \begin{vmatrix} a_{11} & pa_{12} & a_{13} \\ b_{21} & pb_{22} & b_{23} \\ c_{31} & pc_{32} & c_{33} \end{vmatrix}$$

By expanding the determinant along the second column, we get

$$\Delta_2 = pa_{12}\begin{vmatrix} b_{21} & b_{23} \\ c_{31} & c_{33} \end{vmatrix} - pb_{22}\begin{vmatrix} a_{11} & a_{13} \\ c_{31} & c_{33} \end{vmatrix} + pc_{32}\begin{vmatrix} a_{11} & a_{13} \\ b_{21} & b_{23} \end{vmatrix}$$

$$= pa_{12}(b_{21}c_{33} - b_{23}c_{31}) - pb_{22}(a_{11}c_{33} - a_{13}c_{31}) + pc_{32}(a_{11}b_{23} - a_{13}b_{21})$$

$$= p[a_{12}(b_{21}c_{33} - b_{23}c_{31}) - b_{22}(a_{11}c_{33} - a_{13}c_{31}) + c_{32}(a_{11}b_{23} - a_{13}b_{21})]$$

$$\Delta_2 = p\Delta_1$$

Illustration 7

$$\text{Given } \Delta_1 = \begin{vmatrix} 1 & -4 & 7 \\ 2 & 5 & 8 \\ 3 & 6 & -9 \end{vmatrix}, \text{ verify Property 4 for the given determinant}$$

where its first column is multiplied by 2.

Solution

$$\text{Here } \Delta_1 = \begin{vmatrix} 1 & -4 & 7 \\ 2 & 5 & 8 \\ 3 & 6 & -9 \end{vmatrix}$$

Expanding Δ_1 along the first row, we get

$$\Delta_1 = 1(-45 - 48) + 4(-18 - 24) + 7(12 - 15)$$

$$\Delta_1 = -282$$

When the first column of Δ_1 is multiplied by 2, we get

$$\Delta_2 = \begin{vmatrix} 2 & -4 & 7 \\ 4 & 5 & 8 \\ 6 & 6 & -9 \end{vmatrix}$$

Expanding Δ_2 along the first row, we get

$$\Delta_2 = 2(-45 - 48) + 4(-26 - 48) + 7(24 - 30)$$

$$= 2(-45 - 48) + 2 \times 4(-18 - 24) + 2 \times 7(12 - 15)$$

$$= 2\left[(-45 - 48) + 4(-18 - 24) + 7(12 - 15)\right]$$

$$= 2\Delta_1$$

Hence, it is verified.

Property 5

If any one or more elements of a row or a column consist of 'm' terms, then the determinant can be represented as the sum of 'm' determinants.

Proof

Let

$$\Delta = \begin{vmatrix} a_{11} + b_{11} - c_{11} & d_{21} & e_{31} \\ a_{12} + b_{12} - c_{12} & d_{22} & e_{32} \\ a_{13} + b_{13} - c_{13} & d_{23} & e_{33} \end{vmatrix} = \begin{vmatrix} a_{11} & d_{21} & e_{31} \\ a_{12} & d_{22} & e_{32} \\ a_{13} & d_{23} & e_{33} \end{vmatrix}$$

$$+ \begin{vmatrix} b_{11} & d_{21} & e_{31} \\ b_{12} & d_{22} & e_{32} \\ b_{13} & d_{23} & e_{33} \end{vmatrix} - \begin{vmatrix} c_{11} & d_{21} & e_{31} \\ c_{12} & d_{22} & e_{32} \\ c_{13} & d_{23} & e_{33} \end{vmatrix}$$

$$\text{LHS: } \Delta = \begin{vmatrix} a_{11} + b_{11} - c_{11} & d_{21} & e_{31} \\ a_{12} + b_{12} - c_{12} & d_{22} & e_{32} \\ a_{13} + b_{13} - c_{13} & d_{23} & e_{33} \end{vmatrix}$$

By expanding the determinant along the first column, we get

$$\Delta = (a_{11} + b_{11} - c_{11})\begin{vmatrix} d_{22} & e_{32} \\ d_{23} & e_{33} \end{vmatrix} - (a_{12} + b_{12} - c_{12})\begin{vmatrix} d_{21} & e_{31} \\ d_{23} & e_{33} \end{vmatrix}$$

$$+ (a_{13} + b_{13} - c_{13})\begin{vmatrix} d_{21} & e_{31} \\ d_{22} & e_{32} \end{vmatrix}$$

$$= (a_{11} + b_{11} - c_{11})(d_{22}e_{33} - e_{32}d_{23}) - (a_{12} + b_{12} - c_{12})(d_{21}e_{33} - e_{31}d_{23})$$

$$+ (a_{13} + b_{13} - c_{13})(d_{21}e_{32} - e_{31}d_{22})$$

$$= a_{11}(d_{22}e_{33} - e_{32}d_{23}) - a_{12}(d_{22}e_{33} - e_{32}d_{23}) + a_{13}(d_{21}e_{32} - e_{31}d_{22})$$

$$+ b_{11}(d_{22}e_{33} - e_{32}d_{23}) - b_{12}(d_{22}e_{33} - e_{31}d_{23}) + b_{13}(d_{21}e_{32} - e_{31}d_{22})$$

$$- c_{11}(d_{22}e_{33} - e_{32}d_{23}) + c_{12}(d_{21}e_{33} - e_{31}d_{23}) - c_{13}(d_{21}e_{32} - e_{31}d_{22})$$

$$\Delta = \begin{vmatrix} a_{11} & d_{21} & e_{31} \\ a_{12} & d_{22} & e_{32} \\ a_{13} & d_{23} & e_{33} \end{vmatrix} + \begin{vmatrix} b_{11} & d_{21} & e_{31} \\ b_{12} & d_{22} & e_{32} \\ b_{13} & d_{23} & e_{33} \end{vmatrix} - \begin{vmatrix} c_{11} & d_{21} & e_{31} \\ c_{12} & d_{22} & e_{32} \\ c_{13} & d_{23} & e_{33} \end{vmatrix} = \text{R.H.S}$$

Hence, it is verified.

Illustration 8

Given $\begin{vmatrix} a_{11} + 2x & b_{21} & c_{31} \\ a_{12} + 2y & b_{22} & c_{32} \\ a_{13} + 2z & b_{23} & c_{33} \end{vmatrix}$

Solution

$$\begin{vmatrix} a_{11} + 2x & b_{21} & c_{31} \\ a_{12} + 2y & b_{22} & c_{32} \\ a_{13} + 2z & b_{23} & c_{33} \end{vmatrix} = \begin{vmatrix} a_{11} & b_{21} & c_{31} \\ a_{12} & b_{22} & c_{32} \\ a_{13} & b_{23} & c_{33} \end{vmatrix} + \begin{vmatrix} 2x & b_{21} & c_{31} \\ 2y & b_{22} & c_{32} \\ 2z & b_{23} & c_{33} \end{vmatrix}$$

(By using Property 5)

Illustration 9

Given that $\begin{vmatrix} a & b & c \\ a+2x & b+2y & c+2z \\ x & y & z \end{vmatrix} = 0$

Solution

$$\begin{vmatrix} a & b & c \\ a+2x & b+2y & c+2z \\ x & y & z \end{vmatrix} = \begin{vmatrix} a & b & c \\ a & b & c \\ x & y & z \end{vmatrix} + \begin{vmatrix} a & b & c \\ 2x & 2y & 2z \\ x & y & z \end{vmatrix}$$

$$\left(\text{By using Property 5}\right)$$

$$= \begin{vmatrix} a & b & c \\ a & b & c \\ x & y & z \end{vmatrix} + 2\begin{vmatrix} a & b & c \\ x & y & z \\ x & y & z \end{vmatrix}$$

$$(\text{By using Property 4})$$

$$= 0 + 0 = 0 \quad \left(\text{By using Property 3}\right)$$

Hence, it is verified.

Property 6

If each element of any row or column of a determinant is added with equi-multiples of corresponding elements of other rows or columns, then the value of determinants remains the same.

Proof

Let $\Delta_1 = \begin{vmatrix} a_{11} & a_{12} & a_{13} \\ b_{21} & b_{22} & b_{23} \\ c_{31} & c_{32} & c_{33} \end{vmatrix}$ and $\Delta_2 = \begin{vmatrix} a_{11}+pc_{31} & a_{12}+pc_{32} & a_{13}+pc_{33} \\ b_{21} & b_{22} & b_{23} \\ c_{31} & c_{32} & c_{33} \end{vmatrix}$

In Δ_2, the element of the row R_1 is symbolically written as $R_1 \rightarrow R_1 + kR_3$.

From Δ_2, it is clear that the elements of the third row (R_3) are multiplied by a constant p and are added with the corresponding elements of the first row (R_1). Now,

$$\Delta_2 = \begin{vmatrix} a_{11} & a_{12} & a_{13} \\ b_{21} & b_{22} & b_{23} \\ c_{31} & c_{32} & c_{33} \end{vmatrix} + \begin{vmatrix} pc_{31} & pc_{32} & pc_{33} \\ b_{21} & b_{22} & b_{23} \\ c_{31} & c_{32} & c_{33} \end{vmatrix} \quad \text{(By using Property 5)}$$

$$= \begin{vmatrix} a_{11} & a_{12} & a_{13} \\ b_{21} & b_{22} & b_{23} \\ c_{31} & c_{32} & c_{33} \end{vmatrix} + p \begin{vmatrix} c_{31} & c_{32} & c_{33} \\ b_{21} & b_{22} & b_{23} \\ c_{31} & c_{32} & c_{33} \end{vmatrix} \quad \text{(By using Property 4)}$$

$\Delta_2 = \Delta_1 + 0$ (By using Property 2)

$\Delta_2 = \Delta_1$

Illustration 10

Verify Property 6 for the given determinants Δ_1 and Δ_2:

$$\Delta_1 = \begin{vmatrix} 9 & 18 & 27 \\ 2 & 5 & 8 \\ 3 & 6 & 9 \end{vmatrix} \text{ and } \Delta_2 = \begin{vmatrix} 3+9 & 6+18 & 9+27 \\ 2 & 5 & 8 \\ 3 & 6 & 9 \end{vmatrix}$$

Solution

$$\text{Here } \Delta_2 = \begin{vmatrix} 3 & 6 & 9 \\ 2 & 5 & 8 \\ 3 & 6 & 9 \end{vmatrix} + \begin{vmatrix} 6 & 9 & 12 \\ 2 & 5 & 8 \\ 3 & 6 & 9 \end{vmatrix} \quad \text{(By using Property 5)}$$

$$= \begin{vmatrix} 9 & 18 & 27 \\ 2 & 5 & 8 \\ 3 & 6 & 9 \end{vmatrix} + 3 \begin{vmatrix} 3 & 6 & 9 \\ 2 & 5 & 8 \\ 3 & 6 & 9 \end{vmatrix} \quad \text{(By using Property 4)}$$

$\Delta_2 = \Delta_1 + 0$ (By using Property 3)

Hence, $\Delta_2 = \Delta_1$

Property 7

If all the elements on one side of a leading diagonal of a determinant are zero, then the numerical value of that determinant will be equal to the product of the leading diagonal elements.

> **NOTE**
>
> Such a determinant is called triangular determinant.

Proof

$$\text{Let } \Delta = \begin{vmatrix} 1 & 0 & 0 \\ 3 & 2 & 0 \\ 4 & 5 & 3 \end{vmatrix}$$

$$\Delta = 1(2 \times 3 - 5 \times 0) = 1 \times 2 \times 3$$

$\Delta = 6$ (Hence, the determinant of a triangular matrix is the product of the diagonal entries).

Property 8

The product of two determinants of the same order will be a determinant of that order.

Proof

$$\text{Let } \Delta_1 = \begin{vmatrix} a_{11} & b_{21} & c_{31} \\ a_{12} & b_{22} & c_{32} \\ a_{13} & b_{23} & c_{33} \end{vmatrix} \text{ and } \Delta_2 = \begin{vmatrix} p_{11} & q_{21} & r_{31} \\ p_{12} & q_{22} & r_{32} \\ p_{13} & q_{23} & r_{33} \end{vmatrix}$$

Then,

$$\Delta_1 \times \Delta_2 = \begin{vmatrix} a_{11} & b_{21} & c_{31} \\ a_{12} & b_{22} & c_{32} \\ a_{13} & b_{23} & c_{33} \end{vmatrix} \times \begin{vmatrix} p_{11} & q_{21} & r_{31} \\ p_{12} & q_{22} & r_{32} \\ p_{13} & q_{23} & r_{33} \end{vmatrix}$$

$$= \begin{vmatrix} a_{11}p_{11} + b_{21}q_{21} + c_{31}r_{31} & a_{11}p_{12} + b_{21}q_{22} + c_{31}r_{32} & a_{11}p_{13} + b_{21}q_{23} + c_{31}r_{33} \\ a_{12}p_{11} + b_{22}q_{21} + c_{32}r_{31} & a_{12}p_{12} + b_{22}q_{22} + c_{32}r_{32} & a_{12}p_{13} + b_{22}q_{23} + c_{32}r_{33} \\ a_{13}p_{11} + b_{23}q_{21} + c_{33}r_{31} & a_{13}p_{12} + b_{23}q_{22} + c_{33}r_{32} & a_{13}p_{13} + b_{23}q_{23} + c_{33}r_{33} \end{vmatrix}$$

Illustration 11

Prove Property 8 for the given matrices Δ_1 and Δ_2

Given that $\Delta_1 = \begin{vmatrix} 2 & 3 & 4 \\ 5 & 6 & 7 \\ 4 & 8 & 3 \end{vmatrix}$ and $\Delta_2 = \begin{vmatrix} 3 & 4 & 2 \\ 6 & 3 & 7 \\ 2 & 4 & 3 \end{vmatrix}$.

Solution

$$\Delta_1 \times \Delta_2 = \begin{vmatrix} 2 & 3 & 4 \\ 5 & 6 & 7 \\ 4 & 8 & 3 \end{vmatrix} \times \begin{vmatrix} 3 & 4 & 2 \\ 6 & 3 & 7 \\ 2 & 4 & 3 \end{vmatrix}$$

$$= \begin{vmatrix} (2\times3)+(3\times4)+(4\times2) & (2\times6)+(3\times3)+(4\times7) & (2\times2)+(3\times4)+(4\times3) \\ (5\times3)+(6\times4)+(7\times2) & (5\times6)+(6\times3)+(7\times7) & (5\times2)+(6\times4)+(7\times3) \\ (4\times3)+(8\times4)+(3\times2) & (4\times6)+(8\times3)+(3\times7) & (4\times2)+(8\times4)+(3\times3) \end{vmatrix}$$

$$= \begin{vmatrix} 6+12+8 & 12+9+28 & 4+12+21 \\ 15+24+14 & 30+18+49 & 10+24+21 \\ 12+32+6 & 24+24+21 & 8+32+9 \end{vmatrix}$$

$$= \begin{vmatrix} 26 & 49 & 28 \\ 53 & 97 & 55 \\ 50 & 69 & 49 \end{vmatrix}$$

Thus, the order of the determinants Δ_1, Δ_2 and $(\Delta_1 \times \Delta_2)$ are the same, which is equal to 3.

Matrices

In 1860, a French mathematician, Arthur Cayley discovered the matrices which are extensively used in applied mathematics and many branches of engineering.

Definition

The arrangement of elements in the form of rectangular array along 'm' horizontal lines (rows) and 'n' vertical lines (columns) and bounded by the brackets '[ ]' is called an 'm' by 'n' matrix, which is also represented as '$m \times n$' matrix. It is denoted by A with the element in its ith row and jth column as a_{ij}.

$$\text{Thus, } A = \begin{bmatrix} a_{11} & a_{12} & \cdots & a_{1j} & \cdots & a_{1n} \\ a_{21} & a_{22} & \cdots & a_{2j} & \cdots & a_{1n} \\ \vdots & \vdots & \cdots & \vdots & \cdots & \vdots \\ a_{i1} & a_{i2} & \cdots & a_{ij} & \cdots & a_{1n} \\ \vdots & \vdots & \cdots & \vdots & \cdots & \vdots \\ a_{m1} & a_{m2} & \cdots & a_{mj} & \cdots & a_{1n} \end{bmatrix} = \begin{bmatrix} a_{ij} \end{bmatrix}$$

Order of a Matrix

The order of a matrix is given by the number of its rows and columns.
For example,

$A = \begin{bmatrix} 1 & 2 \\ 3 & 4 \end{bmatrix}$ is a matrix of the order 2×2.

$A = \begin{bmatrix} 1 & 2 & 3 \\ 4 & 5 & 6 \end{bmatrix}$ is a matrix of the order 2×3.

Row Matrix

The matrix with elements arranged in only one row is called row matrix.
For example,

$A = \begin{bmatrix} 1 & 2 & 3 & 4 \end{bmatrix}$ is a row matrix with order 1×4.

Column Matrix

The matrix with elements arranged in only one column is called column matrix.
For example,

$A = \begin{bmatrix} 1 \\ 2 \\ 3 \\ 4 \end{bmatrix}$ is a column matrix with order 4×1.

Square Matrix

The matrix having equal numbers of rows and columns is called a square matrix.
For example,

$A = \begin{bmatrix} 1 & 2 \\ 3 & 4 \end{bmatrix}$ and $B = \begin{bmatrix} 1 & 2 & 3 \\ 4 & 5 & 6 \\ 7 & 8 & 9 \end{bmatrix}$ Both A and B are square matrices

with order 2×2 and 3×3, respectively.

Null Matrix

It is a matrix where all the elements are '0'.
   For example,

$$A = \begin{bmatrix} 0 & 0 \\ 0 & 0 \end{bmatrix} \text{ and } B = \begin{bmatrix} 0 & 0 & 0 \\ 0 & 0 & 0 \\ 0 & 0 & 0 \end{bmatrix} \text{ and } C = \begin{bmatrix} 0 & 0 \\ 0 & 0 \\ 0 & 0 \end{bmatrix}$$

Principle Diagonal

The diagonal of the matrix A in this case with elements 1, 5 and 9 is called the leading diagonal or principle diagonal.
   For example,

$$A = \begin{bmatrix} 1 & 2 & 3 \\ 4 & 5 & 6 \\ 7 & 8 & 9 \end{bmatrix}$$

Diagonal Matrix

It is a square matrix where all the elements except those on the principle (leading) diagonal are zero is called diagonal matrix.
   For example,

$$A = \begin{bmatrix} 1 & 0 \\ 0 & 4 \end{bmatrix} \text{ and } B = \begin{bmatrix} 1 & 0 & 0 \\ 0 & 5 & 0 \\ 0 & 0 & 9 \end{bmatrix}$$

Unit Matrix or Identity Matrix

It is a diagonal matrix in which all the elements in the principle (leading) diagonal are unity.
   For example,

$$A = \begin{bmatrix} 1 & 0 \\ 0 & 1 \end{bmatrix} \text{ and } B = \begin{bmatrix} 1 & 0 & 0 \\ 0 & 1 & 0 \\ 0 & 0 & 1 \end{bmatrix}$$

Scalar Matrix

It is a square matrix in which all the elements of its principle (leading) diagonal are equal and all the other elements in it are zero.

For example,

$$A = \begin{bmatrix} 2 & 0 \\ 0 & 2 \end{bmatrix} \text{ and } B = \begin{bmatrix} 3 & 0 & 0 \\ 0 & 3 & 0 \\ 0 & 0 & 3 \end{bmatrix}$$

Upper Triangular Matrix

It is a square matrix in which all the elements below the principle (leading) diagonal are zero.
  For example,

$$A = \begin{bmatrix} 1 & 2 & 3 \\ 0 & 5 & 6 \\ 0 & 0 & 9 \end{bmatrix}$$

Lower Triangular Matrix

It is a square matrix in which all the elements above the principle (leading) diagonal are zero.
  For example,

$$A = \begin{bmatrix} 1 & 0 & 0 \\ 4 & 5 & 0 \\ 7 & 8 & 9 \end{bmatrix}$$

Transpose of a Matrix

The transpose of a matrix 'A' is obtained by interchanging its rows and columns, and it is denoted by 'A^T'.
  For example,

$$\text{If } A = \begin{bmatrix} 1 & 2 & 3 \\ 4 & 5 & 6 \\ 7 & 8 & 9 \end{bmatrix}, \text{ then } A^T = \begin{bmatrix} 1 & 4 & 7 \\ 2 & 5 & 8 \\ 3 & 6 & 9 \end{bmatrix}.$$

Similarly,

$$\text{If } B = \begin{bmatrix} 1 & 2 & 3 \\ 4 & 5 & 6 \end{bmatrix}, \text{ then } B^T = \begin{bmatrix} 1 & 4 \\ 2 & 5 \\ 3 & 6 \end{bmatrix}.$$

Symmetric Matrix

A square matrix 'A' = $[a_{ij}]$ is said to be symmetric when $a_{ij} = a_{ji}$ for all values of i and j.
  In other words, if $A^T = A$, then the matrix A is called symmetric matrix.

For example,

$$\text{If } A = \begin{bmatrix} 1 & 2 & 3 \\ 2 & 4 & 5 \\ 3 & 5 & 6 \end{bmatrix} \text{ then } A^T = \begin{bmatrix} 1 & 2 & 3 \\ 2 & 4 & 5 \\ 3 & 5 & 6 \end{bmatrix}.$$

Skew Symmetric Matrix

A square matrix $'A' = [a_{ij}]$ is said to be skew symmetric when $a_{ij} = -a_{ji}$ for all values of i and j so that all the principle (leading) diagonal elements are zero.
For example,

$$\text{If } A = \begin{bmatrix} 0 & 2 & 3 \\ 2 & 0 & 5 \\ 3 & 5 & 0 \end{bmatrix}, \text{ then skew symmetric of } [A] \text{ is} = \begin{bmatrix} 0 & 2 & -3 \\ -2 & 0 & 5 \\ 3 & -5 & 0 \end{bmatrix}.$$

In other words, if $A^T = -A$, then the matrix A is called skew symmetric matrix.

Singular Matrix

A square matrix $'A'$ is said to be singular matrix, if the determinant $'A'$ is zero.

NOTE

1. If $|A| = 0$, then $[A]$ is a singular matrix.
2. If $|A| \neq 0$, then $[A]$ is a non-singular matrix.

Adjoint of a Matrix

The adjoint of a matrix A is obtained by replacing its each element by its cofactor and then transposing it.
For example,

$$A = \begin{bmatrix} 1 & 4 & 7 \\ 2 & 5 & 8 \\ 3 & 6 & 9 \end{bmatrix}$$

From Illustration 3, we have

$$\text{Cofactor of matrix } A = \begin{bmatrix} -3 & 6 & -3 \\ 6 & -12 & 6 \\ -3 & 6 & -3 \end{bmatrix}$$

Transpose of the above cofactor matrix of $A = \begin{bmatrix} -3 & 6 & -3 \\ 6 & -12 & 6 \\ -3 & 6 & -3 \end{bmatrix} =$

Adjoint of matrix A.

Inverse of a Matrix

The inverse of a non-singular matrix is denoted by A^{-1}:

$$A^{-1} = \frac{\text{adjoint of } [A]}{|A|}$$

For example, let $A = \begin{bmatrix} 4 & 3 \\ 2 & 1 \end{bmatrix}$

$$|A| = \left[(4 \times 1) - (3 \times 2) \right]$$

$$= 4 - 6 = -2 \neq 0 \quad \text{(i.e., non-singular)}$$

Cofactor

$$C_{11} = (-1)^{1+1} [1] = 1$$

$$C_{12} = (-1)^{1+2} [2] = -2$$

$$C_{21} = (-1)^{2+1} [3] = -3$$

$$C_{22} = (-1)^{2+2} [4] = 4$$

Cofactor matrix, $C = \begin{bmatrix} 1 & -2 \\ -3 & 4 \end{bmatrix}$

Adjoint of A = transpose of cofactor matrix, C of $[A]$.
$Adj\ A = C^T$

$$C^T = \begin{bmatrix} 1 & -3 \\ -2 & 4 \end{bmatrix}$$

$$\therefore A^{-1} = \frac{\begin{bmatrix} 1 & -3 \\ -2 & 4 \end{bmatrix}}{-2}$$

$$= \begin{bmatrix} -0.5 & 1.5 \\ 1 & -2 \end{bmatrix}$$

Equality of Matrix

Two matrices A and B are said to be equal, if and only if they are of the same order and each element of A is equal to the corresponding element of B.

Let $A = \begin{bmatrix} 5 & 11 \\ -3 & 3 \end{bmatrix}$ and $B = \begin{bmatrix} 5 & 11 \\ -3 & 3 \end{bmatrix}$. Since A and B are of the same order and have the same elements, they are said to be equal.

Addition of Matrices

If A and B be two matrices of the same order, then their sum $A+B$ is defined as the matrix each element of which is the sum of the corresponding elements of A and B.

Thus,

$$A + B = \begin{bmatrix} p_1 & q_1 \\ p_2 & q_2 \end{bmatrix} + \begin{bmatrix} r_1 & s_1 \\ r_2 & s_2 \end{bmatrix}$$

$$= \begin{bmatrix} p_1 + r_1 & q_1 + s_1 \\ p_2 + r_2 & q_2 + s_2 \end{bmatrix}$$

For example,

$$\text{Let } A = \begin{bmatrix} 11 & 7 \\ 2 & 3 \end{bmatrix} \text{ and } B = \begin{bmatrix} 5 & -9 \\ 6 & 6 \end{bmatrix}$$

$$A + B = \begin{bmatrix} 11 & 7 \\ 2 & 3 \end{bmatrix} + \begin{bmatrix} 5 & -9 \\ 6 & 6 \end{bmatrix}$$

$$= \begin{bmatrix} 16 & -2 \\ 8 & 9 \end{bmatrix}$$

Subtraction of Matrices

If A and B be two matrices of the same order, then their sum $A-B$ is defined as the matrix each element of which is the difference of the corresponding elements of A and B.

Thus,

$$A - B = \begin{bmatrix} p_1 & q_1 \\ p_2 & q_2 \\ p_3 & q_2 \end{bmatrix} - \begin{bmatrix} r_1 & s_1 \\ r_2 & s_2 \\ r_3 & s_3 \end{bmatrix}$$

$$= \begin{bmatrix} p_1 - r_1 & q_1 - s_1 \\ p_2 - r_2 & q_2 - s_2 \\ p_3 - r_3 & q_2 - s_3 \end{bmatrix}$$

For example,

$$\text{Let } A = \begin{bmatrix} 2 & -8 & 5 \\ 1 & 0 & -3 \\ 9 & 1 & 4 \end{bmatrix} \text{ and } B = \begin{bmatrix} 11 & 2 & 5 \\ 3 & 2 & 10 \\ 4 & 19 & 5 \end{bmatrix}$$

$$A - B = \begin{bmatrix} 2 & -8 & 5 \\ 1 & 0 & -3 \\ 9 & 1 & 4 \end{bmatrix} - \begin{bmatrix} 11 & 2 & 5 \\ 3 & 2 & 10 \\ 4 & 19 & 5 \end{bmatrix} = \begin{bmatrix} -9 & -10 & 0 \\ -2 & -2 & -13 \\ 5 & -18 & -1 \end{bmatrix}$$

NOTE

Only matrices of the same order can be added or subtracted.

Multiplication of Matrices

Two matrices can be multiplied only when the number of columns in the first is equal to the number of rows in the second.
    For example,

$$A = \begin{bmatrix} p_1 & q_1 & r_1 \\ p_2 & q_2 & r_2 \\ p_3 & q_2 & r_3 \end{bmatrix}_{3x3} \quad B = \begin{bmatrix} s_1 & t_1 \\ s_2 & t_2 \\ s_3 & t_2 \end{bmatrix}_{3x2}$$

$$A \times B = \begin{bmatrix} p_1 s_1 + q_1 s_2 + r_1 s_3 & p_1 t_1 + q_1 t_2 + r_1 t_3 \\ p_2 s_1 + q_2 s_2 + r_2 s_3 & p_2 t_1 + q_2 t_2 + r_2 t_3 \\ p_3 s_1 + q_3 s_2 + r_3 s_3 & p_3 t_1 + q_3 t_2 + r_3 t_3 \end{bmatrix}_{3x2}$$

For example,

$$\text{Let } A = \begin{bmatrix} 1 & 4 & 3 \\ 7 & 2 & 1 \end{bmatrix} \text{ and } B = \begin{bmatrix} 1 & 5 \\ 2 & 4 \\ 1 & 3 \end{bmatrix}$$

$$A \times B = \begin{bmatrix} 1 & 4 & 3 \\ 7 & 2 & 1 \end{bmatrix} \times \begin{bmatrix} 1 & 5 \\ 2 & 4 \\ 1 & 3 \end{bmatrix} = \begin{bmatrix} 1+8+3 & 5+16+9 \\ 7+4+1 & 35+8+3 \end{bmatrix}$$

$$= \begin{bmatrix} 1+8+3 & 5+16+9 \\ 7+4+1 & 35+8+3 \end{bmatrix} = \begin{bmatrix} 12 & 30 \\ 12 & 46 \end{bmatrix}$$

Conjugate of a Matrix

The conjugate of a matrix A is obtained by replacing its element by the corresponding conjugate and is denoted by $\bar{A}$.

For example,

$$\text{then } \bar{A} = \begin{bmatrix} 1-2i & 3-i4 \\ 5-i6 & 7+i8 \end{bmatrix}$$

Hermitian Matrix

The square matrix A is said to be Hermitian matrix if the transpose of its conjugate is the matrix A itself, i.e., $[\bar{A}]^T = A$.

Skew Hermitian Matrix

The square matrix A is said to be skew Hermitian matrix if the transpose of its conjugate is the negate of matrix A itself, i.e., $[\bar{A}]^T = -A$.

Rank of a Matrix

The rank of a matrix A is the dimension of the largest order of any nonvanishing minor of the matrices (non-zero determinant).

NOTE

1. A matrix is said to be the rank $'r'$ when it has at least one non-zero minor of order $'r'$ and every minor of order higher than $'r'$ vanishes.
2. Rank of $[A^T]$ = Rank of $[A]$.

3. Rank of product of two matrices cannot exceed the rank of either.

4. If a matrix has a non-zero minor of order 'r', then its rank is greater than or equal to 'r'.

5. The rank of a matrix A is denoted by $\rho(A)$.

Illustration 12

Find the rank of matrix $A = \begin{bmatrix} 1 & 2 & 3 \\ 1 & 4 & 2 \\ 2 & 6 & 5 \end{bmatrix}$.

Solution

Method 1

To determine rank of the given matrix, transform matrix to upper triangular form, using elementary row operations

Operate
$R - R_1 \rightarrow$ (Subtract the first row from the second row);
$R_3 - 2R \rightarrow$ (multiply the first row by 2 and subtract it from the third row)

Then, $A = \begin{bmatrix} 1 & 2 & 3 \\ 0 & 2 & -1 \\ 0 & 2 & -1 \end{bmatrix}$

Operate
$R_2/2 \rightarrow$ (divide the second row by 2)

$$A = \begin{bmatrix} 1 & 2 & 3 \\ 0 & 1 & -0.5 \\ 0 & 2 & -1 \end{bmatrix}$$

Operate
$R_3 - 2R_2 \rightarrow$ (multiply the second row by 2 and subtract it from the third row)

$$A = \begin{bmatrix} 1 & 2 & 3 \\ 0 & 1 & -0.5 \\ 0 & 0 & 0 \end{bmatrix}$$

Since, there are two non-zero rows, the rank of the matrix $|A|$ is 2.

Method 2

The determinant of matrix A of order 3 is 0. Find any non-square sub-matrix of order 2.

$$\therefore \text{We have} \begin{bmatrix} 1 & 2 \\ 0 & 2 \end{bmatrix} = 2 \text{ and } \begin{bmatrix} 2 & 3 \\ 2 & -1 \end{bmatrix} = -2 - 6 = -8.$$

Thus, the largest order of the non-zero determinant is 2. Hence, the rank is 2.

The determinants and matrices are very useful in modelling and control of systems. Hence, in the earlier sections, their fundamentals are recalled. Now, before studying state space analysis, the drawbacks of transfer function approach are discussed in the next section.

Definition of Transfer Function

The cause and effect relationship between the output and input of a system is related to each other through a transfer function.

The transfer function of a control system is defined as the ratio of the Laplace transform of the output variable to Laplace transform of the input variable assuming all initial conditions to be zero.

In a Laplace transform, if the input is represented by $R(s)$ and the output is represented by $C(s)$, then the transfer function will be:

$$G(s) = \frac{C(s)}{R(s)} \tag{1.1}$$

Limitations of Transfer Function Approach

1. The classical design methods based on transfer function model is applicable only to linear time-invariant systems.

Proof

Illustration 13

Case 1: Consider the system which is time invariant but nonlinear in nature.

Let, $$y(t) = x(t-1) + 6$$

Applying Laplace transform,

$$L\{(y(t)\} = L\{x(t-1)\} + L\{6\}$$

$$Y(s) = X(s)e^{-s} + \infty$$

Therefore, $L\{6\}$ is undefined and transfer function cannot be determined. Also consider,

$$y(t) = x(t-1) + 6u(t)$$

Applying Laplace transform,

$$L\{(y(t)\} = L\{x(t-1)\} + L\{6\,u(t)\}$$

$$Y(s) = X(s)e^{-s} + \frac{6}{s}$$

$$Y(s) = X(s)\left[e^{-s} + \frac{6}{s\,X(s)}\right]$$

Here, transfer function $G(s) = \dfrac{Y(s)}{X(s)}$ cannot be determined since $X(s)$ is still present in the system.

Case 2: Consider the system which is linear but time variant.

Illustration 14

$$\text{Let, } y(t) = x(4t)$$

Applying Laplace transform,

$$L\{(y(t)\} = L\{x(4t)\}$$

$$Y(s) = \frac{1}{4}X\left(\frac{s}{4}\right)$$

Hence, transfer function $G(s) = \dfrac{Y(s)}{X(s)}$ cannot be determined.

Case 3: Consider the system which is linear time invariant.

Illustration 15

Let, $$y(t) = x(t-5) + x(t+6)$$

Applying Laplace transform,

$$L\{(y(t)\} = L\{x(t-5)\} + L\{x(t+6)\}$$

$$Y(s) = X(s)e^{-5s} + X(s)e^{6s}$$

$$Y(s) = X(s)(e^{-5s} + e^{6s})$$

$$G(s) = \frac{Y(s)}{X(s)} = e^{-5s} + e^{6s}$$

Hence, transfer function approach can be applicable only to linear time-invariant system.

2. The transfer function is derived by assuming zero initial conditions.

Proof

Illustration 16

Consider a system represented by the differential equation

$$\frac{d^2x(t)}{dt^2} + 3\frac{dx(t)}{dt} + 2x(t) = \frac{du(t)}{dt} + u(t) \qquad \text{(i)}$$

Applying Laplace transform with initial conditions, we get

$$\left[s^2X(s) - sx(0) - x(1)\right] + 3\left[sX(s) - x(0)\right] + 2X(s) = [sU(s) - u(0)] + U(s) \qquad \text{(ii)}$$

Case 1: Consider initial conditions = 0

$$\text{i.e., } x(0) = x(1) = 0$$

∴ Then the Equation (1.4) (ii) becomes

$$s^2X(s) + 3sX(s) + 2X(s) = sU(s) + U(s)$$

$$X(s)(s^2 + 3s + 2) = U(s)(s+1)$$

$$\frac{X(s)}{U(s)} = \frac{s+1}{s^2 + 3s + 2}$$

Case 2: Consider the initial conditions ≠ 0

$$\text{i.e., } x(0) = x(1) = 1$$

∴ Then the Equation (1.4) (ii) becomes

$$(s^2 X(s) - s - 1) + 3(sX(s) - 1) + 2X(s) = sU(s) - 1 + U(s)$$

$$s^2 X(s) - s - 1 + 3sX(s) - 3 + 2X(s) = U(s)(s+1) - 1$$

$$s^2 X(s) + X(s)(3s+2) - s - 4 = U(s)(s+1) - 1$$

$$s^2 X(s) + X(s)(3s+2) - s - 3 = U(s)(s+1)$$

$$X(s)(s^2 + 3s + 2) - s - 3 = U(s)(s+1)$$

The extra 's' term tends the system to nonlinearity. Hence, it is assumed that the transfer function is derived under zero initial conditions.

3. The transfer function approach reveals only the system output for a given input and provides no information about the internal behaviour of the system.

Proof

Illustration 17

Consider the block diagram shown in Figure 1.1.

where $G_1(s) = \dfrac{K_1}{(s+1)(s-1)}$ and $G_2(s) = \dfrac{K_2(s-1)}{(s+2)}$

Using block diagram reduction techniques, we can combine the cascaded blocks into one block as shown in Figure 1.2.

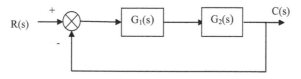

FIGURE 1.1
General block diagram.

FIGURE 1.2
Simplified block diagram.

$$G_1(s)G_2(s) = \left[\frac{K_1}{(s+1)(s-1)}\right]\left[\frac{K_2(s-1)}{(s+2)}\right]$$

$$G_1(s)G_2(s) = \frac{K_1 K_2}{(s+1)(s+2)}$$

$$T.F = \frac{C(s)}{R(s)} = \frac{G_1(s)G_2(s)}{1 + G_1(s)G_2(s)H(s)}$$

$G_1(s)$ and $G_2(s)$ are independently open loop unstable systems. However, when they are cascaded, due to pole-zero cancellation, the system approaches to be stable.

In other words, the transfer function approach does not reveal the information about the internal behaviour of the system.

4. The transfer function approach is restricted to Single Input Single Output (SISO) systems because it becomes highly cumbersome for use in Multi Input Multi Output (MIMO) systems.

Proof

Illustration 18

Case 1: Consider a SISO system shown in Figure 1.3a
Using block diagram reduction technique, the transfer function of the SISO system $C(s)/R(s)$ can be determined using the following steps.

Step 1: Study the system and redraw the diagram in a simplified manner as shown in Figure 1.3b

Step 2: Identify the feedback and feed forward paths separately. In this case, we have a feed forward path with one unity path. Hence, add up the blocks as shown in Figure 1.3c

Step 3: Cascade blocks in series as shown in Figure 1.3d

Step 4: Solve for the feed forward path (Figure 1.3e)

Step 5: Obtain the transfer function $C(s)/R(s)$

$$\frac{C(s)}{R(s)} = G_1(s)G_2(s) + G_2(s) + 1$$

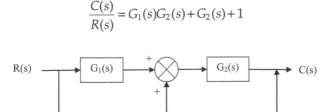

FIGURE 1.3A

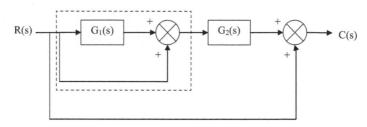

FIGURE 1.3B

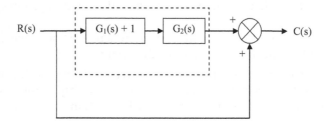

FIGURE 1.3C

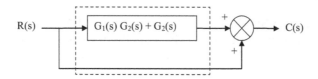

FIGURE 1.3D

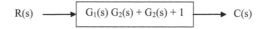

FIGURE 1.3E

Illustration 19

Case 2: Consider an MIMO system shown in Figure 1.4a
  Case 2a: When input = $R_1(s)$ and output = $C_1(s)$ (with $R_2(s) = 0$)

Step 1: Put $R_2(s) = C_2(s) = 0$ and redraw the diagram as shown in Figure 1.4b
Step 2: Cascade the blocks in series and arrange the blocks as in Figure 1.4c
Step 3: Solve for the negative feedback loop as shown in Figure 1.4d

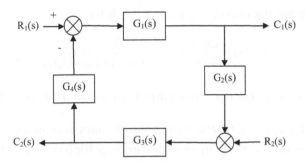

FIGURE 1.4A

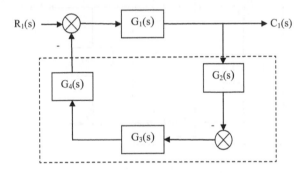

FIGURE 1.4B

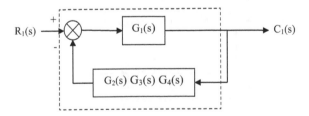

FIGURE 1.4C

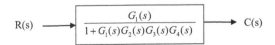

FIGURE 1.4D

Step 4: Obtain the transfer function $C_1(s)/R_2(s)$

$$G_{11}(s) = \frac{C_1(s)}{R_1(s)} = \frac{G_1(s)}{1 + G_1(s)G_2(s)G_3(s)G_4(s)}$$

Case 2b: When input = $R_2(s)$ and output = $C_1(s)$ (with $R_1(s) = 0$)

Step 1: Put $R_1(s) = C_2(s) = 0$ and redraw the diagram as shown in Figure 1.4e
Step 2: Cascade the blocks in series and arrange the blocks as in Figure 1.4f
Step 3: Solve for the negative feedback loop as shown in Figure 1.4g

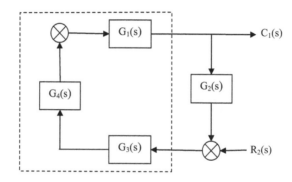

FIGURE 1.4E

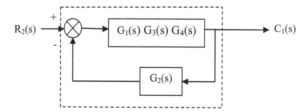

FIGURE 1.4F

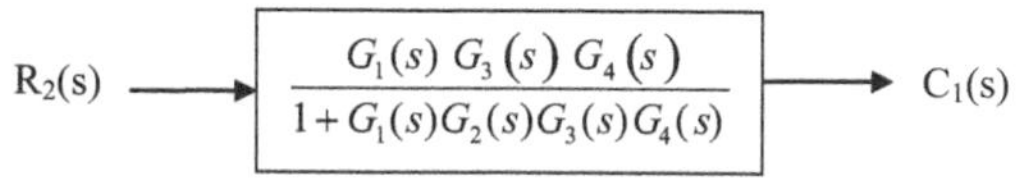

$$R_2(s) \longrightarrow \boxed{\frac{G_1(s)\,G_3(s)\,G_4(s)}{1 + G_1(s)G_2(s)G_3(s)G_4(s)}} \longrightarrow C_1(s)$$

FIGURE 1.4G

Step 4: Obtain the transfer function $C_1(s)/R_2(s)$

$$G_{12}(s) = \frac{C_1(s)}{R_1(s)} = \frac{G_1(s)G_3(s)G_4(s)}{1+G_1(s)G_2(s)G_3(s)G_4(s)}$$

Case 2c: When input $= R_1(s)$ and output $= C_2(s)$ (with $R_2(s) = 0$)

Step 1: Put $R_2(s) = C_1(s) = 0$ and redraw the diagram as shown in Figure 1.4h
Step 2: Cascade the blocks in series and arrange the blocks as in Figure 1.4i
Step 3: Solve for the negative feedback loop as shown in Figure 1.4j
Step 4: Obtain the transfer function $C_2(s)/R_1(s)$

$$G_{21}(s) = \frac{C_2(s)}{R_1(s)} = \frac{-G_1(s)G_3(s)G_4(s)}{1+G_1(s)G_2(s)G_3(s)G_4(s)}$$

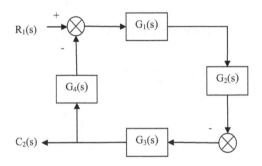

FIGURE 1.4H

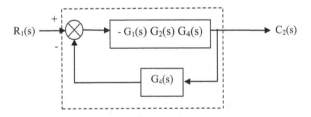

FIGURE 1.4I

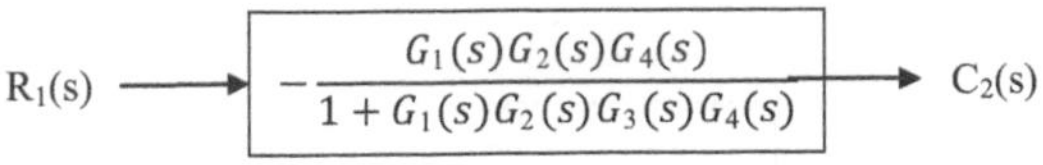

FIGURE 1.4J

Case 2d: When input = $R_2(s)$ and output = $C_2(s)$ (with $R_1(s) = 0$)

Step 1: Put $R_1(s) = C_1(s) = 0$ and redraw the diagram as shown in Figure 1.4k
Step 2: Cascade the blocks in series and arrange the blocks as in Figure 1.4l
Step 3: Solve for the negative feedback loop as shown in Figure 1.4m
Step 4: Obtai the transfer function $C_2(s)/R_1(s)$

$$G_{22}(s) = \frac{C_2(s)}{R_2(s)} = \frac{G_3(s)}{1 + G_1(s)G_2(s)G_3(s)G_4(s)}$$

Now, the transfer function of the MIMO system is given by the transfer function matrix,

$$\begin{bmatrix} C_1(s) \\ C_2(s) \end{bmatrix} = \begin{bmatrix} G_{11}(s) & G_{12}(s) \\ G_{21}(s) & G_{22}(s) \end{bmatrix} \begin{bmatrix} R_1(s) \\ R_2(s) \end{bmatrix}$$

$G_{ij}(s)$ must have $q \times p$ dimensionality and thus has a total of '$q \times p$' elements.

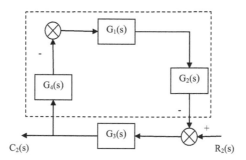

FIGURE 1.4K

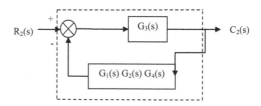

FIGURE 1.4L

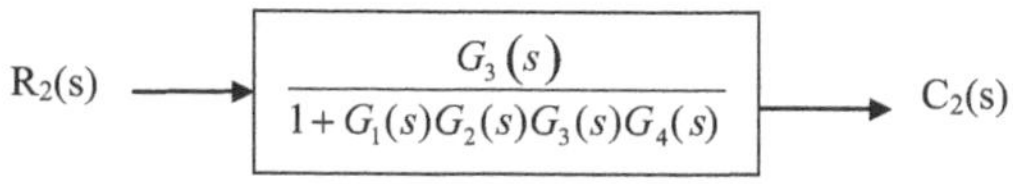

$$R_2(s) \longrightarrow \boxed{\frac{G_3(s)}{1 + G_1(s)G_2(s)G_3(s)G_4(s)}} \longrightarrow C_2(s)$$

FIGURE 1.4M

Therefore, for every input, there are 'q' transfer functions with one for each output. The procedure is repeated many times to find the transfer function. Hence, it is laborious for MIMO systems.

Introduction to State Space Analysis

The state space analysis alleviates the demerits of transfer function approach discussed in the earlier section. In the state space analysis, there are three variables that are useful in the modelling of dynamic systems. They are state variables, input variables, and output variables.

State: It is the smallest set of variables of a dynamic system, the knowledge of whom at $t = t_0$ together with the knowledge of input for $t \geq t_0$, completely establishes the behaviour of the system for any time $t \geq t_0$.

State variable: The variables of a dynamic system that are the necessary ingredients of the smallest set of variables that determine the state of a system are called state variables.

If 'n' variables are required to completely establish the behaviour of the system, then the state variables are given by

$$X(t) = \begin{bmatrix} x_1(t) \\ x_2(t) \\ \vdots \\ x_n(t) \end{bmatrix} \tag{1.2}$$

NOTE

It is handy to choose easily measurable variable as the state variable. However, physical quantities that are neither measureable nor observable can also be chosen as state variables.

Input and Output Variables

In a MIMO system, there may be 'p' number of input variables ($u_1(t), u_2(t)\ldots u_p(t)$) and '$q$' number of output variables ($y_1(t), y_2(t)\ldots y_q(t)$) as shown in Figure 1.5. Thus, the input and output variables can be represented as follows:

$$U(t) = \begin{bmatrix} u_1(t) \\ u_2(t) \\ \vdots \\ u_p(t) \end{bmatrix} \tag{1.3}$$

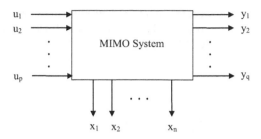

FIGURE 1.5
General block diagram of MIMO system.

$$Y(t) = \begin{bmatrix} y_1(t) \\ y_2(t) \\ \vdots \\ y_q(t) \end{bmatrix} \tag{1.4}$$

State Model

The state model of a MIMO system consists of state equation and output equation. The state equation relates state variables with the input variables as given in equation (1.5):

$$\dot{X}(t) = AX(t) + BU(t) \tag{1.5}$$

The output equation relates output with input and state variables as given in equation (1.6):

$$Y(t) = CX(t) + DU(t) \tag{1.6}$$

where

$$A = \begin{bmatrix} a_{11} & a_{12} & \cdots & a_{1n} \\ a_{21} & a_{22} & \cdots & a_{2n} \\ \vdots & \vdots & & \vdots \\ a_{n1} & a_{n2} & \cdots & a_{nn} \end{bmatrix} \text{ and } B = \begin{bmatrix} b_{11} & b_{12} & \cdots & b_{1p} \\ b_{21} & b_{22} & \cdots & b_{2p} \\ \vdots & \vdots & & \vdots \\ b_{n1} & b_{n2} & \cdots & b_{np} \end{bmatrix}$$

$$C = \begin{bmatrix} c_{11} & c_{12} & \cdots & c_{1n} \\ c_{21} & c_{22} & \cdots & c_{2n} \\ \vdots & \vdots & & \vdots \\ c_{q1} & c_{q2} & \cdots & c_{qn} \end{bmatrix} \text{ and } D = \begin{bmatrix} d_{11} & d_{12} & \cdots & d_{1p} \\ d_{21} & d_{22} & \cdots & d_{2p} \\ \vdots & \vdots & & \vdots \\ d_{q1} & d_{q2} & \cdots & d_{qp} \end{bmatrix} \tag{1.7}$$

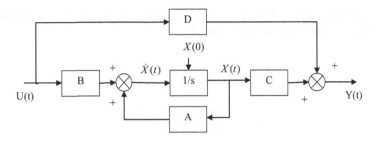

FIGURE 1.6
Block diagram of linear system.

$$X \to n \times 1; Y \to q \times 1; U \to p \times 1; A \to n \times n; B \to n \times p; C \to q \times n; D \to q \times p$$

Thus, the block diagram of a linear system represented in state space is shown in Figure 1.6.

Review of State Models

In this section, the procedure for obtaining state models of dynamic systems represented in different forms, namely, differential equation forms, transfer function form, block diagram form and signal flow graph form, is illustrated with the help of numerical examples.

State Space Model from Differential Equation

Illustration 20

Develop a state space model for a dynamic system with input and output related by the following differential equation:

$$\frac{d^3 e_o}{dt^3} + 5 \frac{d^2 e_o}{dt^2} - 3 \frac{de_o}{dt} + 7 e_o = 2u$$

NOTE

The number of state variables needed to form a complete state model is equal to the order of the given system.

Given

$$\frac{d^3 e_o}{dt^3} + 5 \frac{d^2 e_o}{dt^2} - 3 \frac{de_o}{dt} + 7 e_o = 2u \tag{i}$$

The order of the above differential equation is 3. Hence, let us take three state variables to formulate the state model:

$$\text{Let} \quad x_1 = e_o(t) \Rightarrow \dot{x}_1 = \frac{de_o(t)}{dt} \tag{ii}$$

$$x_2 = \frac{de_o(t)}{dt} \Rightarrow \dot{x}_2 = \frac{d^2 e_o(t)}{dt^2} \tag{iii}$$

$$x_3 = \frac{d^2 e_o(t)}{dt^2} \Rightarrow \dot{x}_3 = \frac{d^3 e_0(t)}{dt^3} \tag{iv}$$

From (i), (ii), (iii) and (iv), we have

$$\dot{x}_1 = \frac{de_o(t)}{dt} = x_2 = 0\,x_1 + 1\,x_2 + 0\,x_3 + 0\,u \tag{v}$$

$$\dot{x}_2 = \frac{d^2 e_o(t)}{dt^2} = x_3 = 0\,x_1 + 0\,x_2 + 1\,x_3 + 0\,u \tag{vi}$$

$$\dot{x}_3 = \frac{d^3 e_o(t)}{dt^3} = 2u - 5\frac{d^2 e_o}{dt^2} + 3\frac{de_o}{dt} - 7e_o = 2u - 5x_3 + 3x_2 - 7x_1 \tag{vii}$$

Equations (v), (vi) and (vii) can be represented in the form of a matrix. Thus, we have

$$\begin{bmatrix} \dot{x}_1 \\ \dot{x}_2 \\ \dot{x}_3 \end{bmatrix} = \begin{bmatrix} 0 & 1 & 0 \\ 0 & 0 & 1 \\ -7 & 3 & -5 \end{bmatrix} \begin{bmatrix} x_1 \\ x_2 \\ x_3 \end{bmatrix} + \begin{bmatrix} 0 \\ 0 \\ 2 \end{bmatrix} u \tag{viii}$$

The output equation Y(t) is given by

$$Y(t) = e_0(t)$$

$$Y(t) = 1\,x_1 + 0\,x_2 + 0\,x_3 + 0\,u$$

$$Y = \begin{bmatrix} 1 & 0 & 0 \end{bmatrix} \begin{bmatrix} x_1 \\ x_2 \\ x_3 \end{bmatrix} + [0]u \tag{ix}$$

Equation (viii) is called state equation and equation (ix) is called output equation. Both equations (viii) and (ix) together form the state model of the given dynamic system.

State Space Model by Transfer Function

Illustration 21

Obtain the state space model of a system whose input–output relationship is given by the transfer function

$$\frac{Y(s)}{U(s)} = \frac{2(s+3)}{(s+1)(s+4)}$$

Given transfer function is

$$\frac{Y(s)}{U(s)} = \frac{2(s+3)}{(s+1)(s+4)}$$

The above transfer function can be rearranged using partial fraction approach. Thus, we have

$$\frac{2(s+3)}{(s+1)(s+4)} = \frac{2s+6}{(s+1)(s+4)} = \frac{A}{(s+1)} + \frac{B}{(s+4)}$$

$$2s+6 = A(s+4) + B(s+1)$$

$$2s+6 = As + 4A + Bs + B$$

$$2s+6 = s(A+B) + (4A+B)$$

Equating 's' terms and constant terms, we get

$$A + B = 2$$

$$4A + B = 6$$

Solving the above two equations for A and B, we get $A = 4/3$ and $B = 2/3$.

$$\therefore \frac{Y(s)}{U(s)} = \frac{2s+6}{(s+1)(s+4)} = \frac{\frac{4}{3}}{(s+1)} + \frac{\frac{2}{3}}{(s+4)}$$

$$Y(s) = \left(\frac{4}{3}\right)\left(\frac{U(s)}{(s+1)}\right) + \left(\frac{2}{3}\right)\left(\frac{U(s)}{(s+4)}\right)$$

$$= \left(\frac{4}{3}\right)X_1(s) + \left(\frac{2}{3}\right)X_2(s)$$

where $\qquad X_1(s) = \dfrac{U(s)}{(s+1)}$ and $X_2(s) = \dfrac{U(s)}{(s+4)}$

We have

$$X_1(s) = \frac{U(s)}{(s+1)} \Rightarrow X_1(s)(s+1) = U(s)$$

$$\dot{x}_1(t) + x_1 = u(t)$$

$$\dot{x}_1(t) = u(t) - x_1 \qquad\qquad\qquad (i)$$

$$\dot{x}_1(t) = 1u(t) - 1x_1 + 0x_2$$

We have

$$X_2(s) = \frac{U(s)}{(s+4)} \Rightarrow X_2(s)(s+4) = U(s)$$

$$\dot{x}_2(t) + 4x_2 = u(t)$$

$$\dot{x}_2(t) = u(t) - 4x_1 \qquad\qquad\qquad (ii)$$

$$\dot{x}_2(t) = 0x_1(t) - 4x_1 + 1u(t)$$

Also we have

$$Y(s) = \tfrac{4}{3}X_1(s) + \tfrac{2}{3}X_2(s)$$

$$y(t) = \tfrac{4}{3}x_1(t) + \tfrac{2}{3}x_2(t) \qquad\qquad\qquad (iii)$$

$$y(t) = \tfrac{4}{3}x_1(t) + \tfrac{2}{3}x_2(t) + 0u(t)$$

From equations (i), (ii) and (iii), the state model of given transfer function can be written as follows:

$$\begin{bmatrix} \dot{x}_1 \\ \dot{x}_2 \end{bmatrix} = \begin{bmatrix} -1 & 0 \\ 0 & -4 \end{bmatrix} \begin{bmatrix} x_1 \\ x_2 \end{bmatrix} + \begin{bmatrix} 1 \\ 1 \end{bmatrix} u$$

$$Y = \begin{bmatrix} \tfrac{4}{3} & \tfrac{2}{3} \end{bmatrix} \begin{bmatrix} x_1 \\ x_2 \end{bmatrix} + [0]u$$

State Space Model of an Electrical System

Illustration 22

Obtain the state space model of an electrical system as shown in Figure 1.7.

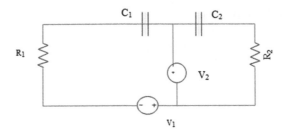

FIGURE 1.7
Block diagram of electrical network.

Solution

In the given electrical system, there are two loops. We can apply Kirchhoff's voltage law and obtain two equations for the two loops. Thus, the given electrical circuit can be redrawn as shown in Figure 1.8.

Applying Kirchhoff's voltage law to the first loop, we get

$$V_2 - V_1 - x_1 R_1 - \frac{1}{C_1} \int x_1 \, dt = 0$$

Differentiating the above equation, we get

$$\dot{V}_2 - \dot{V}_1 - \dot{x}_1 R_1 - \frac{1}{C_1} x_1 = 0$$

(i)

$$\dot{x}_1 = \frac{-x_1}{R_1 C_1} - \frac{\dot{V}_1}{R_1} + \frac{\dot{V}_2}{R_2}$$

Similarly, by applying Kirchhoff's voltage law to the second loop, we get

$$-\frac{1}{C_2} \int x_2 \, dt - x_2 R_2 - V_2 = 0$$

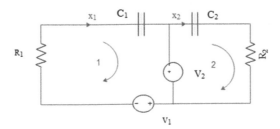

FIGURE 1.8
Block diagram of electrical network.

Differentiating the above equation, we get

$$-\frac{x_2}{C_2} - \dot{x}_2 R_2 - \dot{V}_2 = 0$$

(ii)

$$\dot{x}_2 = \frac{-x_2}{R_2 C_2} - \frac{\dot{V}_2}{R_2}$$

From equations (i) and (ii), we can write the state equation as follows:

$$\begin{bmatrix} \dot{x}_1 \\ \dot{x}_2 \end{bmatrix} = \begin{bmatrix} -\dfrac{1}{R_1 C_1} & 0 \\ 0 & -\dfrac{1}{R_2 C_2} \end{bmatrix} \begin{bmatrix} x_1 \\ x_2 \end{bmatrix} + \begin{bmatrix} -\dfrac{1}{R_1} & \dfrac{1}{R_2} \\ 0 & -\dfrac{1}{R_2} \end{bmatrix} \begin{bmatrix} \dot{V}_1 \\ \dot{V}_2 \end{bmatrix}$$

Similarly, we have
$$y_1 = x_1 = 1x_1 + 0x_2 + 0u$$
$$y_2 = x_2 = 0x_1 + 1x_2 + 0u$$

Thus, the output equation can be written as follows:

$$\begin{bmatrix} y_1 \\ y_2 \end{bmatrix} = \begin{bmatrix} 1 & 0 \\ 0 & 1 \end{bmatrix} \begin{bmatrix} x_1 \\ x_2 \end{bmatrix} + [0]u$$

State Space Model of a Mechanical System

Illustration 23

Consider a mechanical system as shown in Figure 1.9.

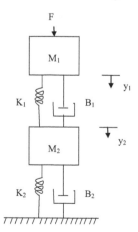

FIGURE 1.9
Block diagram of mechanical system.

F – Applied force in Newton (N)

M_1, M_2 – Mass in kg

K_1, K_2 – Stiffness of the spring in N/m

y_1, y_2 – Displacement in m

B_1, B_2 – Viscous friction coefficient N/ (m/s)

Applying Newton's second law, the force balance equation at M_1 can be written as

$$F = M_1 \frac{d^2 y_1}{dt^2} + B_1 \frac{d(y_1 - y_2)}{dt} + K_1(y_1 - y_2) \tag{i}$$

Similarly, the force balance equation at M_2 can be written as

$$M_2 \frac{d^2 y_2}{dt^2} + B_2 \frac{dy_2}{dt} + K_2 y_2 + B_2 \frac{d(y_2 - y_1)}{dt} + K_2(y_2 - y_1) = 0 \tag{ii}$$

Let us choose state variables as x_1, x_2, x_3 and x_4

$$\text{Let } x_1 = y_1 \Rightarrow \dot{x}_1 = \frac{dy_1}{dt}$$

$$x_2 = y_2 \Rightarrow \dot{x}_2 = \frac{dy_2}{dt}$$

$$x_3 = \frac{dy_1}{dt} \Rightarrow \dot{x}_3 = \frac{d^2 y_1}{dt^2}$$

$$x_4 = \frac{dy_2}{dt} \Rightarrow \dot{x}_4 = \frac{d^2 y_2}{dt^2}$$

Also, $F = u$.

Substituting the above values in equations (i) and (ii), we get

$$u = M_1 \dot{x}_3 + B_1 x_3 - B_1 x_4 + K_1 x_1 - K_1 x_2$$

$$\dot{x}_3 = \frac{-K_1}{M_1} x_1 + \frac{K_1}{M_1} x_2 - \frac{B_1}{M_1} x_3 + \frac{B_1}{M_1} x_4 + \frac{u}{M_1} \tag{iii}$$

Similarly, from equation (ii), we have

$$M_2 \dot{x}_4 + B_2 x_4 - B_1 x_4 - B_1 x_3 + K_1 x_2 + K_2 x_2 - K_1 x_1 = 0$$

$$\dot{x}_4 = \frac{K_1}{M_2} x_1 - \frac{(K_1 + K_2)}{M_2} x_2 + \frac{B_1}{M_2} x_3 - \frac{(B_1 + B_2)}{M_2} x_4 + 0u \tag{iv}$$

We have

$$\dot{x}_1 = \frac{dy_1}{dt} = x_3 = 0x_1 + 0x_2 + 1x_3 + 0u \tag{v}$$

$$\dot{x}_2 = \frac{dy_2}{dt} = x_4 = 0x_1 + 0x_2 + 0x_3 + 1x_4 + 0u \tag{vi}$$

From equations (iii), (iv), (v) and (vi), we can write the state equation in matrix form as

$$
\begin{bmatrix} \dot{x}_1 \\ \dot{x}_2 \\ \dot{x}_3 \\ \dot{x}_4 \end{bmatrix} =
\begin{bmatrix}
0 & 0 & 1 & 0 \\
0 & 0 & 0 & 1 \\
-\dfrac{K_1}{M_1} & \dfrac{K_1}{M_1} & -\dfrac{B_1}{M_1} & \dfrac{B_1}{M_1} \\
\dfrac{K_1}{M_2} & -\dfrac{(K_1+K_2)}{M_2} & \dfrac{B_1}{M_2} & -\dfrac{(B_1+B_2)}{M_2}
\end{bmatrix}
\begin{bmatrix} x_1 \\ x_2 \\ x_3 \\ x_4 \end{bmatrix} +
\begin{bmatrix} 0 \\ 0 \\ \dfrac{1}{M_1} \\ 0 \end{bmatrix} u
$$

The output equation can be obtained as follows:
The displacements y_1 and y_2 are the outputs of the given mechanical system. We have chosen

$$y_1 = x_1 = 1x_1 + 0x_2 + 0x_3 + 0x_4 + 0u$$

$$y_2 = x_2 = 0x_1 + 1x_2 + 0x_3 + 0x_4 + 0u$$

∴The output equation can be written as

$$
\begin{bmatrix} y_1 \\ y_2 \end{bmatrix} =
\begin{bmatrix} 1 & 0 & 0 & 0 \\ 0 & 1 & 0 & 0 \end{bmatrix}
\begin{bmatrix} x_1 \\ x_2 \\ x_3 \\ x_4 \end{bmatrix} + [0]u
$$

State Space Model of an Electro-Mechanical System

Illustration 24

Let us consider a separately excited DC motor operated in armature control mode as shown in Figure 1.10.

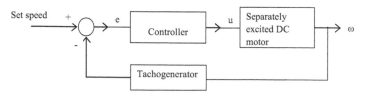

FIGURE 1.10
Speed control of separately excited DC motor (armature control mode).

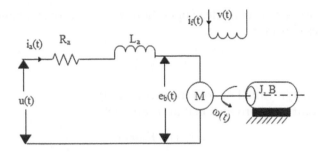

FIGURE 1.11
Simplified model of separately excited DC motor.

$e(t)$ – Error

$u(t)$ – Voltage input to motor in volts

$\omega(t)$ – Angular velocity of the motor shaft in radians/s

The simplified model of separately excited DC motor is shown in Figure 1.11.

$u(t)$ – Voltage input to motor in volts

$v(t)$ – Voltage applied across field coil in volts

R_a – Armature resistance in ohms

L_a – Armature inductance in henrys

$e_b(t)$ – Back emf in volts

$i_a(t)$ – Armature current in amperes

$i_f(t)$ – Field current in amperes

$\omega(t)$ – Angular velocity of motor shaft in radians/s

J – Moment of inertia of motor shaft and load in kg m²

B – Viscous friction coefficient in (N m)/(rad/s)

T_M – Torque developed by the motor in N m

In a DC motor, the air gap flux (Φ) in proportional to field current i_f

$$\therefore \Phi = K_f \, i_f$$

where K_f is a constant.

Also, in a DC motor, the motor torque (T_M) is proportional to air gap flux (Φ) and armature current i_a.

$$\therefore T_M = K_f i_f \, K_a i_a$$

$$T_M = K_f K_a i_f i_a$$

For the armature control mode of operation, 'i_f' is held constant and hence 'Φ' also becomes constant.

$$T_M = K_T i_a \tag{i}$$

where $K_T = K_f K_a i_f$ (called motor torque constant).
  The back emf 'e_b' is proportional to 'ω' and 'Φ'

$$\therefore e_b(t) = K_b \omega(t) \tag{ii}$$

where 'K_b' is called back emf constant.
  Applying Kirchhoff's voltage law to the armature circuit, we get

$$u(t) = R_a i_a(t) + L_a \frac{d i_a(t)}{dt} + e_b(t) = \frac{d i_a(t)}{dt} = -\frac{R_a}{L_a} i_a(t) - \frac{e_b(t)}{L_a} + \frac{u(t)}{L_a}$$

$$\frac{d i_a(t)}{dt} = -\frac{R_a}{L_a} i_a(t) - \frac{K_b \omega(t)}{L_a} + \frac{u(t)}{L_a} \quad \left[\because e_b(t) = K_b \omega(t) \right] \tag{iii}$$

The torque balance equation is given by

$$T_M(t) = J \frac{d\omega(t)}{dt} + B\omega(t) = \frac{d\omega(t)}{dt} = \frac{T_W(t)}{J} - \frac{B}{J}\omega(t) \tag{iv}$$

$$\frac{d\omega(t)}{dt} = \frac{K_T i_a(t)}{J} - \frac{B}{J}\omega(t) \quad \because \left[T_W(t) = K_T i_a(t) \right]$$

Let

$$x_1(t) = \omega(t) \Rightarrow \dot{x}_1(t) = \frac{d\omega(t)}{dt}$$

$$x_2(t) = i_a(t) \Rightarrow \dot{x}_2(t) = \frac{d i_a(t)}{dt}$$

Now, equations (iii) and (iv) can be rewritten as follows:

$$\dot{x}_1(t) = \frac{K_T}{J} x_2(t) - \frac{B}{J} x_1(t) + 0u(t)$$

$$\dot{x}_2(t) = -\frac{R_a}{L_a} x_2(t) - \frac{K_b}{L_a} x_1(t) + \frac{1}{L_a} u(t) \tag{v}$$

$$\begin{bmatrix} \dot{x}_1(t) \\ \dot{x}_2(t) \end{bmatrix} = \begin{bmatrix} -\dfrac{B}{J} & \dfrac{K_T}{J} \\ -\dfrac{K_a}{L_a} & -\dfrac{R_a}{L_a} \end{bmatrix} \begin{bmatrix} x_1 \\ x_2 \end{bmatrix} + \begin{bmatrix} 0 \\ \dfrac{1}{L_a} \end{bmatrix} u(t)$$

We have

$$y(t) = \omega(t)$$

$$y(t) = x_1(t) = 1x_1(t) + 0x_2(t) + 0u(t) \tag{vi}$$

$$y(t) = \begin{bmatrix} 1 & 0 \end{bmatrix} \begin{bmatrix} x_1(t) \\ x_2(t) \end{bmatrix} + [0]u(t)$$

Equations (v) and (vi) describe the state model of a separately excited DC motor operating in the armature control mode.

State Space Model of a Liquid-Level System

Illustration 25

Consider a two-tank non-interacting liquid-level system with level in tank-2 is of interest for output, as shown in Figure 1.12.

For tank-1, the mass balance equation can be written as follows:

$$u_1 - q_1 = A_1 \frac{dh_1}{dt}$$

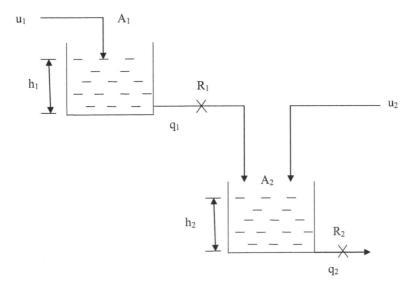

FIGURE 1.12
Two-tank non-interacting liquid-level system.

where
$$q_1 = \frac{h_1}{R_1}$$

$$\therefore u_1 - \frac{h_1}{R_1} = A_1 \frac{dh_1}{dt}$$

$$\frac{dh_1}{dt} = \frac{-h_1}{A_1 R_1} + \frac{u_1}{A_1} \qquad (i)$$

$$\frac{dh_1}{dt} = \frac{-h_1}{A_1 R_1} + 0 h_2 + \frac{u_1}{A_1} + 0 u_2$$

Similarly, for tank-2, the mass balance equation can be written as follows:
For tank-1, the mass balance equation can be written as follows:

$$q_1 + u_2 - q_2 = A_2 \frac{dh_2}{dt}$$

where

$$q_1 = \frac{h_1}{R_1} \text{ and } q_2 = \frac{h_2}{R_2}$$

$$\frac{h_1}{R_1} + u_2 - \frac{h_2}{R_2} = A_2 \frac{dh_2}{dt} \qquad (ii)$$

$$\frac{dh_2}{dt} = \frac{h_1}{A_2 R_1} - \frac{h_2}{A_2 R_2} + 0 u_1 + \frac{u_2}{A_2}$$

From equations (1.14) and (1.15), we can write the state equation as follows:

$$\begin{bmatrix} \dot{h}_1 \\ \dot{h}_2 \end{bmatrix} = \begin{bmatrix} -\dfrac{1}{R_1 A_1} & 0 \\ \dfrac{1}{R_2 A_2} & -\dfrac{1}{R_2 A_2} \end{bmatrix} \begin{bmatrix} h_1 \\ h_2 \end{bmatrix} + \begin{bmatrix} \dfrac{1}{A_1} & 0 \\ 0 & \dfrac{1}{A_2} \end{bmatrix} \begin{bmatrix} u_1 \\ u_2 \end{bmatrix} \qquad (iii)$$

Given that the level 'h_2' alone is of interest. Hence, the output equation can be written as follows:

$$Y = \begin{bmatrix} 0 & 1 \end{bmatrix} \begin{bmatrix} h_1 \\ h_2 \end{bmatrix} \qquad (iv)$$

Equations (iii) and (iv) represent the state model of a two-tank non-interacting system (with level 'h_2' alone being measured).

Non-Uniqueness of State Model

The set of state variables is not unique for the given system. In other words, in the state model given by equations (i) and (ii), the matrices A, B, C and D and the state variables 'x' are not unique for the inter-relationship between U and Y.

$$\dot{X} = AX + BU \qquad \text{(i)}$$

$$Y = CX + DU \qquad \text{(ii)}$$

We have

$$\dot{X} = AX + BU \qquad \text{(iii)}$$

Let $X = MZ$

$$\Rightarrow \dot{X} = M\dot{Z}$$

Substituting $\dot{X} = M\dot{Z}$ in equation (iii), we get

$$M\dot{Z} = AMZ + BU$$
$$\dot{Z} = M^{-1}AMZ + M^{-1}BU \qquad \text{(iv)}$$

Let $M^{-1}AMZ = P$ and $M^{-1}B = Q$.
  Thus, equation (iv) can be rewritten as follows:

$$\dot{Z} = PZ + QU \qquad \text{(v)}$$

Similarly, we have

$$Y = CX + DU \qquad \text{(vi)}$$

Substituting $X = MZ$ in equation (vi), we get

$$Y = CMZ + DU \qquad \text{(vii)}$$

Let $CM = R$.
  Thus, equation (v) can be rewritten as follows:

$$Y = RZ + DU \qquad \text{(viii)}$$

Thus, the new state model is given by

$$\dot{Z} = PZ + QU \qquad \text{(ix)}$$

$$Y = RZ + DU \qquad \text{(x)}$$

Hence, from equations (i) and (ii) and equations (ix) and (x), it is clear that the state model is not unique. The non-uniqueness of state model can be illustrated with the help of a numerical example.

Illustration 26

Develop a state space model for a linear system whose transfer function is given by

$$\frac{Y(s)}{U(s)} = \frac{5}{(s^2 + 4s + 3)}$$

Case 1: State model through block diagram approach:

Solution

Given transfer function is

$$\frac{Y(s)}{U(s)} = \frac{5}{(s^2 + 4s + 3)} = \frac{5}{(s+1)(s+3)}$$

Applying partial fraction and simplifying it, we get

$$\frac{Y(s)}{U(s)} = \frac{\frac{5}{2}}{(s+1)} - \frac{\frac{5}{2}}{(s+3)}$$

The block diagram of $\dfrac{\frac{5}{2}}{(s+1)}$ can be drawn as shown in Figure 1.13a.

The block diagram of $\dfrac{\frac{5}{2}}{(s+3)}$ can be drawn as shown in Figure 1.13b.

Thus, the overall block diagram of combined transfer function can be drawn as shown in Figure 1.13c.
From Figure (1.13c), we have

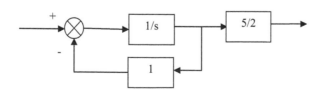

FIGURE 1.13A

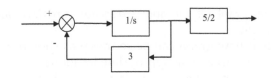

FIGURE 1.13B

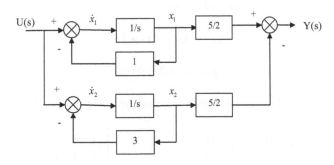

FIGURE 1.13C

$$\dot{x}_1 = u - x_1 = -1x_1 + 0x_2 + 1u$$

$$\dot{x}_2 = u - 3x_2 = 0x_1 - 3x_2 + 1u$$

$$y = \frac{5}{2}x_1 - \frac{5}{2}x_2 + 0u$$

$$\begin{bmatrix} \dot{x}_1 \\ \dot{x}_2 \end{bmatrix} = \begin{bmatrix} -1 & 0 \\ 0 & -3 \end{bmatrix} \begin{bmatrix} x_1 \\ x_2 \end{bmatrix} + \begin{bmatrix} 1 \\ 1 \end{bmatrix} u$$

$$Y = \begin{bmatrix} \frac{5}{2} & -\frac{5}{2} \end{bmatrix} \begin{bmatrix} x_1 \\ x_2 \end{bmatrix} + [0]u$$

Now, let us derive the state model of the same system using signal flow graph approach.

Case 2: State model through signal flow graph approach:

Solution

Given transfer function is

$$\frac{Y(s)}{U(s)} = \frac{5}{(s^2 + 4s + 3)}$$

The above transfer function can be rewritten as follows:

Thus, we have $\dfrac{Y(s)}{U(s)} = \dfrac{\frac{5}{s^2}}{1 + \frac{4}{s} + \frac{3}{s^2}}$

The signal flow graph of the above transfer function can be drawn as shown in Figure 1.14.

The above signal flow graph can be redrawn as shown in Figure 1.15 by identifying the state variables at appropriate locations.

From the above signal flow graph, it is clear that there are two state variables.

Thus, we have $\dot{x}_1 = x_2 = 0x_1 + 1x_2 + 0u$

$$\dot{x}_2 = -3x_1 - 4x_2 + 1u$$

$$y = 5x_1 + 0x_2 + 0u$$

Thus, the state model can be represented as follows:

$$\begin{bmatrix} \dot{x}_1 \\ \dot{x}_2 \end{bmatrix} = \begin{bmatrix} 0 & 1 \\ -3 & -4 \end{bmatrix} \begin{bmatrix} x_1 \\ x_2 \end{bmatrix} + \begin{bmatrix} 1 \\ 1 \end{bmatrix} u$$

$$Y = \begin{bmatrix} 5 & 0 \end{bmatrix} \begin{bmatrix} x_1 \\ x_2 \end{bmatrix} + [0] usssw$$

Thus, from equations (xi) and (xii) and equations (xiii) and (xiv), it is clear that the state model for the given system is not unique.

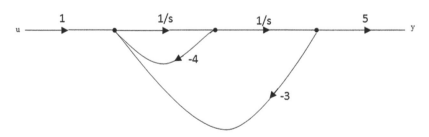

FIGURE 1.14

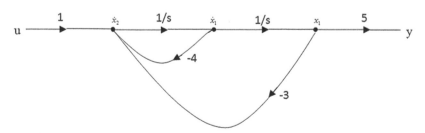

FIGURE 1.15

2

State Space Approach

This chapter on 'State Space Approach' presents the details of topics namely: Role of eigen values and eigen vectors, free response or unforced response or homogeneous state equations, state transition matrix and its properties, forced response or non-homogeneous state equation, pictorial representation of state model using state diagram, minimal realization, non-minimal realization and balanced realization. The eigen values and eigen vectors determine the dynamic response of the given system. The eigen values also turn out to be the poles of the system transfer function. The procedures for obtaining the homogeneous and non-homogeneous state equations are illustrated with numerical examples. Manipulating the given state models into minimal and non-minimal realization forms is also presented. Balanced realization which plays a major role in model reduction is also illustrated with a numerical example.

Role of Eigen Values and Eigen Vectors

Eigen values: The eigen values of a $n \times n$ matrix A are the roots of the characteristic equation given below:

$$| \lambda I - A | = 0$$

Here, 'A' is any square matrix of order 'n'. 'I' is the nth order unit matrix. The eigen values are also called 'characteristic roots'.

NOTE

1. When 'A' represents the dynamic matrix of a linear system, the eigen values determine the dynamic response of the system.
2. The eigen values also turn out to be poles of the corresponding transfer function.
3. Eigen values of a matrix A are invariant under equivalent transformation, for any non-singular matrix P, i.e., $| \lambda I - A | = | \lambda I - P^{-1} A P |$.

Properties of eigen values

i. The sum of the eigen values of a matrix is the sum of the elements of the principal diagonal.

ii. The product of eigen values of a matrix is equal to its determinant.

iii. If 'λ' is an eigen value of matrix A, then $1/\lambda$ is the eigen value of A^{-1}.

iv. If 'λ' is an eigen value of an orthogonal matrix, then $1/\lambda$ is also its eigen value.

v. The eigen values of a lower or upper triangular matrix are the elements on its principal diagonal.

Illustration 1

Find the eigen values of a 3×3 matrix given by $A = \begin{bmatrix} -7 & -4 & -8 \\ 2 & -1 & 2 \\ 5 & 4 & 6 \end{bmatrix}$.

Solution

Given

$$A = \begin{bmatrix} -7 & -4 & -8 \\ 2 & -1 & 2 \\ 5 & 4 & 6 \end{bmatrix}$$

$$[\lambda I - A] = \begin{bmatrix} \lambda & 0 & 0 \\ 0 & \lambda & 0 \\ 0 & 0 & \lambda \end{bmatrix} - \begin{bmatrix} -7 & -4 & -8 \\ 2 & -1 & 2 \\ 5 & 4 & 6 \end{bmatrix}$$

$$= \begin{bmatrix} \lambda+7 & 4 & 8 \\ -2 & \lambda+1 & -2 \\ -5 & -4 & \lambda-6 \end{bmatrix}$$

$$= (\lambda+7)[(\lambda+1)(\lambda-6)-8] - 4[-2(\lambda-6)-10] + 8[8-(-5(\lambda+1))]$$

$$= \lambda^3 - 5\lambda^2 - 14\lambda + 7\lambda^2 - 35\lambda - 98 + 8\lambda - 8 + 40\lambda + 104$$

$$= \lambda^3 + 2\lambda^2 - \lambda - 2$$

The eigen values are $\lambda = -1, 1, -2$

Eigen vectors: The eigen vector of a square matrix A is a non-zero column vector X, provided there exists a scalar λ (eigen value) such that $AX = \lambda X$.

How to Find Eigen Vectors?

Case 1: (Distinct eigen values)
If the eigen values of A are all distinct, then the eigen vector corresponding to λ_i may be obtained by taking cofactors of matrix $(\lambda_i I - A)$ along any row.
Let 'm_i' be the eigen vector corresponding to λ_i, then m_i is given as follows:

$$m_i = \begin{bmatrix} C_{k1} \\ C_{k2} \\ \vdots \\ C_{kn} \end{bmatrix} \quad \text{where, } k = 1,2...n.$$

$C_{k1}, C_{k2}... C_{kn}$ are cofactors of matrix $(\lambda_i I - A)$ along kth row.

Case 2a: (Repeated eigen values)
If the matrix has repeated eigen values with multiplicity 'q', same procedure is followed for the repeated eigen value. For the remaining $(q-1)$ eigen vectors, the following procedure is used.
Let 'm_i' be the eigen vector corresponding to λ_i, then m_i is given as follows:

$$m_i = \begin{bmatrix} \dfrac{1}{\lfloor P} \dfrac{d^P}{d\lambda_i^P} C_{k1} \\[2ex] \dfrac{1}{\lfloor P} \dfrac{d^P}{d\lambda_i^P} C_{k2} \\[2ex] \vdots \\[2ex] \dfrac{1}{\lfloor P} \dfrac{d^P}{d\lambda_i^P} C_{kn} \end{bmatrix} \quad \text{where, } p = 1,2...(q-1)$$

Case 2b: (Repeated eigen values, when cofactors tend to zero)
If the matrix has repeated eigen values with multiplicity 'q', same procedure is followed for the repeated eigen value. If all the cofactors tend to zero, follow the procedure given below:

i. Reduce the given matrix using simple row or column transformation.
ii. After reduction, introduce independent variables and assign every independent variable to each state variable.
iii. Solve for each state variable. Obtain the eigen vectors.

Illustration 2: (Distinct eigen values)

Find the eigen vectors of the matrix $A = \begin{bmatrix} -7 & -4 & -8 \\ 2 & -1 & 2 \\ 5 & 4 & 6 \end{bmatrix}$

Solution

Given

$$A = \begin{bmatrix} -7 & -4 & -8 \\ 2 & -1 & 2 \\ 5 & 4 & 6 \end{bmatrix}$$

From the results obtained in Illustration 1, the eigen values are $\lambda = -1, 1, -2$ which are distinct in nature.

Now, to find eigen vector corresponding to $\lambda_1 = -1$

$$[\lambda_1 I - A] = \begin{bmatrix} -1 & 0 & 0 \\ 0 & -1 & 0 \\ 0 & 0 & -1 \end{bmatrix} - \begin{bmatrix} -7 & -4 & -8 \\ 2 & -1 & 2 \\ 5 & 4 & 6 \end{bmatrix} = \begin{bmatrix} 6 & 4 & 8 \\ -2 & 0 & -2 \\ -5 & -4 & -7 \end{bmatrix}$$

Cofactor of $[\lambda_1 I - A]$ along its first row can be found as follows:

$$C_{11} = (-1)^{1+1} \begin{vmatrix} 0 & -2 \\ -4 & -7 \end{vmatrix} = -8$$

$$C_{12} = (-1)^{1+2} \begin{vmatrix} -2 & -2 \\ -5 & -7 \end{vmatrix} = -(14 - 10) = -4$$

$$C_{13} = (-1)^{1+3} \begin{vmatrix} -2 & 0 \\ -5 & -4 \end{vmatrix} = 8 - 0 = 8$$

Thus, the eigen vector corresponding to $\lambda_1 = -1$ can be obtained by arranging C_{11}, C_{12} and C_{13} in the form of column matrix C_1:

$$\therefore C_1 = \begin{bmatrix} C_{11} \\ C_{12} \\ C_{13} \end{bmatrix} = \begin{bmatrix} -8 \\ -4 \\ 8 \end{bmatrix}$$

Now, to find eigen vector corresponding to $\lambda_2 = 1$

$$[\lambda_2 I - A] = \begin{bmatrix} 1 & 0 & 0 \\ 0 & 1 & 0 \\ 0 & 0 & 1 \end{bmatrix} - \begin{bmatrix} -7 & -4 & -8 \\ 2 & -1 & 2 \\ 5 & 4 & 6 \end{bmatrix} = \begin{bmatrix} 8 & 4 & 8 \\ -2 & 2 & -2 \\ -5 & -4 & -5 \end{bmatrix}$$

Cofactor of $[\lambda_2 I - A]$ along its first row can be found as follows:

$$C_{11} = (-1)^{1+1} \begin{vmatrix} 2 & -2 \\ -4 & -5 \end{vmatrix} = -10 - 8 = -18$$

$$C_{12} = (-1)^{1+2} \begin{vmatrix} -2 & -2 \\ -5 & -5 \end{vmatrix} = 10 - 10 = 0$$

$$C_{13} = (-1)^{1+3} \begin{vmatrix} -2 & 2 \\ -5 & -4 \end{vmatrix} = 8 + 10 = 18$$

Thus, the eigen vector corresponding to $\lambda_2 = 1$ can be obtained by arranging C_{11}, C_{12} and C_{13} in the form of column matrix C_2:

$$\therefore C_2 = \begin{bmatrix} -18 \\ 0 \\ 18 \end{bmatrix}$$

Now, to find eigen vectors corresponding to $\lambda_3 = -2$

$$[\lambda_3 I - A] = \begin{bmatrix} -2 & 0 & 0 \\ 0 & -2 & 0 \\ 0 & 0 & -2 \end{bmatrix} - \begin{bmatrix} -7 & -4 & -8 \\ 2 & -1 & 2 \\ 5 & 4 & 6 \end{bmatrix} = \begin{bmatrix} 5 & 4 & 8 \\ -2 & -1 & -2 \\ -5 & -4 & -8 \end{bmatrix}$$

Cofactor of $[\lambda_3 I - A]$ along its first row can be found as follows

$$C_{11} = (-1)^{1+1} \begin{vmatrix} -1 & -2 \\ -4 & -8 \end{vmatrix} = 8 - 8 = 0$$

$$C_{12} = (-1)^{1+2} \begin{vmatrix} -2 & -2 \\ -5 & -8 \end{vmatrix} = -(16 - 10) = -6$$

$$C_{13} = (-1)^{1+3} \begin{vmatrix} -2 & -1 \\ -5 & -4 \end{vmatrix} = 8 - 5 = 3$$

Thus, the eigen vector corresponding to $\lambda_3 = -2$ can be obtained by arranging C_{11}, C_{12} and C_{13} in the form of column matrix C_3

$$\therefore C_3 = \begin{bmatrix} 0 \\ -6 \\ 3 \end{bmatrix}$$

∴ Eigen values = –1, 1, –2

$$\text{and} \quad \text{Eigen vectors} = \begin{bmatrix} -8 \\ -4 \\ 8 \end{bmatrix}, \begin{bmatrix} -18 \\ 0 \\ 18 \end{bmatrix}, \begin{bmatrix} 0 \\ -6 \\ 3 \end{bmatrix}$$

Illustration 3

Case 2: (Repeated eigen values)

Find the eigen values and eigen vectors of the matrix $A = \begin{bmatrix} 2 & 0 & 0 \\ -1 & -1 & 9 \\ 0 & -1 & 5 \end{bmatrix}$.

Solution

Given

$$A = \begin{bmatrix} 2 & 0 & 0 \\ -1 & -1 & 9 \\ 0 & -1 & 5 \end{bmatrix}$$

$$[\lambda I - A] = \begin{bmatrix} \lambda & 0 & 0 \\ 0 & \lambda & 0 \\ 0 & 0 & \lambda \end{bmatrix} - \begin{bmatrix} 2 & 0 & 0 \\ -1 & -1 & 9 \\ 0 & -1 & 5 \end{bmatrix} = \begin{bmatrix} \lambda-2 & 0 & 0 \\ 1 & \lambda+1 & -9 \\ 0 & 1 & \lambda-5 \end{bmatrix}$$

$|\lambda I - A| = (\lambda - 2)[(\lambda + 1)(\lambda - 5) + 9]$

$\qquad = (\lambda - 5)\,[\lambda^2 - 5\lambda + 1\lambda - 5 + 9]$

$\qquad = \lambda^3 - 6\lambda^2 + 12\lambda - 8$

$\lambda = 2, 2, 2.$

Now, to find the eigen vector corresponding to the eigen value $\lambda_1 = 2$, the following steps are followed.

$$[\lambda_1 I - A] = [\lambda_1] \begin{bmatrix} 1 & 0 & 0 \\ 0 & 1 & 0 \\ 0 & 0 & 1 \end{bmatrix} - \begin{bmatrix} 2 & 0 & 0 \\ -1 & -1 & 9 \\ 0 & -1 & 5 \end{bmatrix}$$

$$= \begin{bmatrix} \lambda_1 - 2 & 0 & 0 \\ 1 & \lambda_1 + 1 & -9 \\ 0 & 1 & \lambda_1 - 5 \end{bmatrix}$$

Cofactor of $[\lambda_1 I - A]$ along its first row can be found as follows:

$$C_{11} = (-1)^{1+1} \begin{vmatrix} \lambda_1 + 1 & -9 \\ 1 & \lambda_1 - 5 \end{vmatrix}$$

$$= (\lambda_1 + 1)(\lambda_1 - 5) = \lambda_1^2 - 5\lambda_1 + 1\lambda_1 - 5 + 9 = \lambda_1^2 - 4\lambda_1 + 4$$

$$C_{12} = (-1)^{1+2} \begin{vmatrix} 1 & -9 \\ 0 & \lambda - 5 \end{vmatrix} = \lambda - 5$$

$$C_{13} = (-1)^{1+3} \begin{vmatrix} 1 & \lambda_1 + 1 \\ 0 & 1 \end{vmatrix} = 1$$

On substituting $\lambda_1 = 2$

$$\therefore C_1 = \begin{bmatrix} 0 \\ -3 \\ 1 \end{bmatrix}$$

Since, $\lambda_1 = \lambda_2 = \lambda_3$, the matrix C_2 can be written as follows:

$$C_2 = \begin{bmatrix} \dfrac{1}{\lfloor P} \dfrac{d^P}{d\lambda_i^P} C_{k1} \\ \dfrac{1}{\lfloor P} \dfrac{d^P}{d\lambda_i^P} C_{k2} \\ \vdots \\ \dfrac{1}{\lfloor P} \dfrac{d^P}{d\lambda_i^P} C_{kn} \end{bmatrix}$$

$$C_{11} = \lambda_2^2 - 4\lambda_2 + 4$$

$$C_{12} = \lambda_2 - 5$$

$$C_{13} = 1$$

$$\frac{d}{d\lambda_2} C_{11} = \frac{d}{d\lambda_2}(\lambda_2^2 - 4\lambda_2 + 4) = 2\lambda_2 - 4$$

$$\frac{d}{d\lambda_2} C_{12} = \frac{d}{d\lambda_2}(\lambda_2 - 5) = 1$$

$$\frac{d}{d\lambda_2} C_{13} = \frac{d}{d\lambda_2}(1) = 0$$

$$\therefore C_2 = \begin{bmatrix} 0 \\ 1 \\ 0 \end{bmatrix}$$

$$C_3 = \begin{bmatrix} \dfrac{1}{\lfloor P} \dfrac{d^P}{d\lambda_i^P} C_{k1} \\[2ex] \dfrac{1}{\lfloor P} \dfrac{d^P}{d\lambda_i^P} C_{k2} \\[1ex] \vdots \\[1ex] \dfrac{1}{\lfloor P} \dfrac{d^P}{d\lambda_i^P} C_{kn} \end{bmatrix}$$

$$C_{11} = 2\lambda_2 - 4$$

$$C_{12} = 1$$

$$C_{13} = 0$$

$$\frac{d}{d\lambda_2} C_{11} = \frac{d}{d\lambda_2} (2\lambda_2 - 4) = 2$$

$$\frac{d}{d\lambda_2} C_{12} = \frac{d}{d\lambda_2} (1) = 0$$

$$\frac{d}{d\lambda_2} C_{13} = \frac{d}{d\lambda_2} (0) = 0$$

$$\therefore C_3 = \begin{bmatrix} 2 \\ 0 \\ 0 \end{bmatrix}$$

$\therefore$ Eigen values = 2, 2, 2

$$\text{and eigen vectors} = \begin{bmatrix} 0 \\ -3 \\ 1 \end{bmatrix}, \begin{bmatrix} 0 \\ 1 \\ 0 \end{bmatrix}, \begin{bmatrix} 2 \\ 0 \\ 0 \end{bmatrix}.$$

Case 3: (Repeated eigen values when cofactors tend to zero)

Find the eigen values and eigen vectors of the matrix $A = \begin{bmatrix} 7 & -2 & -14 \\ 5 & -4 & -7 \\ 5 & -1 & -10 \end{bmatrix}$.

Solution

Given

$$A = \begin{bmatrix} 7 & -2 & -14 \\ 5 & -4 & -7 \\ 5 & -1 & -10 \end{bmatrix}$$

$$[\lambda I - A] = \begin{bmatrix} \lambda & 0 & 0 \\ 0 & \lambda & 0 \\ 0 & 0 & \lambda \end{bmatrix} - \begin{bmatrix} 7 & -2 & -14 \\ 5 & -4 & -7 \\ 5 & -1 & -10 \end{bmatrix} = \begin{bmatrix} \lambda-7 & 2 & 14 \\ -5 & \lambda+4 & 7 \\ -5 & 1 & \lambda+10 \end{bmatrix}$$

$$= (\lambda-7)[(\lambda+4)(\lambda+10)-7] - 2[-5(\lambda+10)-(-5)(7)] - 14[-5-(-5(\lambda+4))]$$

$$= (\lambda-7)[(\lambda+4)(\lambda+10)-7] - 2[-5\lambda-50+35] + [-5+5\lambda+20]$$

$$= \lambda3 + 7\lambda2 + 15\lambda + 9$$

$$= (\lambda+1)(\lambda+3)(\lambda+3)$$

Therefore, the eigen values are –1, –3, –3

Now, to find the eigen vector corresponding to the eigen value $\lambda_1 = -1$

$$[\lambda_1 I - A] = \begin{bmatrix} -1 & 0 & 0 \\ 0 & -1 & 0 \\ 0 & 0 & -1 \end{bmatrix} - \begin{bmatrix} 7 & -2 & -14 \\ 5 & -4 & -7 \\ 5 & -1 & -10 \end{bmatrix} = \begin{bmatrix} -8 & 2 & 14 \\ -5 & 3 & 7 \\ -5 & 1 & 9 \end{bmatrix}$$

Cofactor of $[\lambda_1 I{-}A]$ along its first row can be found as follows:

$$C_{11} = (-1)^{1+1} \begin{vmatrix} 3 & 7 \\ 1 & 9 \end{vmatrix} = 27 - 7 = 20$$

$$C_{12} = (-1)^{1+2} \begin{vmatrix} -5 & 7 \\ -5 & 9 \end{vmatrix} = -(-45+35) = 10$$

$$C_{13} = (-1)^{1+3} \begin{vmatrix} -5 & 3 \\ -5 & 1 \end{vmatrix} = -5+15 = 10$$

Thus, the eigen vector corresponding to $\lambda_1 = -1$ can be obtained by arranging C_{11}, C_{12} and C_{13} in the form of column matrix C_1.

$$\therefore C_1 = \begin{bmatrix} 20 \\ 10 \\ 10 \end{bmatrix} = \begin{bmatrix} 2 \\ 1 \\ 1 \end{bmatrix}$$

Now to find eigen vectors for the repeating eigen values $\lambda_2 = \lambda_3 = -3$,

$$[\lambda_2 I - A] = \begin{bmatrix} \lambda & 0 & 0 \\ 0 & \lambda & 0 \\ 0 & 0 & \lambda \end{bmatrix} - \begin{bmatrix} 7 & -2 & -14 \\ 5 & -4 & -7 \\ 5 & -1 & -10 \end{bmatrix} = \begin{bmatrix} \lambda-7 & 2 & 14 \\ -5 & \lambda+4 & 7 \\ -5 & 1 & \lambda+10 \end{bmatrix}$$

Cofactor of $[\lambda_2 I - A]$ along its first row can be found as follows:

$$C_{21} = (-1)^{1+1} \begin{vmatrix} \lambda+4 & 7 \\ 1 & \lambda+10 \end{vmatrix} = \lambda^2 + 14\lambda + 33$$

$$C_{22} = (-1)^{1+2} \begin{vmatrix} -5 & 7 \\ -5 & \lambda+10 \end{vmatrix} = -5\lambda - 15$$

$$C_{23} = (-1)^{1+3} \begin{vmatrix} -5 & \lambda+4 \\ -5 & 1 \end{vmatrix} = 5\lambda + 15$$

On substituting the $\lambda_2 = -3$ in the above equations, the cofactors tend to zero. Hence, a different approach is followed, to find the eigen vectors.
   We have

$$[\lambda_2 I - A] = \begin{bmatrix} -3 & 0 & 0 \\ 0 & -3 & 0 \\ 0 & 0 & -3 \end{bmatrix} - \begin{bmatrix} 7 & -2 & -14 \\ 5 & -4 & -7 \\ 5 & -1 & -10 \end{bmatrix} = \begin{bmatrix} -10 & 2 & 14 \\ -5 & 1 & 7 \\ -5 & 1 & 7 \end{bmatrix}$$

This matrix can be reduced to $\begin{bmatrix} 1 & -0.2 & -1.4 \\ 0 & 0 & 0 \\ 0 & 0 & 0 \end{bmatrix}$ using row reduction

techniques.
   Now, to find the eigen vectors of repeated roots,

$$\begin{bmatrix} 1 & -0.2 & -1.4 \\ 0 & 0 & 0 \\ 0 & 0 & 0 \end{bmatrix} \begin{bmatrix} v_1 \\ v_2 \\ v_3 \end{bmatrix} = \begin{bmatrix} 0 \\ 0 \\ 0 \end{bmatrix}$$

We get the equation, $v_1 - 0.2v_2 - 1.4v_3 = 0$
   Set s and t as free variables to v_2 and v_3, respectively, and solve for v_1.

$$\begin{bmatrix} 0.2s + 1.4t \\ s \\ t \end{bmatrix} = s \begin{bmatrix} 0.2 \\ 1 \\ 0 \end{bmatrix} + t \begin{bmatrix} 1.4 \\ 0 \\ 1 \end{bmatrix}$$

The vector v_2 and v_3 are the eigen vectors of the repeated eigen values.
Eigen values $= -1, -3, -3$.

$$\text{Eigen vectors} = \begin{bmatrix} 2 \\ 1 \\ 1 \end{bmatrix}, \begin{bmatrix} 0.2 \\ 1 \\ 0 \end{bmatrix}, \begin{bmatrix} 1.4 \\ 0 \\ 1 \end{bmatrix}.$$

Free and Forced Responses

Free response or unforced response or homogeneous state equation
Consider the state equation given by equation (2.1)

$$\dot{X}(t) = AX(t) + Bu(t) \tag{2.1}$$

In the absence of input, equation (2.1) can be rewritten as follows:

$$\dot{X}(t) = AX(t) \tag{2.2}$$

Equation (2.2) is called free response or unforced or homogeneous state equation whose solution can be obtained as follows:
Let us assume a solution $X(t)$ of the form

$$x(t) = e^{At} K \tag{2.3}$$

where e^{At} is the matrix exponential function

$$e^{At} = I + At + \frac{1}{\lfloor 2} A^2 t^2 + \cdots + \frac{1}{\lfloor K} A^K t^K$$

$$= \sum_{i=0}^{\infty} \frac{1}{\lfloor i} A^i t^i \text{ and } K \text{ is a suitably chosen constant vector.}$$

To verify whether the assumed solution is the true solution, differentiate equation (2.3)

$$\dot{X}(t) = \frac{d}{dt} \left[e^{At} K \right] \text{(By using matrix exponential property)}$$

$$\dot{X}(t) = \frac{d}{dt} \left[e^{At} \right] K = A e^{At} K$$

$$\dot{X}(t) = Ax(t)$$

In order to evaluate the constant vector 'K', substitute $t = t_0$ in equation (2.3)

$$x(t_0) = e^{At_0} K$$

$$K = \left[e^{At_0} \right]^{-1} x(t_0)$$

$$= \left[e^{-At_0} \right] x(t_0) \quad \text{(By using matrix exponential property)}$$

Substituting the value of 'K' in equation (2.3), we get

$$x(t) = e^{At} \cdot e^{-At_0} \cdot x(t_0)$$

$$= e^{A(t-t_0)} x(t_0)$$

If the initial time $t_0 = 0$, i.e., initial state X^0 is known at time $t = 0$, then

$$x(t) = e^{At} x(t_0) \tag{2.4}$$

where $e^{At} = I + At + \dfrac{1}{\lfloor 2} A^2 t^2 + \cdots + \dfrac{1}{\lfloor K} A^K t^K$

From equation (2.4), it is observed that the initial state $x(0) \underline{\Delta} x^0$ at $t = 0$ is driven to a state $x(t)$ at time 't'.

This transition in state is carried out by the matrix exponential e^{At}. Due to this, e^{At} is known as 'state transition matrix' and it is denoted by $\Phi(t)$.

Properties of State Transition Matrix

   i. $\dfrac{d}{dt} \varphi(t) = A.\varphi(t); \quad \varphi(0) = I$

   ii. $[\varphi(t_2 - t_1)][\varphi(t_1 - t_0)] = \varphi(t_2 - t_0)$ for any t_0, t_1 and t_2

   iii. $\varphi^{-1}(t) = \varphi(-t)$

   iv. $\varphi(t)$ is a non-singular matrix for all 't'.

Evaluation of State Transition Matrix

From equation (2.2), we have the unforced/homogeneous equation when $u(t) = 0$ as

$$\dot{x}(t) = A x(t)$$

Taking Laplace on both sides, we get

$$sX(s) - X^0 = AX(s)$$

where $X(s) = L[x(t)]$ and $X^0 = X(0)$

Solving for X(s), we get

$$sX(s) - AX(s) = X^0$$

$$X(s)[sI - A] = X^0$$

$$X(s) = [sI - A]^{-1} X^0$$

Taking inverse Laplace transform

$$x(t) = L^{-1}[sI - A]^{-1} x^0 \qquad (2.5)$$

This is the response of free or homogeneous or unforced response.
Comparing equations (2.2) and (2.5), we get

$$e^{At} = \varphi(t) = L^{-1}[sI - A]^{-1} \qquad (2.6)$$

This is called state transition matrix.
Forced response or non-homogeneous state equation
We have the state equation given by

$$\dot{x}(t) = Ax(t) + Bu(t) \qquad (2.7)$$

$$\dot{x}(t) - Ax(t) = Bu(t)$$

When u(t) is present, then it is called as forced system or non-homogeneous system.
Premultiplying the above equation by e^{-At}, then we get

$$e^{-At}[\dot{x}(t) - Ax(t)] = e^{-At} Bu(t) \qquad (2.8)$$

But, $\dfrac{d}{dt}[e^{-At} x(t)] = e^{-At} \dfrac{d}{dt}(x(t)) + \dfrac{d}{dt}(e^{-At})x(t)$

$$= e^{-At}\dot{x}(t) - e^{-At} Ax(t)$$

$$= e^{-At}[\dot{x}(t) - Ax(t)] \qquad (2.9)$$

From equations (2.8) and (2.9), we have

$$\frac{d}{dt}[e^{-At} x(t)] = e^{-At} Bu(t)$$

Integrating both sides, we get

$$e^{-At}x(t) - x(0) = \int_0^t e^{-At}Bu(t)\,dt$$

$$e^{-At}x(t) = \left[x(0) + \int_0^t e^{-At}Bu(t)\,dt \right]$$

Premultiplying both sides by e^{At}, we have

$$x(t) = e^{At}x(0) + \int_0^t e^{A(t-\tau)}Bu(\tau)\,d\tau$$

When initial conditions are known at $t = 0$

$$x(t) = e^{At}x(0) + \int_0^t e^{A(t-\tau)}Bu(\tau)\,d\tau$$

$$= \varphi(t)x(0) + \int_0^t \varphi(t-\tau)Bu(\tau)\,d\tau \quad \text{where } \varphi(t) = e^{At}$$

If the initial state is known at $t = t_0$, rather than $t = 0$, then

$$x(t) = e^{A(t-t_0)}x(t_0) + \int_{t_0}^t e^{A(t-\tau)}Bu(\tau)\,d\tau$$

$$x(t) = \varphi(t-t_0)x(t_0) + \int_{t_0}^t \varphi(t-\tau)Bu(\tau)\,d\tau \qquad (2.10)$$

where $\varphi(t) = e^{At}$

Equation (2.10) is the solution for equation (2.7) and is called 'state transition equation'.

Free response computation

The value of state transition matrix and the expression for free response can be obtained by three methods, namely,

   1. Laplace inverse approach
      We have $X(t) = AX(t)$, where solution is given by
      $X(t) = e^{At}x(0)$
      $X(t) = L^{-1}[(sI-A)^{-1}]\,x(0)$ $(\because e^{At} = L^{-1}[(sI-A)^{-1}])$

2. Similarity transformation approach

$e^{At} = Pe^{\wedge t}P^{-1}$ where $\wedge$ = diagonal matrix

3. Cayley Hamilton approach

$e^{At} = \beta_0 I + \beta_1 A + \beta_1 A^2$.

Laplace inverse approach

The Laplace inverse approach is a straightforward method.

For a given 'A' matrix, use the following procedural steps to calculate $x(t)$.

Step 1: Find the matrix $[sI-A]$

Step 2: Find the inverse of matrix $[sI-A]$

Step 3: Find the laplace inverse of matrix $[sI-A]^{-1}$

Step 4: The free response can be obtained by using the formula

$$x(t) = L^{-1}[sI - A]^{-1}x(0)$$

Similarity transformation approach

For a given 'A' matrix, use the following procedural steps to calculate $x(t)$.

Step 1: Find the eigen values

Step 2: Find the cofactor matrix of the matrix $[\lambda I-A]$, and hence the eigen vectors

Step 3: Find the modal matrix $P = \begin{bmatrix} C_{11} & C_{21} \\ C_{12} & C_{22} \end{bmatrix}$

Step 4: Find P^{-1}

Step 5: Find $e^{At} = [P][e^{\lambda t}][P^{-1}]$

Step 6: Find $x(t) = [e^{At}]x(0)$.

Cayley Hamilton approach

For a given 'A' matrix, use the following procedural steps to calculate $x(t)$ using Cayley Hamilton approach.

Step 1: Find $[\lambda I]$, $[\lambda I-A]$ $|\lambda I-A|$ and hence, its eigen values.

Step 2: Find $\begin{aligned} g(\lambda_1) &= \beta_0 + \beta_1 \lambda_1 \\ g(\lambda_2) &= \beta_0 + \beta_1 \lambda_2 \end{aligned}$

Step 3: Find $\begin{aligned} f(\lambda_1) &= e^{\lambda_1 t} \\ f(\lambda_2) &= e^{\lambda_2 t} \end{aligned}$

Step 4: Find β_0 and β_1 from steps 2 and 3

Step 5: Find $e^{At} = \beta_0 I + \beta_1 A$

Step 6: Find $x(t) = [e^{At}]x(0)$

Illustration 4: (Free response)

A system is represented by $\dot{x} = Ax$, where $A = \begin{bmatrix} 3 & 4 \\ 2 & 1 \end{bmatrix}$. The value of

$x(0) = \begin{bmatrix} 1 \\ 0 \end{bmatrix}$. Find the values of e^{At} and $x(t)$ using (A) Laplace inverse

approach, (B) similarity transformation approach and (C) Cayley Hamilton approach.

Solution

Given

$$A = \begin{bmatrix} 3 & 4 \\ 2 & 1 \end{bmatrix}; \quad x(0) = \begin{bmatrix} 1 \\ 0 \end{bmatrix}$$

Case 1: Laplace inverse approach

Step 1: Find the matrix sI

$$I = \begin{bmatrix} 1 & 0 \\ 0 & 1 \end{bmatrix}$$

$$sI = \begin{bmatrix} s & 0 \\ 0 & s \end{bmatrix}$$

Step 2: Find the matrix $[sI–A]$

$$[sI - A] = \begin{bmatrix} s & 0 \\ 0 & s \end{bmatrix} - \begin{bmatrix} 3 & 4 \\ 2 & 1 \end{bmatrix}$$

$$= \begin{bmatrix} s-3 & -4 \\ -2 & s-1 \end{bmatrix}$$

Step 3: Find the inverse of the matrix $[sI–A]$

$$[sI - A]^{-1} = \frac{Adj(sI - A)}{|sI - A|}$$

$$|sI - A| = (s-3)(s-1) - 8$$

$$= s^2 - s - 3s + 3 - 8 = s^2 - 4s - 5$$

$$= (s-5)(s+1)$$

$$Adj(sI - A) = \begin{bmatrix} s-1 & 4 \\ 2 & s-3 \end{bmatrix}$$

$$[sI - A]^{-1} = \frac{Adj(sI - A)}{|sI - A|} = \frac{\begin{bmatrix} s-1 & 4 \\ 2 & s-3 \end{bmatrix}}{(s-5)(s+1)}$$

$$= \begin{bmatrix} \dfrac{s-1}{(s-5)(s+1)} & \dfrac{4}{(s-5)(s+1)} \\ \dfrac{2}{(s-5)(s+1)} & \dfrac{s-3}{(s-5)(s+1)} \end{bmatrix}$$

Step 4: Find the laplace inverse of $[sI-A]^{-1}$

$$e^{At} = L^{-1}[sI - A]^{-1}$$

$$= L^{-1} \begin{bmatrix} \dfrac{s-1}{(s-5)(s+1)} & \dfrac{4}{(s-5)(s+1)} \\ \dfrac{2}{(s-5)(s+1)} & \dfrac{s-3}{(s-5)(s+1)} \end{bmatrix}$$

Finding the partial fraction

$$\Rightarrow \frac{s-1}{(s-5)(s+1)} = \frac{A}{(s-5)} + \frac{B}{(s+1)}$$

$$s-1 = A(s+1) + B(s-5)$$

When $s = 5$	When $s = -1$
$5-1 = A(5+1) + 0$	$-1-1 = A(0) + B(-1-5)$
$4 = 6A$	$-2 = -6B$
$A = 4/6$	$B = 2/6$

$$\Rightarrow \frac{4}{(s-5)(s+1)} = \frac{A}{(s-5)} + \frac{B}{(s+1)}$$

$$4 = A(s+1) + B(s-5)$$

When $s = 5$	When $s = -1$
$4 = A(5+1) + 0$	$4 = A(0) + B(-1-5)$
$4 = 6A$	$4 = -6B$
$A = 4/6$	$B = -4/6$

$$\Rightarrow \frac{2}{(s-5)(s+1)} = \frac{A}{(s-5)} + \frac{B}{(s+1)}$$

$$2 = A(s+1) + B(s-5)$$

When $s = 5$ | When $s = -1$
$2 = A(5+1) + 0$ | $2 = A(0) + B(-1-5)$
$2 = 6A$ | $2 = -6B$
$A = 2/6$ | $B = -2/6$

$$\Rightarrow \frac{s-3}{(s-5)(s+1)} = \frac{A}{(s-5)} + \frac{B}{(s+1)}$$

$$s-3 = A(s+1) + B(s-5)$$

When $s = 5$ | When $s = -1$
$5-3 = A(5+1) + 0$ | $-1-3 = A(0) + B(-1-5)$
$2 = 6A$ | $-4 = -6B$
$A = 2/6$ | $B = 4/6$

$$\therefore e^{At} = L^{-1} \begin{bmatrix} \dfrac{4/6}{(s-5)} + \dfrac{2/6}{(s+1)} & \dfrac{4/6}{(s-5)} - \dfrac{4/6}{(s+1)} \\ \dfrac{2/6}{(s-5)} - \dfrac{2/6}{(s+1)} & \dfrac{2/6}{(s-5)} + \dfrac{4/6}{(s+1)} \end{bmatrix}$$

$$e^{At} = \begin{bmatrix} \dfrac{4}{6}e^{5t} + \dfrac{2}{6}e^{-t} & \dfrac{4}{6}e^{5t} - \dfrac{4}{6}e^{-t} \\ \dfrac{2}{6}e^{5t} - \dfrac{2}{6}e^{-t} & \dfrac{2}{6}e^{5t} + \dfrac{4}{6}e^{-t} \end{bmatrix}$$

$$\ldots \therefore e^{At} = \begin{bmatrix} \dfrac{2}{3}e^{5t} + \dfrac{1}{3}e^{-t} & \dfrac{2}{3}e^{5t} - \dfrac{2}{3}e^{-t} \\ \dfrac{1}{3}e^{5t} - \dfrac{1}{3}e^{-t} & \dfrac{1}{3}e^{5t} + \dfrac{2}{3}e^{-t} \end{bmatrix}$$

Step 5: Find the free response $x(t)$

$$x(t) = e^{At}x(0)$$

$$x(t) = e^{At} = \begin{bmatrix} \dfrac{2}{3}e^{5t} + \dfrac{1}{3}e^{-t} & \dfrac{2}{3}e^{5t} - \dfrac{2}{3}e^{-t} \\[3mm] \dfrac{1}{3}e^{5t} - \dfrac{1}{3}e^{-t} & \dfrac{1}{3}e^{5t} + \dfrac{2}{3}e^{-t} \end{bmatrix} \begin{bmatrix} 1 \\ 0 \end{bmatrix}$$

$$x(t) = \begin{bmatrix} \dfrac{2}{3}e^{5t} + \dfrac{1}{3}e^{-t} \\[3mm] \dfrac{1}{3}e^{5t} - \dfrac{1}{3}e^{-t} \end{bmatrix}$$

Case 2: Similarity transformation approach

Given

$$A = \begin{bmatrix} 3 & 4 \\ 2 & 1 \end{bmatrix}; \quad x(0) = \begin{bmatrix} 1 \\ 0 \end{bmatrix}$$

Step 1: Find the eigen values

$$I = \begin{bmatrix} 1 & 0 \\ 0 & 1 \end{bmatrix}$$

$$\lambda I = \begin{bmatrix} \lambda & 0 \\ 0 & \lambda \end{bmatrix}$$

$$\Rightarrow [\lambda I - A] = \begin{bmatrix} \lambda & 0 \\ 0 & \lambda \end{bmatrix} - \begin{bmatrix} 3 & 4 \\ 2 & 1 \end{bmatrix}$$

$$= \begin{bmatrix} \lambda - 3 & -4 \\ -2 & \lambda - 1 \end{bmatrix}$$

$$|\lambda I - A| = (\lambda - 3)(\lambda - 1) - 8$$

$$= \lambda^2 - \lambda - 3\lambda + 3 - 8 = \lambda^2 - 4\lambda - 5$$

$$= (\lambda - 5)(\lambda + 1)$$

The eigen values are $\lambda_1 = 5$, $\lambda_2 = -1$.

Step 2: Find the cofactor matrix of matrix $[\lambda I - A]$

For $\lambda = \lambda_1 = 5$

$$[\lambda_1 I - A] = \begin{bmatrix} 5 & 0 \\ 0 & 5 \end{bmatrix} - \begin{bmatrix} 3 & 4 \\ 2 & 1 \end{bmatrix} = \begin{bmatrix} 2 & -4 \\ -2 & 4 \end{bmatrix}$$

$$C_{11} = (-1)^{1+1}(4) = 4$$

$$C_{12} = (-1)^{1+2}(-2) = 2$$

$$p_1 = \begin{bmatrix} C_{11} \\ C_{12} \end{bmatrix} = \begin{bmatrix} 4 \\ 2 \end{bmatrix}$$

For $\lambda = \lambda_2 = -1$

$$[\lambda_2 I - A] = \begin{bmatrix} -1 & 0 \\ 0 & -1 \end{bmatrix} - \begin{bmatrix} 3 & 4 \\ 2 & 1 \end{bmatrix} = \begin{bmatrix} -4 & -4 \\ -2 & -2 \end{bmatrix}$$

$$C_{11} = (-1)^{1+1}(-2) = -2$$

$$C_{12} = (-1)^{1+2}(-2) = 2$$

$$p_2 = \begin{bmatrix} C_{11} \\ C_{12} \end{bmatrix} = \begin{bmatrix} -2 \\ 2 \end{bmatrix}$$

$\therefore$ Eigen vectors corresponding to $\lambda_1 = \begin{bmatrix} 4 \\ 2 \end{bmatrix}$

$\therefore$ Eigen vectors corresponding to $\lambda_2 = \begin{bmatrix} -2 \\ 2 \end{bmatrix}$

Step 3: Find the matrix $P = \begin{bmatrix} p_1 & p_2 \end{bmatrix}$

$$P = \begin{bmatrix} 4 & -2 \\ 2 & 2 \end{bmatrix}$$

Step 4: Find P^{-1}

$$[P]^{-1} = \frac{1}{12} \begin{bmatrix} 2 & 2 \\ -2 & 4 \end{bmatrix}$$

Step 5: Find e^{At}

$$e^{At} = [P][e^{\lambda t}][P^{-1}]$$

$$= \begin{bmatrix} 4 & -2 \\ 2 & 2 \end{bmatrix} \begin{bmatrix} e^{5t} & 0 \\ 0 & e^{-t} \end{bmatrix} \begin{bmatrix} \dfrac{2}{12} & \dfrac{2}{12} \\ \dfrac{-2}{12} & \dfrac{4}{12} \end{bmatrix}$$

$$= \begin{bmatrix} 4e^{5t} & -2e^{-t} \\ 2e^{5t} & 2e^{-t} \end{bmatrix} \begin{bmatrix} \dfrac{2}{12} & \dfrac{2}{12} \\ \dfrac{-2}{12} & \dfrac{4}{12} \end{bmatrix}$$

$$e^{At} = \begin{bmatrix} \dfrac{8}{12}e^{-5t} + \dfrac{4}{12}e^{-t} & \dfrac{8}{12}e^{-5t} - \dfrac{8}{12}e^{-t} \\ \dfrac{4}{12}e^{-5t} - \dfrac{4}{12}e^{-t} & \dfrac{4}{12}e^{-5t} + \dfrac{8}{12}e^{-t} \end{bmatrix}$$

$$\therefore e^{At} = \begin{bmatrix} \dfrac{2}{3}e^{-5t} + \dfrac{1}{3}e^{-t} & \dfrac{2}{3}e^{-5t} - \dfrac{2}{3}e^{-t} \\ \dfrac{1}{3}e^{-5t} - \dfrac{1}{3}e^{-t} & \dfrac{1}{3}e^{-5t} + \dfrac{2}{3}e^{-t} \end{bmatrix}$$

Step 6: Find the free response $x(t)$

$$x(t) = e^{At}x(0)$$

$$x(t) = e^{At} = \begin{bmatrix} \dfrac{2}{3}e^{-5t} + \dfrac{1}{3}e^{-t} & \dfrac{2}{3}e^{-5t} - \dfrac{2}{3}e^{-t} \\ \dfrac{1}{3}e^{-5t} - \dfrac{1}{3}e^{-t} & \dfrac{1}{3}e^{-5t} + \dfrac{2}{3}e^{-t} \end{bmatrix} \begin{bmatrix} 1 \\ 0 \end{bmatrix}$$

$$\Rightarrow x(t) = \begin{bmatrix} \dfrac{2}{3}e^{-5t} + \dfrac{1}{3}e^{-t} \\ \dfrac{1}{3}e^{-5t} - \dfrac{1}{3}e^{-t} \end{bmatrix}$$

Case 3: Cayley Hamilton approach

Given

$$A = \begin{bmatrix} 3 & 4 \\ 2 & 1 \end{bmatrix}; \quad x(0) = \begin{bmatrix} 1 \\ 0 \end{bmatrix}$$

Step 1: Find the eigen values

$$[\lambda I - A] = \begin{bmatrix} \lambda & 0 \\ 0 & \lambda \end{bmatrix} - \begin{bmatrix} 3 & 4 \\ 2 & 1 \end{bmatrix}$$

$$= \begin{bmatrix} \lambda - 3 & -4 \\ -2 & \lambda - 1 \end{bmatrix}$$

$$[\lambda I - A]^{-1} = \frac{Adj(\lambda I - A)}{|\lambda I - A|}$$

$$|\lambda I - A| = (\lambda - 3)(\lambda - 1) - 8$$

$$= \lambda^2 - \lambda - 3\lambda + 3 - 8 \ = \lambda^2 - 4\lambda - 5$$

$$= (\lambda - 5)(\lambda + 1)$$

The eigen values are $\lambda_1 = 5, \lambda_2 = -1$

Step 2: Find
$$g(\lambda_1) = \beta_0 + \beta_1 \lambda_1$$
$$g(\lambda_2) = \beta_0 + \beta_1 \lambda_2$$

∵ The matrix A is second order, and hence, the polynomial $g(\lambda)$ will be of the form, $g(\lambda) = \beta_0 + \beta_1 \lambda$

For $\lambda_1 = 5$, we have

$$g(\lambda_1) = \beta_0 + \beta_1 \lambda_1$$

$$g(5) = \beta_0 + 5\beta_1$$

For $\lambda_2 = -1$, we have

$$g(\lambda_2) = \beta_0 + \beta_1 \lambda_2$$

$$g(-1) = \beta_0 + \beta_1(-1)$$

$$g(-1) = \beta_0 - \beta_1$$

Step 3: Find $f(\lambda_1) = e^{\lambda_1 t}$ and $f(\lambda_2) = e^{\lambda_2 t}$

The coefficients $g(\lambda) = \beta_0 + \beta_1 \lambda$ are evaluated using the following relations:

$$f(\lambda_1) = e^{\lambda_1 t} = g(\lambda_1) = \beta_0 + \beta_1 \lambda_1$$

$$f(\lambda_2) = e^{\lambda_2 t} = g(\lambda_2) = \beta_0 + \beta_1 \lambda_2$$

For $\lambda_1 = 5$, we have

$$f(\lambda_1) = e^{5t}$$

$$g(\lambda_1) = \beta_0 + 5\beta_1$$

$$f(\lambda_1) = g(\lambda_1)$$

$$e^{5t} = \beta_0 + 5\beta_1 \qquad \text{(i)}$$

For $\lambda_1 = -1$, we have

$$f(\lambda_2) = e^{-t}$$

$$g(\lambda_2) = \beta_0 + \beta_1$$

$$f(\lambda_2) = g(\lambda_2)$$

$$e^{-t} = \beta_0 - \beta_1 \qquad \text{(ii)}$$

By solving equations (i) and (ii), we get

$$6\beta_1 = e^{5t} - e^{-t} \qquad \text{(iii)}$$

$$\beta_1 = (e^{5t} - e^{-t})/6 \qquad \text{(iv)}$$

Substituting (iv) in (i), we get

$$\beta_0 = \left(e^{5t} + 5e^{-t}\right)/6 \qquad \text{(v)}$$

Step 5: Find $e^{At} = \beta_0 I + \beta_1 A$

$$e^{At} = \left[\left(e^{5t} + 5e^{-t} \right)/6 \right] \begin{bmatrix} 1 & 0 \\ 0 & 1 \end{bmatrix} + \left[\left(e^{5t} - e^{-t} \right)/6 \right] \begin{bmatrix} 3 & 4 \\ 2 & 1 \end{bmatrix}$$

$$= \begin{bmatrix} \dfrac{1}{6}e^{-5t} + \dfrac{5}{6}e^{-t} & 0 \\ 0 & \dfrac{1}{6}e^{-5t} + \dfrac{5}{6}e^{-t} \end{bmatrix} + \begin{bmatrix} \dfrac{3}{6}e^{-5t} - \dfrac{3}{6}e^{-t} & \dfrac{4}{6}e^{-5t} - \dfrac{4}{6}e^{-t} \\ \dfrac{2}{6}e^{-5t} - \dfrac{2}{6}e^{-t} & \dfrac{1}{6}e^{-5t} - \dfrac{1}{6}e^{-t} \end{bmatrix}$$

$$= \begin{bmatrix} \dfrac{4}{6}e^{-5t} + \dfrac{2}{6}e^{-t} & \dfrac{4}{6}e^{-5t} - \dfrac{4}{6}e^{-t} \\ \dfrac{2}{6}e^{-5t} - \dfrac{2}{6}e^{-t} & \dfrac{2}{6}e^{-5t} + \dfrac{4}{6}e^{-t} \end{bmatrix}$$

$$\therefore e^{At} = \begin{bmatrix} \dfrac{2}{3}e^{-5t} + \dfrac{1}{3}e^{-t} & \dfrac{2}{3}e^{-5t} - \dfrac{2}{3}e^{-t} \\ \dfrac{1}{3}e^{-5t} - \dfrac{1}{3}e^{-t} & \dfrac{1}{3}e^{-5t} + \dfrac{2}{3}e^{-t} \end{bmatrix}$$

Step 6: Find the free response $x(t)$

$$x(t) = e^{At} x(0)$$

$$x(t) = e^{At} = \begin{bmatrix} \dfrac{2}{3}e^{-5t} + \dfrac{1}{3}e^{-t} & \dfrac{2}{3}e^{-5t} - \dfrac{2}{3}e^{-t} \\ \dfrac{1}{3}e^{-5t} - \dfrac{1}{3}e^{-t} & \dfrac{1}{3}e^{-5t} + \dfrac{2}{3}e^{-t} \end{bmatrix} \begin{bmatrix} 1 \\ 0 \end{bmatrix}$$

$$\Rightarrow x(t) = \begin{bmatrix} \dfrac{2}{3}e^{-5t} + \dfrac{1}{3}e^{-t} \\ \dfrac{1}{3}e^{-5t} - \dfrac{1}{3}e^{-t} \end{bmatrix}$$

Illustration 5: (Free response)

A system is represented by $\dot{x} = Ax$, where $A = \begin{bmatrix} 0 & 1 & 0 \\ 0 & 0 & 1 \\ -6 & -11 & -6 \end{bmatrix}$.

The value of $x(0) = \begin{bmatrix} 0 \\ 1 \\ 0 \end{bmatrix}$.

Find the values of e^{At} and $x(t)$ using the following methods

A. Laplace inverse approach
B. Similarity transformation approach
C. Cayley Hamilton approach

Solution

Given

$$A = \begin{bmatrix} 0 & 1 & 0 \\ 0 & 0 & 1 \\ -6 & -11 & -6 \end{bmatrix}; \quad x(0) = \begin{bmatrix} 0 \\ 1 \\ 0 \end{bmatrix}$$

Case 1: Laplace inverse approach

Step 1: Find the matrix $[sI]$ and hence $[sI-A]$

$$I = \begin{bmatrix} 1 & 0 & 0 \\ 0 & 1 & 0 \\ 0 & 0 & 1 \end{bmatrix}$$

$$sI = \begin{bmatrix} s & 0 & 0 \\ 0 & s & 0 \\ 0 & 0 & s \end{bmatrix}$$

$$[sI - A] = \begin{bmatrix} s & 0 & 0 \\ 0 & s & 0 \\ 0 & 0 & s \end{bmatrix} - \begin{bmatrix} 0 & 1 & 0 \\ 0 & 0 & 1 \\ -6 & -11 & -6 \end{bmatrix} = \begin{bmatrix} s & -1 & 0 \\ 0 & s & -1 \\ 6 & 11 & s+6 \end{bmatrix}$$

Step 2: Find the inverse of the matrix $[sI-A]$

$$[sI - A]^{-1} = \frac{Adj(sI - A)}{|sI - A|}$$

$$|sI - A| = s[s(s+6)+11]+1(0+6)$$

$$= s^3 + 6s^2 + 11s + 6$$

$$= (s+1)(s+2)(s+3)$$

$Adj(sI–A)$ = Transpose of cofactor of $(sI–A)$
   Cofactor matrix of $(sI–A)$ can be found as follows:

$$C_{11} = (-1)^{1+1} \begin{vmatrix} s & -1 \\ 11 & s+6 \end{vmatrix} \qquad C_{12} = (-1)^{1+2} \begin{vmatrix} 0 & -1 \\ 6 & s+6 \end{vmatrix} \qquad C_{13} = (-1)^{1+3} \begin{vmatrix} 0 & s \\ 6 & 11 \end{vmatrix}$$

$$= s^2 + 6s + 11 \qquad\qquad = -6 \qquad\qquad = -6s$$

$$C_{21} = (-1)^{2+1} \begin{vmatrix} -1 & 0 \\ 11 & s+6 \end{vmatrix} \qquad C_{22} = (-1)^{2+2} \begin{vmatrix} s & 0 \\ 6 & s+6 \end{vmatrix} \qquad C_{23} = (-1)^{2+3} \begin{vmatrix} s & -1 \\ 6 & 11 \end{vmatrix}$$

$$= s+6 \qquad\qquad = s(s+6) \qquad\qquad = -11s - 6$$

$$C_{31} = (-1)^{3+1} \begin{vmatrix} -1 & 0 \\ s & -1 \end{vmatrix} \qquad C_{32} = (-1)^{3+2} \begin{vmatrix} s & 0 \\ 0 & -1 \end{vmatrix} \qquad C_{33} = (-1)^{3+3} \begin{vmatrix} s & -1 \\ 0 & s \end{vmatrix}$$

$$= 1 \qquad\qquad = s \qquad\qquad = s^2$$

$\therefore$ Cofactor matrix is given by

$$\begin{bmatrix} C_{11} & C_{12} & C_{13} \\ C_{21} & C_{22} & C_{23} \\ C_{31} & C_{32} & C_{33} \end{bmatrix} = \begin{bmatrix} s^2 + 6s + 11 & -6 & -6s \\ s+6 & s(s+6) & -11s - 6 \\ 1 & s & s^2 \end{bmatrix}$$

$$\therefore \quad [sI - A]^{-1} = \frac{Adj(sI - A)}{|sI - A|} = \frac{\begin{bmatrix} s^2 + 6s + 11 & s+6 & 1 \\ -6 & s(s+6) & s \\ -6s & -11s - 6 & s^2 \end{bmatrix}}{(s+1)(s+2)(s+3)}$$

$$\therefore \quad [sI - A]^{-1} = \begin{bmatrix} \dfrac{s^2 + 6s + 11}{(s+1)(s+2)(s+3)} & \dfrac{s+6}{(s+1)(s+2)(s+3)} & \dfrac{1}{(s+1)(s+2)(s+3)} \\[3mm] \dfrac{-6}{(s+1)(s+2)(s+3)} & \dfrac{s(s+6)}{(s+1)(s+2)(s+3)} & \dfrac{s}{(s+1)(s+2)(s+3)} \\[3mm] \dfrac{-6s}{(s+1)(s+2)(s+3)} & \dfrac{-11s - 6}{(s+1)(s+2)(s+3)} & \dfrac{s^2}{(s+1)(s+2)(s+3)} \end{bmatrix}$$

Step 3: Find the laplace inverse of $[sI–A]^{-1}$

$$e^{At} = L^{-1}[sI-A]^{:}$$

$$\therefore e^{At} = L^{-1} \begin{bmatrix} \dfrac{s^2+6s+11}{(s+1)(s+2)(s+3)} & \dfrac{s+6}{(s+1)(s+2)(s+3)} & \dfrac{1}{(s+1)(s+2)(s+3)} \\[4mm] \dfrac{-6}{(s+1)(s+2)(s+3)} & \dfrac{s(s+6)}{(s+1)(s+2)(s+3)} & \dfrac{s}{(s+1)(s+2)(s+3)} \\[4mm] \dfrac{-6s}{(s+1)(s+2)(s+3)} & \dfrac{-11s-6}{(s+1)(s+2)(s+3)} & \dfrac{s^2}{(s+1)(s+2)(s+3)} \end{bmatrix}$$

Finding the partial fraction

$$\Rightarrow \frac{s^2+6s+11}{(s+1)(s+2)(s+3)} = \frac{A}{(s+1)} + \frac{B}{(s+2)} + \frac{C}{(s+3)}$$

$$s^2+6s+11 = A(s+2)(s+3) + B(s+1)(s+3) + C(s+1)(s+2)$$

When $s=-1$	When $s=-2$	When $s=-3$
$1-6+11 = A(-1+2)(-1+3)$	$4-12+11 = B(-2+1)(-2+3)$	$9-18+11 = C(-3+1)(-3+2)$
$-5+11 = A(1)(2)$	$3 = B(-1)(1)$	$-9+11 = C(-2)(-1)$
$6 = 2A$	$B = 3$	$2 = 2C$
$A = 3$		$C = 1$

$$\Rightarrow \frac{s+6}{(s+1)(s+2)(s+3)} = \frac{A}{(s+1)} + \frac{B}{(s+2)} + \frac{C}{(s+3)}$$

$$s+6 = A(s+2)(s+3) + B(s+1)(s+3) + C(s+1)(s+2)$$

When $s=-1$	When $s=-2$	When $s=-3$
$-1+6 = A(-1+2)(-1+3)$	$-2+6 = B(-2+1)(-2+3)$	$-3+6 = C(-3+1)(-3+2)$
$5 = A(1)(2)$	$4 = B(-1)(1)$	$3 = C(-2)(-1)$
$5/2 = A$	$B = -4$	$3/2 = C$
$A = 2.5$		$C = 1.5$

$$\Rightarrow \frac{1}{(s+1)(s+2)(s+3)} = \frac{A}{(s+1)} + \frac{B}{(s+2)} + \frac{C}{(s+3)}$$

$$1 = A(s+2)(s+3) + B(s+1)(s+3) + C(s+1)(s+2)$$

When $s = -1$	When $s = -2$	When $s = -3$
$1 = A(-1+2)(-1+3)$	$1 = B(-2+1)(-2+3)$	$1 = C(-3+1)(-3+2)$
$1 = A(1)(2)$	$1 = B(-1)(1)$	$1 = C(-2)(-1)$
$1/2 = A$	$B = -1$	$1/2 = C$
$A = 0.5$		$C = 0.5$

$$\Rightarrow \frac{-6}{(s+1)(s+2)(s+3)} = \frac{A}{(s+1)} + \frac{B}{(s+2)} + \frac{C}{(s+3)}$$

$$-6 = A(s+2)(s+3) + B(s+1)(s+3) + C(s+1)(s+2)$$

When $s = -1$	When $s = -2$	When $s = -3$
$-6 = A(-1+2)(-1+3)$	$-6 = B(-2+1)(-2+3)$	$-6 = C(-3+1)(-3+2)$
$-6 = A(1)(2)$	$-6 = B(-1)(1)$	$-6 = C(-2)(-1)$
$-6/2 = A$	$B = 6$	$-6 = 2C$
$A = -3$		$C = -3$

$$\Rightarrow \frac{s(s+6)}{(s+1)(s+2)(s+3)} = \frac{A}{(s+1)} + \frac{B}{(s+2)} + \frac{C}{(s+3)}$$

$$s(s+6) = A(s+2)(s+3) + B(s+1)(s+3) + C(s+1)(s+2)$$

When $s = -1$	When $s = -2$	When $s = -3$
$-1(-1+6) = A(-1+2)(-1+3)$	$-2(-2+6) = B(-2+1)(-2+3)$	$-3(-3+6) = C(-3+1)(-3+2)$
$-1(5) = A(1)(2)$	$-2(4) = B(-1)(1)$	$-3(3) = C(-2)(-1)$
$-5/2 = A$	$-8 = -B$	$-9 = 2C$
$A = -2.5$	$B = 8$	$C = -4.5$

$$\Rightarrow \frac{s}{(s+1)(s+2)(s+3)} = \frac{A}{(s+1)} + \frac{B}{(s+2)} + \frac{C}{(s+3)}$$

$$s = A(s+2)(s+3) + B(s+1)(s+3) + C(s+1)(s+2)$$

When $s = -1$	When $s = -2$	When $s = -3$
$-1 = A(-1+2)(-1+3)$	$-2 = B(-2+1)(-2+3)$	$-3 = C(-3+1)(-3+2)$
$-1 = A(1)(2)$	$-2 = B(-1)(1)$	$-3 = C(-2)(-1)$
$-1/2 = A$	$-2 = -B$	$-3 = 2C$
$A = -0.5$	$B = 2$	$C = -1.5$

$$\Rightarrow \frac{-6s}{(s+1)(s+2)(s+3)} = \frac{A}{(s+1)} + \frac{B}{(s+2)} + \frac{C}{(s+3)}$$

$$-6s = A(s+2)(s+3) + B(s+1)(s+3) + C(s+1)(s+2)$$

When $s = -1$	When $s = -2$	When $s = -3$
$-6(-1) = A(-1+2)(-1+3)$	$-6(-2) = B(-2+1)(-2+3)$	$-6(-3) = C(-3+1)(-3+2)$
$6 = A(1)(2)$	$12 = B(-1)(1)$	$18 = C(-2)(-1)$
$6/2 = A$	$12 = -B$	$18 = 2C$
$A = 3$	$B = -12$	$C = 9$

$$\Rightarrow \frac{-11s-6}{(s+1)(s+2)(s+3)} = \frac{A}{(s+1)} + \frac{B}{(s+2)} + \frac{C}{(s+3)}$$

$$-11s - 6 = A(s+2)(s+3) + B(s+1)(s+3) + C(s+1)(s+2)$$

When $s = -1$	When $s = -2$	When $s = -3$
$-11(-1)-6 = A(-1+2)(-1+3)$	$-11(-2)-6 = B(-2+1)(-2+3)$	$-11(-3)-6 = C(-3+1)(-3+2)$
$11-6 = A(1)(2)$	$22-6 = B(-1)(1)$	$33-6 = C(-2)(-1)$
$5 = 2A$	$16 = -B$	$27 = 2C$
$A = 2.5$	$B = -16$	$C = 13.5$

$$\Rightarrow \frac{s^2}{(s+1)(s+2)(s+3)} = \frac{A}{(s+1)} + \frac{B}{(s+2)} + \frac{C}{(s+3)}$$

$$s^2 = A(s+2)(s+3) + B(s+1)(s+3) + C(s+1)(s+2)$$

When $s = -1$	When $s = -2$	When $s = -3$
$(-1)^2 = A(-1+2)(-1+3)$	$(-2)^2 = B(-2+1)(-2+3)$	$(-3)^2 = C(-3+1)(-3+2)$
$1 = A(1)(2)$	$4 = B(-1)(1)$	$9 = C(-2)(-1)$
$1 = 2A$	$4 = -B$	$9 = 2C$
$A = 0.5$	$B = -4$	$C = 4.5$

$$\therefore \; e^{At} = \begin{bmatrix} 3e^{-t} + 3e^{-2t} + e^{-3t} & 2.5e^{-t} - 4e^{-2t} + 1.5e^{-3t} & 0.5e^{-t} - 1e^{-2t} + 0.5e^{-3t} \\ -3e^{-t} + 6e^{-2t} - 3e^{-3t} & -2.5e^{-t} + 8e^{-2t} - 4.5e^{-3t} & -0.5e^{-t} + 2e^{-2t} - 1.5e^{-3t} \\ 3e^{-t} - 12e^{-2t} + 9e^{-3t} & 2.5e^{-t} - 16e^{-2t} + 13.5e^{-3t} & 0.5e^{-t} - 4e^{-2t} + 4.5e^{-3t} \end{bmatrix}$$

Step 4: Find the free response $x(t)$

$$x(t) = e^{At}x(0)$$

$$x(t) = \begin{bmatrix} 3e^{-t} - 3e^{-2t} + e^{-3t} & 2.5e^{-t} - 4e^{-2t} + 1.5e^{-3t} & 0.5e^{-t} - 1e^{-2t} + 0.5e^{-3t} \\ -3e^{-t} + 6e^{-2t} - 3e^{-3t} & -2.5e^{-t} + 8e^{-2t} - 4.5e^{-3t} & -0.5e^{-t} + 2e^{-2t} - 1.5e^{-3t} \\ 3e^{-t} - 12e^{-2t} + 9e^{-3t} & 2.5e^{-t} - 16e^{-2t} + 13.5e^{-3t} & 0.5e^{-t} - 4e^{-2t} + 4.5e^{-3t} \end{bmatrix}$$

$$\times \begin{bmatrix} 0 \\ 1 \\ 0 \end{bmatrix}$$

$$x(t) = \begin{bmatrix} 2.5e^{-t} - 4e^{-2t} + 1.5e^{-3t} \\ -2.5e^{-t} + 8e^{-2t} - 4.5e^{-3t} \\ 2.5e^{-t} - 16e^{-2t} + 13.5e^{-3t} \end{bmatrix}$$

Case 2: Similarity transformation approach
Given

$$A = \begin{bmatrix} 0 & 1 & 0 \\ 0 & 0 & 1 \\ -6 & -11 & -6 \end{bmatrix}; \quad x(0) = \begin{bmatrix} 0 \\ 1 \\ 0 \end{bmatrix}$$

Step 1: Find the eigen values

$$\lambda I = \begin{bmatrix} \lambda & 0 & 0 \\ 0 & \lambda & 0 \\ 0 & 0 & \lambda \end{bmatrix}$$

Step 2: Find the matrix $[\lambda I - A]$

$$[\lambda I - A] = \begin{bmatrix} \lambda & 0 & 0 \\ 0 & \lambda & 0 \\ 0 & 0 & \lambda \end{bmatrix} - \begin{bmatrix} 0 & 1 & 0 \\ 0 & 0 & 1 \\ -6 & -11 & -6 \end{bmatrix}$$

$$= \begin{bmatrix} \lambda & -1 & 0 \\ 0 & \lambda & -1 \\ 6 & 11 & \lambda+6 \end{bmatrix}$$

The characteristic equation is $|\lambda I - A| = \lambda^3 + 6\lambda^2 + 11\lambda + 6$

$$= (\lambda + 1)(\lambda + 2)(\lambda + 3)$$

∴ The eigen values are $\lambda_1 = -1$, $\lambda_2 = -2$, $\lambda_3 = -3$

Step 2: Find the cofactor matrix of matrix $[\lambda I - A]$

Method 1

For $\lambda = \lambda_1 = -1$

$$[\lambda_1 I - A] = \begin{bmatrix} -1 & 0 & 0 \\ 0 & -1 & 0 \\ 0 & 0 & -1 \end{bmatrix} - \begin{bmatrix} 0 & 1 & 0 \\ 0 & 0 & 1 \\ -6 & -11 & -6 \end{bmatrix} = \begin{bmatrix} -1 & -1 & 0 \\ 0 & -1 & -1 \\ 6 & 11 & 5 \end{bmatrix}$$

$$C_{11} = (-1)^{1+1} \begin{vmatrix} -1 & -1 \\ 11 & 5 \end{vmatrix} \quad C_{12} = (-1)^{1+2} \begin{vmatrix} 0 & -1 \\ 6 & 5 \end{vmatrix} \quad C_{13} = (-1)^{1+3} \begin{vmatrix} 0 & -1 \\ 6 & 11 \end{vmatrix}$$

$$= 6 \qquad\qquad\qquad = -6 \qquad\qquad\qquad = 6$$

$$p_1 = \begin{bmatrix} C_{11} \\ C_{12} \\ C_{13} \end{bmatrix} = \begin{bmatrix} 6 \\ -6 \\ 6 \end{bmatrix} = \begin{bmatrix} 1 \\ -1 \\ 1 \end{bmatrix}$$

For $\lambda = \lambda_2 = -2$

$$[\lambda_2 I - A] = \begin{bmatrix} -2 & 0 & 0 \\ 0 & -2 & 0 \\ 0 & 0 & -2 \end{bmatrix} - \begin{bmatrix} 0 & 1 & 0 \\ 0 & 0 & 1 \\ -6 & -11 & -6 \end{bmatrix} = \begin{bmatrix} -2 & -1 & 0 \\ 0 & -2 & -1 \\ 6 & 11 & 4 \end{bmatrix}$$

$$C_{11} = (-1)^{1+2} \begin{vmatrix} -2 & -1 \\ 11 & 4 \end{vmatrix} \quad C_{12} = (-1)^{1+2} \begin{vmatrix} 0 & 1 \\ 6 & 4 \end{vmatrix} \quad C_{13} = (-1)^{1+3} \begin{vmatrix} 0 & -2 \\ 6 & 11 \end{vmatrix}$$

$$= 3 \qquad\qquad\qquad = -6 \qquad\qquad\qquad = 12$$

$$p_2 = \begin{bmatrix} C_{11} \\ C_{12} \\ C_{13} \end{bmatrix} = \begin{bmatrix} 3 \\ -6 \\ 12 \end{bmatrix} = \begin{bmatrix} 1 \\ -2 \\ 4 \end{bmatrix}$$

For $\lambda = \lambda_3 = -3$

$$[\lambda_3 I - A] = \begin{bmatrix} -3 & 0 & 0 \\ 0 & -3 & 0 \\ 0 & 0 & -3 \end{bmatrix} - \begin{bmatrix} 0 & 1 & 0 \\ 0 & 0 & 1 \\ -6 & -11 & -6 \end{bmatrix} = \begin{bmatrix} -3 & -1 & 0 \\ 0 & -3 & -1 \\ 6 & 11 & 3 \end{bmatrix}$$

$$C_{11} = (-1)^{1+2} \begin{vmatrix} -3 & -1 \\ 11 & 3 \end{vmatrix} \qquad C_{12} = (-1)^{1+2} \begin{vmatrix} 0 & 1 \\ 6 & 3 \end{vmatrix} \qquad C_{13} = (-1)^{1+3} \begin{vmatrix} 0 & -3 \\ 6 & 11 \end{vmatrix}$$

$$= 2 \qquad\qquad\qquad = -6 \qquad\qquad\qquad = 18$$

$$p_3 = \begin{bmatrix} C_{11} \\ C_{12} \\ C_{13} \end{bmatrix} = \begin{bmatrix} 2 \\ -6 \\ 18 \end{bmatrix} = \begin{bmatrix} 1 \\ -3 \\ 9 \end{bmatrix}$$

Step 3: Find the eigen vectors

$$\text{Eigen vectors corresponding to } \lambda_1 = \begin{bmatrix} 1 \\ -1 \\ 1 \end{bmatrix}$$

$$\text{Eigen vectors corresponding to } \lambda_2 = \begin{bmatrix} 1 \\ -2 \\ 4 \end{bmatrix}$$

$$\text{Eigen vectors corresponding to } \lambda_3 = \begin{bmatrix} 1 \\ -3 \\ 9 \end{bmatrix}$$

Step 4: Find the modal matrix $P = \begin{bmatrix} p_1 & p_2 & p_3 \end{bmatrix}$

$$P = \begin{bmatrix} 1 & 1 & 1 \\ -1 & -2 & -3 \\ 1 & 4 & 9 \end{bmatrix}$$

Method 2

$$\text{Modal matrix } P = \begin{bmatrix} 1 & 1 & 1 \\ \lambda_1 & \lambda_2 & \lambda_3 \\ \lambda_1^2 & \lambda_2^2 & \lambda_3^2 \end{bmatrix} = \begin{bmatrix} 1 & 1 & 1 \\ -1 & -2 & -3 \\ 1 & 4 & 9 \end{bmatrix}$$

Step 5: Find P^{-1}

$$[P]^{-1} = \frac{Adj(P)}{|P|}$$

$$|P| = \begin{vmatrix} 1 & 1 & 1 \\ -1 & -2 & -3 \\ 1 & 4 & 9 \end{vmatrix} = 1(-18+12)-1(-9+3)+(-4+2) = -2$$

$$Adj\ P = \begin{bmatrix} -6 & -5 & -1 \\ 6 & 8 & 2 \\ -2 & -3 & -1 \end{bmatrix}$$

$$[P]^{-1} = \frac{-1}{2}\begin{bmatrix} -6 & -5 & -1 \\ 6 & 8 & 2 \\ -2 & -3 & -1 \end{bmatrix} = \begin{bmatrix} 3 & 2.5 & 0.5 \\ -3 & -4 & -1 \\ 1 & 1.5 & 0.5 \end{bmatrix}$$

Step 6: Find e^{At}

$$e^{At} = Pe^{\lambda t}P^{-1}$$

$$= \begin{bmatrix} 1 & 1 & 1 \\ -1 & -2 & -3 \\ 1 & 4 & 9 \end{bmatrix} \begin{bmatrix} e^{-t} & 0 & 0 \\ 0 & e^{-2t} & 0 \\ 0 & 0 & e^{-3t} \end{bmatrix} \begin{bmatrix} 3 & 2.5 & 0.5 \\ -3 & -4 & -1 \\ 1 & 1.5 & 0.5 \end{bmatrix}$$

$$= \begin{bmatrix} e^{-t} & e^{-2t} & e^{-3t} \\ -e^{-t} & -2e^{-2t} & -3e^{-3t} \\ e^{-t} & 4e^{-2t} & 9e^{-3t} \end{bmatrix} \begin{bmatrix} 3 & 2.5 & 0.5 \\ -3 & -4 & -1 \\ 1 & 1.5 & 0.5 \end{bmatrix}$$

$$\therefore\ e^{At} = \begin{bmatrix} 3e^{-t}-3e^{-2t}+e^{-3t} & 2.5e^{-t}-4e^{-2t}+1.5e^{-3t} & 0.5e^{-t}-1e^{-2t}+0.5e^{-3t} \\ -3e^{-t}+6e^{-2t}-3e^{-3t} & -2.5e^{-t}+8e^{-2t}-4.5e^{-3t} & -0.5e^{-t}+2e^{-2t}-1.5e^{-3t} \\ 3e^{-t}-12e^{-2t}+9e^{-3t} & 2.5e^{-t}-16e^{-2t}+13.5e^{-3t} & 0.5e^{-t}-4e^{-2t}+4.5e^{-3t} \end{bmatrix}$$

Step 7: Find the free response $x(t)$

$$x(t) = e^{At}x(0)$$

$$x(t) = \begin{bmatrix} 3e^{-t} - 3e^{-2t} + e^{-3t} & 2.5e^{-t} - 4e^{-2t} + 1.5e^{-3t} & 0.5e^{-t} - 1e^{-2t} + 0.5e^{-3t} \\ -3e^{-t} + 6e^{-2t} - 3e^{-3t} & -2.5e^{-t} + 8e^{-2t} - 4.5e^{-3t} & -0.5e^{-t} + 2e^{-2t} - 1.5e^{-3t} \\ 3e^{-t} - 12e^{-2t} + 9e^{-3t} & 2.5e^{-t} - 16e^{-2t} + 13.5e^{-3t} & 0.5e^{-t} - 4e^{-2t} + 4.5e^{-3t} \end{bmatrix}$$

$$\times \begin{bmatrix} 0 \\ 1 \\ 0 \end{bmatrix}$$

$$\Rightarrow x(t) = \begin{bmatrix} 2.5e^{-t} - 4e^{-2t} + 1.5e^{-3t} \\ -2.5e^{-t} + 8e^{-2t} - 4.5e^{-3t} \\ 2.5e^{-t} - 16e^{-2t} + 13.5e^{-3t} \end{bmatrix}$$

Case 3: Cayley Hamilton approach
Given

$$A = \begin{bmatrix} 0 & 1 & 0 \\ 0 & 0 & 1 \\ -6 & -11 & -6 \end{bmatrix}; \quad x(0) = \begin{bmatrix} 0 \\ 1 \\ 0 \end{bmatrix}$$

Step 1: Find the matrix $[\lambda I]$, $[\lambda I{-}A]$, $|\lambda I{-}A|$ and hence the eigen values

$$\lambda I = \begin{bmatrix} \lambda & 0 & 0 \\ 0 & \lambda & 0 \\ 0 & 0 & \lambda \end{bmatrix}$$

$$[\lambda I - A] = \begin{bmatrix} \lambda & 0 & 0 \\ 0 & \lambda & 0 \\ 0 & 0 & \lambda \end{bmatrix} - \begin{bmatrix} 0 & 1 & 0 \\ 0 & 0 & 1 \\ -6 & -11 & -6 \end{bmatrix}$$

$$= \begin{bmatrix} \lambda & -1 & 0 \\ 0 & \lambda & -1 \\ 6 & 11 & \lambda+6 \end{bmatrix}$$

$$|\lambda - A| = \lambda^3 + 6\lambda^2 + 11\lambda + 6$$

The characteristic polynomial of matrix 'A' is $\lambda^3 + 6\lambda^2 + 11\lambda + 6 = 0$
The roots of characteristic polynomial are $\lambda_1 = -1$, $\lambda_2 = -2$, $\lambda_3 = -3$

$$g(\lambda_1) = \beta_0 + \beta_1\lambda_1 + \beta_2\lambda_1^2$$

Step 2: Find $g(\lambda_2) = \beta_0 + \beta_1\lambda_2 + \beta_2\lambda_2^2$

$$g(\lambda_3) = \beta_0 + \beta_1\lambda_3 + \beta_2\lambda_3^2$$

The matrix A is third order; hence, the polynomial $g(\lambda)$ will be of the form

$$g(\lambda) = \beta_0 + \beta_1\lambda + \beta_2\lambda^2$$

$g(\lambda_1) = \beta_0 + \beta_1\lambda_1 + \beta_2\lambda_1^2$	$g(\lambda_2) = \beta_0 + \beta_1\lambda_2 + \beta_2\lambda_2^2$	$g(\lambda_3) = \beta_0 + \beta_1\lambda_3 + \beta_2\lambda_3^2$
$g(-1) = \beta_0 + \beta_1(-1) + \beta_2(-1)^2$	$g(-2) = \beta_0 + \beta_1(-2) + \beta_2(-2)^2$	$g(-3) = \beta_0 + \beta_1(-3) + \beta_2(-3)^2$
$g(-1) = \beta_0 - \beta_1 + \beta_2$	$g(-2) = \beta_0 - 2\beta_1 + 4\beta_2$	$g(-3) = \beta_0 - 3\beta_1 + 9\beta_2$

Step 3: Find $f(\lambda_1) = e^{\lambda_1 t}; f(\lambda_2) = e^{\lambda_2 t}; f(\lambda_3) = e^{\lambda_3 t}$

The coefficients $g(\lambda) = \beta_0 + \beta_1\lambda + \beta_2\lambda^2$ are evaluated using the following relations:

$$f(\lambda_1) = e^{\lambda_1 t} = g(\lambda_1) = \beta_0 + \beta_1\lambda_1 + \beta_2\lambda_1^2$$

$$f(\lambda_2) = e^{\lambda_2 t} = g(\lambda_2) = \beta_0 + \beta_1\lambda_2 + \beta_2\lambda_2^2$$

$$f(\lambda_3) = e^{\lambda_3 t} = g(\lambda_3) = \beta_0 + \beta_1\lambda_3 + \beta_2\lambda_3^2$$

For $\lambda_1 = -1$	For $\lambda_2 = -2$	For $\lambda_3 = -3$
$f(\lambda_1) = e^{-t}$	$f(\lambda_2) = e^{-2t}$	$f(\lambda_3) = e^{-3t}$
$g(\lambda_1) = \beta_0 - \beta_1 + \beta_2$	$g(\lambda_2) = \beta_0 - 2\beta_1 + 4\beta_2$	$g(\lambda_3) = \beta_0 - 3\beta_1 + 9\beta_2$
$f(\lambda_1) = g(\lambda_1)$	$f(\lambda_2) = g(\lambda_2)$	$f(\lambda_3) = g(\lambda_3)$
$e^{-t} = \beta_0 - \beta_1 + \beta_2$ (i)	$e^{-2t} = \beta_0 - 2\beta_1 + 4\beta_2$ (ii)	$e^{-3t} = \beta_0 - 3\beta_1 + 9\beta_2$ (iii)

Step 4: Find β_0, β_1 and β_2

By solving equations (i) and (ii), we get

$$2e^{-t} - e^{-2t} = \beta_0 - 2\beta_2 \tag{iv}$$

By solving equations (i) and (iii), we get

$$3e^{-t} - e^{-3t} = 2\beta_0 - 6\beta_2 \tag{v}$$

By solving equations (iv) and (v), we get

$$\beta_2 = 0.5e^{-t} - e^{-2t} + 0.5e^{-3t} \tag{vi}$$

Substituting (vi) in (iv), we get

$$\beta_0 = 3e^{-t} - 3e^{-2t} + e^{-3t} \tag{vii}$$

Substituting (vi) and (vii) in (i), we get

$$\beta_1 = 2.5e^{-t} - 4e^{-2t} + 1.5e^{-3t} \tag{viii}$$

Step 5: Find $e^{At} = \beta_0 I + \beta_1 A + \beta_2 A^2$

$$e^{At} = (3e^{-t} - 3e^{-2t} + e^{-3t})\begin{bmatrix} 1 & 0 & 0 \\ 0 & 1 & 0 \\ 0 & 0 & 1 \end{bmatrix} + (2.5e^{-t} - 4e^{-2t} + 1.5e^{-3t})\begin{bmatrix} 0 & 1 & 0 \\ 0 & 0 & 1 \\ -6 & -11 & -6 \end{bmatrix}$$

$$+ (0.5e^{-t} - e^{-2t} + 0.5e^{-3t})\begin{bmatrix} 0 & 0 & 1 \\ -6 & -11 & -6 \\ 36 & 60 & 25 \end{bmatrix} e^{At}$$

$$= \begin{bmatrix} 3e^{-t} - 3e^{-2t} + e^{-3t} & 0 & 0 \\ 0 & 3e^{-t} - 3e^{-2t} + e^{-3t} & 0 \\ 0 & 0 & 3e^{-t} - 3e^{-2t} + e^{-3t} \end{bmatrix}$$

$$+ \begin{bmatrix} 0 & 2.5e^{-t} - 4e^{-2t} + 1.5e^{-3t} & 0 \\ 0 & 0 & 2.5e^{-t} - 4e^{-2t} + 1.5e^{-3t} \\ 3e^{-t} - 12e^{-2t} + 9e^{-3t} & 2.5e^{-t} - 16e^{-2t} + 13.5e^{-3t} & 0.5e^{-t} - 4e^{-2t} + 4.5e^{-3t} \end{bmatrix}$$

$$+ \begin{bmatrix} 0 & 0 & 0.5e^{-t} - 1e^{-2t} + 0.5e^{-3t} \\ -3e^{-t} + 6e^{-2t} - 3e^{-3t} & -5.5e^{-t} + 11e^{-2t} - 5.5e^{-3t} & -3e^{-t} + 6e^{-2t} - 3e^{-3t} \\ 18e^{-t} - 36e^{-2t} + 18e^{-3t} & 30e^{-t} - 60e^{-2t} + 30e^{-3t} & 12.5e^{-t} - 25e^{-2t} + 12.5e^{-3t} \end{bmatrix}$$

$$\therefore e^{At} = \begin{bmatrix} 3e^{-t} + 3e^{-2t} + e^{-3t} & 2.5e^{-t} - 4e^{-2t} + 1.5e^{-3t} & 0.5e^{-t} - 1e^{-2t} + 0.5e^{-3t} \\ -3e^{-t} + 6e^{-2t} - 3e^{-3t} & -2.5e^{-t} + 8e^{-2t} - 4.5e^{-3t} & -0.5e^{-t} + 2e^{-2t} - 1.5e^{-3t} \\ 3e^{-t} - 12e^{-2t} + 9e^{-3t} & 2.5e^{-t} - 16e^{-2t} + 13.5e^{-3t} & 0.5e^{-t} - 4e^{-2t} + 4.5e^{-3t} \end{bmatrix}$$

Step 6: Find the free response $x(t)$

$$x(t) = e^{At}x(0)$$

$$x(t) = \begin{bmatrix} 3e^{-t} - 3e^{-2t} + e^{-3t} & 2.5e^{-t} - 4e^{-2t} + 1.5e^{-3t} & 0.5e^{-t} - 1e^{-2t} + 0.5e^{-3t} \\ -3e^{-t} + 6e^{-2t} - 3e^{-3t} & -2.5e^{-t} + 8e^{-2t} - 4.5e^{-3t} & -0.5e^{-t} + 2e^{-2t} - 1.5e^{-3t} \\ 3e^{-t} - 12e^{-2t} + 9e^{-3t} & 2.5e^{-t} - 16e^{-2t} + 13.5e^{-3t} & 0.5e^{-t} - 4e^{-2t} + 4.5e^{-3t} \end{bmatrix} \begin{bmatrix} 0 \\ 1 \\ 0 \end{bmatrix}$$

$$\Rightarrow x(t) = \begin{bmatrix} 2.5e^{-t} - 4e^{-2t} + 1.5e^{-3t} \\ -2.5e^{-t} + 8e^{-2t} - 4.5e^{-3t} \\ 2.5e^{-t} - 16e^{-2t} + 13.5e^{-3t} \end{bmatrix}$$

Steps to find the forced response:
Step 1: Find the matrix [sI], [sI–A] and |sI–A|
Step 2: Find the matrix [sI–A]$^{-1}$
Step 3: Find Laplace inverse, $L^{-1}[sI-A]^{-1} = e^{At}$ or $\Phi(t)$

Step 4: Find x(t) using the formula $x(t) = e^{At}x(0) + \int_0^t e^{A(t-\tau)}Bu(\tau)d\tau$.

Illustration 6: (Forced response)

Find the forced response of the system represented by a state model of the

form $\dot{x} = Ax + Bu$ and $y = Cx$ where $A = \begin{bmatrix} 0 & 1 \\ -2 & -3 \end{bmatrix}$; $B = \begin{bmatrix} 0 \\ 10 \end{bmatrix}$ and $C =$

$\begin{bmatrix} 1 & 0 \end{bmatrix}$. The system is excited with unit step input. Assume that the initial conditions are zero.

Solution

Given

$$A = \begin{bmatrix} 0 & 1 \\ -2 & -3 \end{bmatrix}; B = \begin{bmatrix} 0 \\ 10 \end{bmatrix} \text{ and } C = \begin{bmatrix} 1 & 0 \end{bmatrix}$$

Given that the initial conditions are zero, i.e., x(0) = 0

$$\therefore x(t) = \int_0^t e^{A(t-\tau)}Bu(\tau)d\tau$$

Step 1: Find [sI–A]

$$[sI - A] = \begin{bmatrix} s & 0 \\ 0 & s \end{bmatrix} - \begin{bmatrix} 0 & 1 \\ -2 & -3 \end{bmatrix}$$

$$= \begin{bmatrix} s & -1 \\ 2 & s+3 \end{bmatrix}$$

$$\therefore |sI - A| = s(s+3) - (-1)(2)$$

$$= s^2 + 3s + 2$$

$$= (s+1)(s+2)$$

Step 2: Find the inverse of the matrix $[sI-A]$

$$[sI - A]^{-1} = \frac{Adj(sI - A)}{|sI - A|}$$

$Adj(sI–A)$ = Transpose of cofactor of $(sI–A)$
Cofactor matrix of $(sI–A)$

$$C_{11} = (-1)^{1+1}|s+3| = s+3$$

$$C_{12} = (-1)^{1+2}|2| = -2$$

$$C_{21} = (-1)^{2+1}|-1| = 1$$

$$C_{22} = (-1)^{2+2}|s| = s$$

$$\therefore Adj(sI - A) = \begin{bmatrix} s+3 & 1 \\ -2 & s \end{bmatrix}$$

$$(sI - A)^{-1} = \frac{Adj(sI - A)}{|sI - A|} = \frac{\begin{vmatrix} s+3 & 1 \\ -2 & s \end{vmatrix}}{(s+1)(s+2)}$$

$$= \frac{1}{(s+1)(s+2)}\begin{bmatrix} s+3 & 1 \\ -2 & s \end{bmatrix}$$

$$= \begin{bmatrix} \dfrac{s+3}{(s+1)(s+2)} & \dfrac{1}{(s+1)(s+2)} \\ \dfrac{-2}{(s+1)(s+2)} & \dfrac{s}{(s+1)(s+2)} \end{bmatrix}$$

Step 3: Find the laplace inverse of $[sI–A]^{-1}$

$$e^{At} = L^{-1}\begin{bmatrix} \dfrac{s+3}{(s+1)(s+2)} & \dfrac{1}{(s+1)(s+2)} \\ \dfrac{-2}{(s+1)(s+2)} & \dfrac{s}{(s+1)(s+2)} \end{bmatrix}$$

By taking partial fractions, we get

$$\Rightarrow \frac{s+3}{(s+1)(s+2)} = \frac{A}{(s+1)} + \frac{B}{(s+2)}$$

$$s+3 = A(s+2) + B(s+1)$$

When $s = -1$
$-1+3 = A(-1+2) + B(-1+1)$
$2 = A(1) + B(0)$
$A = 2$

When $s = -2$
$-2+3 = A(-2+2) + B(-2+1)$
$1 = A(0) + B(-1)$
$B = -1$

$$\Rightarrow \frac{s}{(s+1)(s+2)} = \frac{A}{(s+1)} + \frac{B}{(s+2)}$$

$$s = A(s+2) + B(s+1)$$

When $s = -1$
$-1 = A(-1+2) + B(-1+1)$
$-1 = A(1) + B(0)$
$A = -1$

When $s = -2$
$-2 = A(-2+2) + B(-2+1)$
$-2 = A(0) + B(-1)$
$B = 2$

$$L^{-1}[sI - A]^{-1} = \begin{bmatrix} \dfrac{2}{(s+1)} - \dfrac{1}{(s+2)} & \dfrac{1}{(s+1)} - \dfrac{1}{(s+2)} \\ \dfrac{-2}{(s+1)} + \dfrac{2}{(s+2)} & \dfrac{-1}{(s+1)} + \dfrac{2}{(s+2)} \end{bmatrix}$$

$$e^{At} = \begin{bmatrix} 2e^{-t} - 1e^{-2t} & e^{-t} - e^{-2t} \\ -2e^{-t} + 2e^{-2t} & -e^{-t} + 2e^{-2t} \end{bmatrix}$$

Step 4: Find the forced response $x(t)$

$$x(t) = e^{At}x(0) + \int_0^t e^{A(t-\tau)}Bu(\tau)d\tau$$

Given $x(0) = 0$, $u = 1/s$, $B = \begin{bmatrix} 0 \\ 10 \end{bmatrix}$

$$x(t) = 0 + \int_0^t e^{A(t-\tau)} B \, d\tau$$

$$= 0 + \int_0^t \begin{bmatrix} 2e^{-t} - 1e^{-2t} & e^{-t} - e^{-2t} \\ -2e^{-t} + 2e^{-2t} & -e^{-t} + 2e^{-2t} \end{bmatrix} \begin{bmatrix} 0 \\ 10 \end{bmatrix}$$

$$= 0 + \int_0^t \begin{bmatrix} 10(e^{-t} - e^{-2t}) \\ 10(e^{-t} + 2e^{-2t}) \end{bmatrix} d\tau$$

$$= \int_0^t \begin{bmatrix} 10(e^{-(t-\tau)} - e^{-2(t-\tau)}) \\ 10(e^{-(t-\tau)} + 2e^{-2(t-\tau)}) \end{bmatrix} d\tau$$

$$\therefore x(t) = \begin{bmatrix} 5 - 10e^{-t} + 5e^{-2t} \\ 10e^{-t} - 10e^{-2t} \end{bmatrix}$$

$$y(t) = Cx(t)$$

$$y(t) = 5 - 10e^{-t} + 5e^{-2t}$$

Alternate method:

Illustration 7: (Forced response)

Solution

$$x(t) = [\Phi(t) \, x(0)] + L^{-1}[\Phi(s).B.u(s)]$$

$$x(t) = \underbrace{L^{-1}[\Phi(t)] \, [x(0)]}_{PartA} + \underbrace{L^{-1}[\Phi(s).B.u(s)]}_{PartB}$$

Given $x(0) = 0$, $u = 1/s$, $B = \begin{bmatrix} 0 \\ 10 \end{bmatrix}$

Since, $x(0) = 0$, Part $A = 0$
Part B can be found as follows:

$$\Phi(s) = \begin{bmatrix} \dfrac{s+3}{(s+1)(s+2)} & \dfrac{1}{(s+1)(s+2)} \\ \dfrac{-2}{(s+1)(s+2)} & \dfrac{s}{(s+1)(s+2)} \end{bmatrix}$$

$$\Phi(s)Bu(s) = \begin{bmatrix} \dfrac{s+3}{(s+1)(s+2)} & \dfrac{1}{(s+1)(s+2)} \\ \dfrac{-2}{(s+1)(s+2)} & \dfrac{s}{(s+1)(s+2)} \end{bmatrix} \begin{bmatrix} 0 \\ 10 \end{bmatrix} [1/s]$$

$$= \begin{bmatrix} \dfrac{10}{s(s+1)(s+2)} \\ \dfrac{10s}{s(s+1)(s+2)} \end{bmatrix}$$

We know that

$$x(t) = 0 + L^{-1}[\Phi(s).B.u(s)]$$

Taking partial fraction

$$\Rightarrow \frac{10}{s(s+1)(s+2)} = \frac{A}{s} + \frac{B}{(s+1)} + \frac{C}{(s+2)}$$

$$10 = A(s+1)(s+2) + Bs(s+2) + Cs(s+1)$$

When $s = 0$	When $s = -1$	When $s = -2$
$10 = A(1)(2) + 0 + 0$	$10 = 0 - B(1) - C(0)$	$10 = 0 + 0 - 2C(-1)$
$10 = 2A$	$10 = -B$	$10 = 2C$
$A = 5$	$B = -10$	$C = 5$

$$\Rightarrow \frac{10}{(s+1)(s+2)} = \frac{A}{(s+1)} + \frac{B}{(s+2)}$$

$$10 = A(s+2) + Bs(s+1)$$

When $s = -2$	When $s = -1$
$10 = 0 + B(-2+1)$	$10 = A(-1+2)$
$10 = -B$	$10 = A$
$B = -10$	

Taking Laplace inverse

$$x(t) = 0 + L^{-1}[\Phi(s).B.u(s)] = \begin{bmatrix} \dfrac{5}{s} - \dfrac{10}{s+1} + \dfrac{5}{s+2} \\ \dfrac{10}{s+1} - \dfrac{10}{s+2} \end{bmatrix} = \begin{bmatrix} 5 - 10e^{-t} + 5e^{-2t} \\ 10e^{-t} - 10e^{-2t} \end{bmatrix}$$

$$y(t) = [C][x(t)]$$

$$y(t) = \begin{bmatrix} 1 & 0 \end{bmatrix} \begin{bmatrix} 5 - 10e^{-t} + 5e^{-2t} \\ 10e^{-t} - 10e^{-2t} \end{bmatrix}$$

$$y(t) = 5 - 10e^{-t} + 5e^{-2t}$$

State diagram

A state diagram is a pictorial representation of a given state model. Let us consider the state model represented by the two equations, namely, state equation and output equation as given below

$$\dot{X}(t) = AX(t) + BU(t)$$

$$Y(t) = CX(t) + DU(t)$$

The above state model can be represented pictorially as shown in Figure 2.1. This representation is called state diagram.

Illustration 8

Let us consider a second-order system represented by the following state model

$$\dot{X}_1(t) = a_{11}x_1(t) + a_{12}x_2(t) + b_1u(t)$$

$$\dot{X}_2(t) = a_{21}x_1(t) + a_{22}x_2(t) + b_2u(t)$$

$$y(t) = c_1x_1(t) + c_2x_2(t)$$

Draw the corresponding state diagram.

Solution

The state diagram of the given state model is drawn as shown in Figure 2.2.

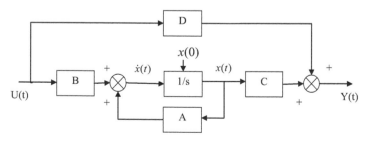

FIGURE 2.1
State diagram representing the state model.

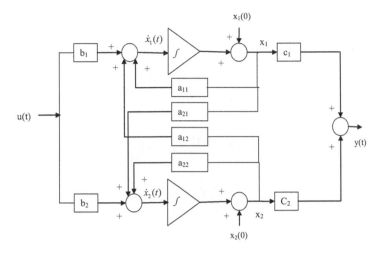

FIGURE 2.2
State diagram representing the given state model.

Illustration 9

Obtain the state diagram of a system whose input–output relationship is given by the transfer function.

$$\frac{Y(s)}{U(s)} = \frac{3(s+4)}{(s+3)(s+5)}$$

Solution

Given transfer function is

$$\frac{Y(s)}{U(s)} = \frac{3(s+4)}{(s+3)(s+5)}$$

The above transfer function can be rearranged using partial fraction approach. Thus, we have

$$\frac{3(s+4)}{(s+3)(s+5)} = \frac{3s+12}{(s+3)(s+5)} = \frac{A}{(s+3)} + \frac{B}{(s+5)}$$

$$3s+12 = A(s+5) + B(s+3)$$

$$3s+12 = As + 5A + Bs + 3B$$

$$3s+12 = s(A+B) = (5A+3B)$$

Equating 's' terms and constant terms, we get

$$A+B = 3$$

$$5A + 3B = 12$$

Solving the above two equations for A and B, we get, $A = 4/3$ and $B = 2/3$.

$$\therefore \frac{Y(s)}{U(s)} = \frac{3s+12}{(s+3)(s+5)} = \frac{3/2}{(s+3)} + \frac{3/2}{(s+5)}$$

$$Y(s) = \left(\frac{3}{2}\right)\left(\frac{U(s)}{(s+3)}\right) + \left(\frac{3}{2}\right)\left(\frac{U(s)}{(s+5)}\right)$$

$$Y(s) = \left(\frac{3}{2}\right)X_1(s) + \left(\frac{3}{2}\right)X_2(s)$$

where $X_1(s) = \dfrac{U(s)}{(s+3)}$ and $X_2(s) = \dfrac{U(s)}{(s+5)}$

We have

$$X_1(s) = \frac{U(s)}{(s+3)} \Rightarrow X_1(s)(s+3) = U(s)$$

$$\dot{x}_1(t) + 3x_1 = u(t)$$

$$\dot{x}_1(t) = u(t) - 3x_1$$

$$\dot{x}_1(t) = 1u(t) - 3x_1 + 0x_2 \qquad \text{(i)}$$

We have

$$X_2(s) = \frac{U(s)}{(s+5)} \Rightarrow X_2(s)(s+5) = U(s)$$

$$\dot{x}_2(t) + 5x_2 = u(t)$$

$$\dot{x}_2(t) = u(t) - 5x_2 \qquad \text{(ii)}$$

$$\dot{x}_2(t) = 0x_1(t) - 5x_2 + 1u(t)$$

Also we have

$$Y(s) = \frac{3}{2}X_1(s) + \frac{3}{2}X_2(s)$$

$$y(t) = \frac{3}{2}x_1(t) + \frac{3}{2}x_2(t) + 0u(t) \qquad \text{(iii)}$$

Thus, the state diagram can be drawn as shown in Figure 2.3

From equations (i), (ii) and (iii), the state model of given transfer function can be written as follows:

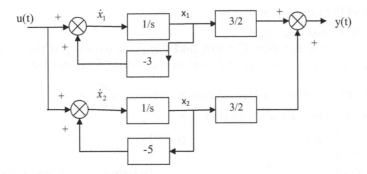

FIGURE 2.3
State diagram representing the given transfer function.

$$
\begin{bmatrix} \dot{x}_1 \\ \dot{x}_2 \end{bmatrix} = \begin{bmatrix} -3 & 0 \\ 0 & -5 \end{bmatrix} \begin{bmatrix} x_1 \\ x_2 \end{bmatrix} + \begin{bmatrix} 1 \\ 1 \end{bmatrix} u
$$

$$
Y = \begin{bmatrix} \tfrac{3}{2} & \tfrac{3}{2} \end{bmatrix} \begin{bmatrix} x_1 \\ x_2 \end{bmatrix} + [0] u
$$

Minimal Realization

An n-dimensional realization $\{A, B, C, D\}$ is called minimal realization or irreducible realization if there is no other realization of dimensions lower than 'n' is realizable. The procedure for the computation of minimal realization for the given transfer function matrices is discussed below.

Minimal Realization Using Transfer Function Matrix

A given state model can be manipulated to obtain minimal realization using the following steps.

1. Find the lowest common multiple (LCM) of the transfer function matrix $d(s)$
2. Find the degree of $d(s) = r$ and roots $\lambda_1, \lambda_2, \ldots \lambda_r$
3. Expand the transfer function model $H(s)$ into partial fractions

$$
H(s) = \frac{n(s)}{d(s)} = \sum_{i=1}^{r} \frac{R_i}{(s - \lambda_i)} \text{ where } R_i \text{ is the residue matrix}
$$

4. Find R_i, $i = 1,2,3\ldots r$ where 'r' is the degree of $d(s)$

5. Find the rank (ρ_i) of R_i where $i = 1,2,3...r$
6. Find C_i and B_i using R_i
   i.e., $R_i = C_i B_i$, where $C_i = p \times \rho_i$ and $B_i = \rho_i \times m$. (where p is the number of inputs, ρ_i is the rank of R_i and m is the number of outputs).

7. Find the order (m) of realization 'n' where $n = \sum_{i=1}^{r} \rho_i$
8. Obtain the realization of $H(s)$

$$
A = \begin{bmatrix} \lambda_1 I_{\rho 1} & & & 0 \\ & \lambda_2 I_{\rho 2} & & \\ & & \ddots & \\ 0 & & & \lambda_r I_{\rho r} \end{bmatrix} \text{ where } \lambda_1, \lambda_2 \dots \lambda_r \text{ are roots } d(s).
$$

$I_{\rho 1}, I_{\rho 1} \dots I_{\rho r}$ are identity matrices of order $\rho_1, \rho_2, \dots \rho_r$, respectively, where '$r$' is the degree of $d(s)$.

$$
B = \begin{bmatrix} B_1 \\ B_2 \\ \vdots \\ B_r \end{bmatrix}; \quad C = \begin{bmatrix} C_1 & C_2 & \cdots & C_r \end{bmatrix}
$$

Illustration 10

Find the minimal realization of a system, whose transfer function matrix is

$$
\text{given by } H(s) = \begin{bmatrix} \dfrac{2}{(s-1)^2} & \dfrac{1}{(s-1)} \\ \dfrac{-6}{(s-1)(s+3)} & \dfrac{1}{(s+3)} \end{bmatrix}
$$

Solution
Given

$$
H(s) = \begin{bmatrix} \dfrac{2}{(s-1)^2} & \dfrac{1}{(s-1)} \\ \dfrac{-6}{(s-1)(s+3)} & \dfrac{1}{(s+3)} \end{bmatrix}
$$

1. Find the LCM of the transfer function matrix $d(s)$
   $d(s) = (s-1)(s+3)$
2. Find the degree of $d(s) = r$ and roots $\lambda_1, \lambda_2, \dots \lambda_r$
   Degree $r = 2$; roots $\lambda_1 = 1$, $\lambda_2 = -3$
3. Expand the transfer function model $H(s)$

$$H(s) = \sum_{i=1}^{r} \frac{R_i}{(s - \lambda_i)} = \frac{R_1}{s - \lambda_1} + \frac{R_2}{s - \lambda_2}$$

$$\Rightarrow H(s) = \frac{R_1}{s-1} + \frac{R_2}{s+3}$$

4. Find R_i, $i = 1,2,3...r$

R_1 and R_2 can be written as numerator coefficients of transfer function matrix, which contain $(s-1)$ and $(s+3)$ terms.

Here, R_1 is a residue matrix of the pole $(s-1)$, hence consider only the numerator (residue) containing the $(s-1)$ terms in $H(s)$.

$$\text{where } H(s) = \begin{bmatrix} \dfrac{2}{(s-1)^2} & \dfrac{1}{(s-1)} \\ \dfrac{-6}{(s-1)(s+3)} & \dfrac{1}{(s+3)} \end{bmatrix}$$

$$\text{Accordingly, } R_1 = \begin{bmatrix} 2 & 1 \\ -6 & 0 \end{bmatrix}$$

Similarly, R_2 is a residue matrix of the pole $(s+3)$

$$\therefore R_2 = \begin{bmatrix} 0 & 0 \\ -6 & 1 \end{bmatrix}$$

5. Find the rank of $R_i = \rho_i$, $i = 1,2,3...r$.

$$R_1 = \begin{vmatrix} 2 & 1 \\ -6 & 0 \end{vmatrix} = 0 + 6 \neq 0 \therefore \text{Rank of } R_1, \text{ i.e., } \rho_1 = 2$$

$$R_2 = \begin{vmatrix} 0 & 0 \\ -6 & 1 \end{vmatrix} = 0 \therefore \text{Rank of } R_2, \text{ i.e., } \rho_1 = 1$$

6. Find C_i and B_i using R_i, i.e., $R_i = C_i B_i$

$$R_1 = \begin{bmatrix} 2 & 1 \\ -6 & 0 \end{bmatrix} = C_1 B_1 = \begin{bmatrix} 2 & 0 \\ -6 & 0 \end{bmatrix} \begin{bmatrix} 1 & 0 \\ 0 & 1 \end{bmatrix}$$

$$R_2 = \begin{bmatrix} 0 & 0 \\ -6 & 1 \end{bmatrix} = C_2 B_2 = \begin{bmatrix} 0 \\ 1 \end{bmatrix} \begin{bmatrix} -6 & 1 \end{bmatrix}$$

7. Find the order of realization 'n' where $n = n = \sum_{i=1}^{r} \rho_i = \rho_1 + \rho_2$

$\therefore n = 2 + 1 = 3$

$\therefore$ Number of state variables $= 3$

8. Obtain the minimal realization of $H(s)$

$$A = \begin{bmatrix} \lambda_1 I_{\rho 1} & \\ & \lambda_2 I_{\rho 2} \end{bmatrix}; \quad B = \begin{bmatrix} B_1 \\ B_2 \end{bmatrix}; \quad C = \begin{bmatrix} C_1 & C_2 \end{bmatrix}$$

$$\lambda_1 I_{\rho 1} = (1)\begin{bmatrix} 1 & 0 \\ 0 & 1 \end{bmatrix} = \begin{bmatrix} 1 & 0 \\ 0 & 1 \end{bmatrix}$$

$$\lambda_2 I_{\rho 2} = (-3)\begin{bmatrix} 1 \end{bmatrix} = -3$$

$$\therefore A = \begin{bmatrix} 1 & 0 & 0 \\ 0 & 1 & 0 \\ 0 & 0 & -3 \end{bmatrix}; \quad B = \begin{bmatrix} 1 & 0 \\ 0 & 1 \\ -6 & 1 \end{bmatrix}; \quad C = \begin{bmatrix} 2 & 0 & 0 \\ -6 & 0 & 1 \end{bmatrix}$$

$\therefore$ State model is given by

$$\begin{bmatrix} \dot{x}_1(t) \\ \dot{x}_2(t) \\ \dot{x}_3(t) \end{bmatrix} = \begin{bmatrix} 1 & 0 & 0 \\ 0 & 1 & 0 \\ 0 & 0 & -3 \end{bmatrix} \begin{bmatrix} x_1(t) \\ x_2(t) \\ x_3(t) \end{bmatrix} + \begin{bmatrix} 1 & 0 \\ 0 & 1 \\ -6 & 1 \end{bmatrix} \begin{bmatrix} u_1 \\ u_2 \end{bmatrix}$$

$$\begin{bmatrix} B_1 \\ B_2 \end{bmatrix} = \begin{bmatrix} 2 & 0 & 0 \\ -6 & 0 & 1 \end{bmatrix} \begin{bmatrix} x_1(t) \\ x_2(t) \\ x_3(t) \end{bmatrix}$$

Non-Minimal Realization

A state space model is said to be non-minimal model when its transfer function matrix $H(s)$ has a lower order than the dimension of the state space, i.e., some eigen values of matrix A do not appear in the transfer function matrix.

Non-Minimal Realization Using Transfer Function Matrix

A given state model can be manipulated to obtain non-minimal realization using the following steps.

1. Find the transfer functions of the MIMO system as $G_{11}(s)$, $G_{12}(s)$... $G_{nm}(s)$ separately.
2. Convert them into their first companion form.
3. Find the elements of $[A]$, $[B]$ and $[C]$ matrices.

Illustration 11

Find the non-minimal realization of a system, whose transfer function matrix is given by

$$G(s) = \begin{bmatrix} \dfrac{s^2 + 2}{s^3 + 6s^2 - 4s - 1} & \dfrac{3s^2 - 7s + 2}{s^3 + 6s^2 - 4s - 1} \\ \dfrac{s^2 - 2s + 1}{s^3 + 6s^2 - 4s - 1} & \dfrac{-s^2 + 2s + 7}{s^3 + 6s^2 - 4s - 1} \end{bmatrix}$$

Solution
Given

$$G(s) = \begin{bmatrix} \dfrac{s^2 + 2}{s^3 + 6s^2 - 4s - 1} & \dfrac{3s^2 - 7s + 2}{s^3 + 6s^2 - 4s - 1} \\ \dfrac{s^2 - 2s + 1}{s^3 + 6s^2 - 4s - 1} & \dfrac{-s^2 + 2s + 7}{s^3 + 6s^2 - 4s - 1} \end{bmatrix}$$

Here,

$$\Rightarrow G_{11}(s) = \frac{s^2 + 2}{s^3 + 6s^2 - 4s - 1} = \frac{\beta_0 s^n + \beta_1 s^{n-1} + \beta_2 s^{n-2} + \ldots + \beta_n}{s^n + \alpha_1 s^{n-1} + \alpha_2 s^{n-2} + \alpha_n} \qquad (i)$$

Rewriting $G_{11}(s)$, we have

$$G_{11}(s) = \frac{0s^3 + s^2 + 0s + 2}{s^3 + 6s^2 - 4s - 1} \qquad (ii)$$

From (i) and (ii), we have

$$\alpha_1 = 6;\ \alpha_2 = -4;\ \alpha_3 = -1 \text{ and } \beta_0 = 0;\ \beta_1 = 1;\ \beta_2 = 0;\ \beta_3 = 2 \qquad (iii)$$

We know that for single-input single-output (SISO) system of order 3, the first companion form can be written as follows:

$$A = \begin{bmatrix} 0 & 1 & 0 \\ 0 & 0 & 1 \\ -\alpha_n & -\alpha_{n-1} & -\alpha_{n-2} \end{bmatrix};\quad B = \begin{bmatrix} 0 \\ 0 \\ 1 \end{bmatrix}$$

$$C = \begin{bmatrix} \beta_n - \alpha_n \beta_0 & \beta_{n-1} - \alpha_{n-1}\beta_0 & \beta_{n-2} - \alpha_{n-2}\beta_0 \end{bmatrix} \qquad (iv)$$

$$D = \beta_0$$

From (iii) and (iv), we have

$$a_{11} = \beta_n - \alpha_n \beta_0 = \beta_3 - \alpha_3 \beta_0 = 2 - (-1)0 = 2$$

$$b_{11} = \beta_{n-1} - \alpha_{n-1} \beta_0 = \beta_2 - \alpha_2 \beta_0 = 0 - (-4)0 = 0$$

$$c_{11} = \beta_{n-2} - \alpha_{n-2} \beta_0 = \beta_1 - \alpha_1 \beta_0 = 1 - (6)0 = 1$$

$$\Rightarrow G_{12}(s) = \frac{3s^2 - 7s + 2}{s^3 + 6s^2 - 4s - 1} = \frac{\beta_0 s^n + \beta_1 s^{n-1} + \beta_2 s^{n-2} + \cdots + \beta_n}{s^n + \alpha_1 s^{n-1} + \alpha_2 s^{n-2} + \alpha_n} \qquad \text{(v)}$$

Rewriting $G_{12}(s)$, we have

$$G_{12}(s) = \frac{0s^3 + 3s^2 - 7s + 2}{s^3 + 6s^2 - 4s - 1} \qquad \text{(vi)}$$

From (v) and (vi), we have

$$\alpha_1 = 6; \; \alpha_2 = -4; \; \alpha_3 = -1 \text{ and } \beta_0 = 0; \; \beta_1 = 3; \; \beta_2 = -7; \; \beta_3 = 2 \qquad \text{(vii)}$$

From (vii), we have

$$a_{12} = \beta_n - \alpha_n \beta_0 = \beta_3 - \alpha_3 \beta_0 = 2 - (-1)0 = 2$$

$$b_{12} = \beta_{n-1} - \alpha_{n-1} \beta_0 = \beta_2 - \alpha_2 \beta_0 = -7 - (-4)0 = 0$$

$$c_{12} = \beta_{n-2} - \alpha_{n-2} \beta_0 = 3 - (6)0 = 1$$

$$\Rightarrow G_{21}(s) = \frac{s^2 - 2s + 1}{s^3 + 6s^2 - 4s - 1} = \frac{\beta_0 s^n + \beta_1 s^{n-1} + \beta_2 s^{n-2} + \cdots + \beta_n}{s^n + \alpha_1 s^{n-1} + \alpha_2 s^{n-2} + \alpha_n} \qquad \text{(viii)}$$

Rewriting $G_{21}(s)$, we have

$$G_{21}(s) = \frac{0s^3 + s^2 - 2s + 1}{s^3 + 6s^2 - 4s - 1} \qquad \text{(ix)}$$

From (viii) and (ix), we have

$$\alpha_1 = 6; \; \alpha_2 = -4; \; \alpha_3 = -1 \text{ and } \beta_0 = 0; \; \beta_1 = 1; \; \beta_2 = -2; \; \beta_3 = 1 \qquad \text{(x)}$$

From (x), we have

$$a_{21} = \beta_n - \alpha_n \beta_0 = \beta_3 - \alpha_3 \beta_0 = 1 - (-1)0 = 2$$

$$b_{21} = \beta_{n-1} - \alpha_{n-1} \beta_0 = \beta_2 - \alpha_2 \beta_0 = -2 - (-4)0 = 0$$

$$c_{21} = \beta_{n-2} - \alpha_{n-2} \beta_0 = \beta_1 - \alpha_1 \beta_0 = 1 - (6)0 = 1$$

$$\Rightarrow G_{22}(s) = \frac{-s^2 + 2s + 7}{s^3 + 6s^2 - 4s - 1} = \frac{\beta_0 s^n + \beta_1 s^{n-1} + \beta_2 s^{n-2} + \cdots + \beta_n}{s^n + \alpha_1 s^{n-1} + \alpha_2 s^{n-2} + \alpha_n} \tag{xi}$$

Rewriting $G_{21}(s)$, we have

$$G_{22}(s) = \frac{0s^3 - s^2 + 2s + 7}{s^3 + 6s^2 - 4s - 1} \tag{xii}$$

From (xi) and (xii), we have

$$\alpha_1 = 6;\ \alpha_2 = -4;\ \alpha_3 = -1 \text{ and } \beta_0 = 0;\ \beta_1 = -1;\ \beta_2 = 2;\ \beta_3 = 7 \tag{xiii}$$

From (xiii),

$$a_{22} = \beta_n - \alpha_n \beta_0 = \beta_3 - \alpha_3 \beta_0 = 7 - (-1)0 = 2$$
$$b_{22} = \beta_{n-1} - \alpha_{n-1}\beta_0 = \beta_2 - \alpha_2 \beta_0 = 2 - (-4)0 = 0$$
$$c_{22} = \beta_{n-2} - \alpha_{n-2}\beta_0 = \beta_1 - \alpha_1 \beta_0 = -1 - (6)0 = 1$$

Now,

$$A = \begin{bmatrix} 0 & 1 & 0 \\ 0 & 0 & 1 \\ -\alpha_n & -\alpha_{n-1} & -\alpha_{n-2} \end{bmatrix};\quad B = \begin{bmatrix} 0 \\ 0 \\ 1 \end{bmatrix};$$

$$C = \begin{bmatrix} \beta_n - \alpha_n \beta_0 & \beta_{n-1} - \alpha_{n-1}\beta_0 & \beta_{n-2} - \alpha_{n-2}\beta_0 \end{bmatrix} \text{ and } D = \beta_0$$

$$A = \begin{bmatrix} 0 & 0 & 1 & 0 & 0 & 0 \\ 0 & 0 & 0 & 1 & 0 & 0 \\ 0 & 0 & 0 & 0 & 1 & 0 \\ 0 & 0 & 0 & 0 & 0 & 1 \\ 1 & 0 & 4 & 0 & -6 & 0 \\ 0 & 1 & 0 & 4 & 0 & -6 \end{bmatrix} \text{ and } B = \begin{bmatrix} 0 & 0 \\ 0 & 0 \\ 0 & 0 \\ 0 & 0 \\ 1 & 0 \\ 0 & 1 \end{bmatrix}$$

$$C = \begin{bmatrix} 2 & 2 & 0 & -7 & 1 & 3 \\ 1 & 7 & -7 & 2 & 1 & -1 \end{bmatrix};\ D = 0$$

Balanced Realization

In minimal realization of stable continuous time systems, there exists a coordinate transformation 'T' such that controllability and observability grammians are equal and diagonal. The diagonal entries of the transformed controllability and observability grammians will be Hankel singular values. Such a realization is called balanced realization. This approach plays a major role in model reduction.

The various steps involved in obtaining balanced realization are as follows:

Step 1: Compute controllability grammian P and observability grammian Q such that it satisfies the Lyapunov equation

$$AP + PA^T + BB^T = 0$$

$$A^T Q + QA + C^T C = 0$$

Step 2: Compute Hankel singular values of 'Σ' that are equal to the positive square root of eigen values of the product of gramians PQ

$$\sigma_i = \sqrt{\lambda_i(PQ)}$$

Step 3: Determine 'Σ' = diag($\sigma_1, \sigma_2 \ldots \sigma_n$) where $\sigma_1 > \sigma_2 > \ldots \sigma_n$

Step 4: Determine the lower triangular Cholesky factors of the grammians L_c and L_o

Step 5: Perform singular value decomposition (SVD) of the product of $L_o^T L_c$

Step 6: Find transformation matrix 'T' where

$$T = L_c V S^{-1/2}$$

$$T^{-1} = S^{-1/2} U^T L_o^T$$

Step 7: Balanced realization ($\tilde{A}, \tilde{B}, \tilde{C}$) is determined as follows:

$$\tilde{A} = T^{-1} A^T; \quad \tilde{B} = T^{-1} B; \quad \tilde{C} = CT$$

The balanced state equations are given by

$$\dot{x} = \begin{bmatrix} \tilde{A} \end{bmatrix} \begin{bmatrix} x_1 \\ x_2 \end{bmatrix} + \begin{bmatrix} \tilde{B} \end{bmatrix} u; \quad y = \begin{bmatrix} \tilde{C} \end{bmatrix} \begin{bmatrix} x_1 \\ x_2 \end{bmatrix}$$

Illustration 12

Obtain the balanced realization of the given continuous system represented in state space

$$\dot{x} = \begin{bmatrix} 0 & 1 \\ -2 & -3 \end{bmatrix} \begin{bmatrix} x_1 \\ x_2 \end{bmatrix} + \begin{bmatrix} 0 \\ 1 \end{bmatrix} u; \quad y = \begin{bmatrix} 1 & 0 \end{bmatrix} \begin{bmatrix} x_1 \\ x_2 \end{bmatrix}$$

Solution

Given

$$A = \begin{bmatrix} 0 & 1 \\ -2 & -3 \end{bmatrix}; \quad B = \begin{bmatrix} 0 \\ 1 \end{bmatrix}; \quad C = \begin{bmatrix} 1 & 0 \end{bmatrix}$$

Step 1: Find P and Q such that the Lyapunov equations are satisfied

We have $AP + PA^T + BB^T = 0$

$$\begin{bmatrix} 0 & 1 \\ -2 & -3 \end{bmatrix} \begin{bmatrix} p_{11} & p_{12} \\ p_{21} & p_{22} \end{bmatrix} + \begin{bmatrix} p_{11} & p_{12} \\ p_{21} & p_{22} \end{bmatrix} \begin{bmatrix} 0 & -2 \\ 1 & -3 \end{bmatrix} + \begin{bmatrix} 0 \\ 1 \end{bmatrix} \begin{bmatrix} 0 & 1 \end{bmatrix} = 0$$

By expanding, we get

$$0p_{11} + 1p_{12} + 1p_{21} + 0p_{22} = 0$$

$$-2p_{11} - 3p_{12} + 0p_{21} + 1p_{22} = 0$$

$$-2p_{11} + 0p_{12} - 3p_{21} + 1p_{22} = 0$$

$$0p_{11} - 2p_{12} - 2p_{21} - 6p_{22} = -1$$

Solving for $p_{11}, p_{12}, p_{21}\ p_{22}$ using Cramer's rule, we get

$$P = \begin{bmatrix} 1/12 & 0 \\ 0 & 1/6 \end{bmatrix}$$

Similarly, we have $A^T Q + QA + C^T C = 0$

$$\begin{bmatrix} 0 & -2 \\ 1 & -3 \end{bmatrix} \begin{bmatrix} Q_{11} & Q_{12} \\ Q_{21} & Q_{22} \end{bmatrix} + \begin{bmatrix} Q_{11} & Q_{12} \\ Q_{21} & Q_{22} \end{bmatrix} \begin{bmatrix} 0 & 1 \\ -2 & -3 \end{bmatrix} + \begin{bmatrix} 1 \\ 0 \end{bmatrix} \begin{bmatrix} 1 & 0 \end{bmatrix} = 0$$

By expanding, we get

$$0Q_{11} - 2Q_{12} - 2Q_{21} + 0Q_{22} = -1$$

$$1Q_{11} - 3Q_{12} + 0Q_{21} - 2Q_{22} = 0$$

$$1Q_{11} + 0Q_{12} - 3Q_{21} - 2Q_{22} = 0$$

$$0Q_{11} + 1Q_{12} + 1Q_{21} - 6Q_{22} = 0$$

Solving for $Q_{11}, Q_{12}, Q_{21}, Q_{22}$ using Cramer's rule, we get

$$Q = \begin{bmatrix} 11/12 & 1/4 \\ 14 & 1/12 \end{bmatrix}$$

Step 2: Compute $\sigma_i = \sqrt{\lambda_i(PQ)}$, where $i = 1, 2$

$$[P][Q] = \begin{bmatrix} 1/12 & 0 \\ 0 & 1/6 \end{bmatrix} \begin{bmatrix} 11/12 & 1/4 \\ 14 & 1/12 \end{bmatrix} = \begin{bmatrix} 11/144 & 1/48 \\ 1/24 & 1/72 \end{bmatrix}$$

$$|\lambda I - PQ| = gives \begin{vmatrix} \lambda - \frac{11}{144} & -\frac{1}{48} \\ -\frac{1}{24} & \lambda - \frac{1}{72} \end{vmatrix} = 0$$

Solving for 'λ', we get $\lambda = 0.088,\ 2.198 \times 10^{-3}$

$$\sigma_1 = \sqrt{0.088} = 0.29665$$

$$\sigma_2 = \sqrt{2.198 \times 10^{-3}} = 0.04688$$

Step 3: Find 'Σ' = diag(σ_1, σ_2)

$$\Sigma = \begin{bmatrix} 0.29665 & 0 \\ 0 & 0.04688 \end{bmatrix}$$

Step 4: Determine L_C and L_O

$$L_c = \begin{bmatrix} 0.2887 & 0 \\ 0 & 0.4082 \end{bmatrix} \quad L_o = \begin{bmatrix} 0.9574 & 0 \\ 0.2611 & 0.1231 \end{bmatrix}$$

Step 5: Perform SVD of the product of $L_o^T L_c$

$$U = \begin{bmatrix} 0.9980 & -0.0625 \\ 0.0625 & 0.9980 \end{bmatrix}; \quad S = \begin{bmatrix} 0.2968 & 0 \\ 0 & 0.468 \end{bmatrix};$$

$$V = \begin{bmatrix} 0.9294 & -0.3690 \\ 0.3690 & 0.9294 \end{bmatrix}$$

Step 6: Find transformation matrix 'T' and T^{-1}

$$T = L_c V S^{-1/2} = \begin{bmatrix} 0.2887 & 0 \\ 0 & 0.4082 \end{bmatrix} \begin{bmatrix} 0.9294 & -0.3690 \\ 0.3690 & 0.9294 \end{bmatrix}$$

$$\times \begin{bmatrix} 0.2968 & 0 \\ 0 & 0.468 \end{bmatrix}^{-1/2}$$

$$T = \begin{bmatrix} 0.4925 & -0.4925 \\ 0.2766 & 1.7540 \end{bmatrix}$$

$$T^{-1} = S^{-1/2} U^T L_o^T = \begin{bmatrix} 0.2968 & 0 \\ 0 & 0.468 \end{bmatrix}^{-1/2} \begin{bmatrix} 0.9980 & -0.0625 \\ 0.0625 & 0.9980 \end{bmatrix}^T$$

$$\times \begin{bmatrix} 0.9574 & 0 \\ 0.2611 & 0.1231 \end{bmatrix}^T$$

$$T^{-1} = \begin{bmatrix} 1.7540 & 0.4925 \\ -0.2766 & 0.4925 \end{bmatrix}$$

$$T^{-1} = \begin{bmatrix} 1.7540 & 0.4925 \\ -0.2766 & 0.4925 \end{bmatrix}$$

Step 7: Balanced realization $(\tilde{A}, \tilde{B}, \tilde{C})$ is determined as follows:

$$\tilde{A} = T^{-1} A^T = \begin{bmatrix} -0.4086 & 0.9701 \\ -0.9701 & -2.591 \end{bmatrix}$$

$$\tilde{B} = T^{-1} B = \begin{bmatrix} 0.4925 \\ 0.4925 \end{bmatrix}$$

$$\tilde{C} = C T = \begin{bmatrix} 0.4925 & -0.4925 \end{bmatrix}$$

Thus, the balanced realization state equations can be written as follows:

$$\dot{x} = \begin{bmatrix} \tilde{A} \end{bmatrix} \begin{bmatrix} x_1 \\ x_2 \end{bmatrix} + \begin{bmatrix} \tilde{B} \end{bmatrix} u$$

$$\dot{x} = \begin{bmatrix} -0.4086 & 0.901 \\ -0.9701 & -2.591 \end{bmatrix} \begin{bmatrix} x_1 \\ x_2 \end{bmatrix} + \begin{bmatrix} 0.4925 \\ 0.4925 \end{bmatrix} u$$

$$y = \begin{bmatrix} \tilde{C} \end{bmatrix} \begin{bmatrix} x_1 \\ x_2 \end{bmatrix}$$

$$y = \begin{bmatrix} 0.4925 & -0.4925 \end{bmatrix} \begin{bmatrix} x_1 \\ x_2 \end{bmatrix}$$

3

State Feedback Control and State Estimator

This chapter on 'State Feedback Control and State Estimator' deals with the concepts of state controllability, output controllability, stabilizability, complete observability, Kalman's tests for controllability and observability, detectability, state space representation in various canonical forms (controllable canonical form, observable canonical from, diagonal canonical form and Jordan canonical form), state feedback control using pole placement technique, determination of state feedback gain matrix (using transformation matrix approach, direct substitution method and Ackermann's formula), state estimation (observer), full order/reduced order/minimum order observers, mathematical model of state observer and determination of state observer gain matrix (using transformation matrix approach, direct substitution method and Ackermann's formula). The concepts are also illustrated with suitable numerical examples.

The concepts of controllability and observability are important in the control system design in state space. The Kalman's test helps us to perform the controllability and observability tests. The canonical form of representation of state model results in decoupling of various components which makes it convenient for carrying out analysis. The advantage of control system design using state feedback is that, any inner parameter can be used as feedback and also the closed loop pole can be located at any desired place. In the pole placement technique, it is assumed that all the state variables are measureable. In real world, all the state variables are not measurable. The state estimation (observation) technique helps to estimate the unmeasured state variable.

Concept of Controllability and Observability

Most of the engineering applications are interactive in nature. It is essential to determine the mode of interaction and the parameters forcing the interactions in the system. Also, the concept of controllability and observability plays an important role in the control system design. By using the controllability and observability properties, the structured features of a dynamic system can be analysed.

Controllability

a) State Controllability

For a linear system (represented in the form of a state model given by equations (3.1 & 3.2), if there exists an input $U_{[0,t_1]}$ which transfers the initial state $x(0)$ to any final state in a final interval of time $t_0 \leq t \leq t_1$, then the state $x(0)$ is said to be controllable. If every state is controllable, then the system is said to be completely controllable.

$$\dot{X}(t) = AX + BU \tag{3.1}$$

$$Y(t) = CX + DU \tag{3.2}$$

We have seen earlier, the solution of equation 3.1 as

$$X(t) = e^{At}x(0) + \int_0^t e^{A(t-\tau)}Bu(\tau)d\tau$$

Applying the definition of complete state controllability, we have

$$X(t_1) = 0 = e^{At_1}x(0) + \int_0^{t_1} e^{A(t_1-\tau)}Bu(\tau)d\tau$$

$$x(0) = -\int_0^{t_1} e^{-A\tau}Bu(\tau)d\tau$$

$$\Rightarrow x(0) = -\sum_{k=0}^{n-1} A^k B \int_0^{t_1} \alpha_k(\tau)u(\tau)d\tau \quad \left(\because e^{-At} = \sum_{k=0}^{n-1} \alpha_k(\tau)A^k \right)$$

Let, $\int_0^{t_1} \alpha_k(\tau)u(\tau)d\tau = \beta_k$

$$\Rightarrow x(0) = -\sum_{k=0}^{n-1} A^k B \beta_k$$

$$\Rightarrow x(0) = -\begin{bmatrix} B & AB & A^2B & \cdots & A^{n-1}B \end{bmatrix} \begin{bmatrix} \beta_0 \\ \beta_1 \\ \beta_2 \\ \vdots \\ \beta_{n-1} \end{bmatrix} \tag{3.3}$$

Now, as per Kalman's controllability test, for the system to be completely state controllable, the equation (3.1) has to be satisfied. This implies that the rank of matrix $\begin{bmatrix} B & AB & A^2B & \cdots & A^{n-1}B \end{bmatrix}$ shall be 'n'. The matrix $\begin{bmatrix} B & AB & A^2B & \cdots & A^{n-1}B \end{bmatrix}$ is called controllability matrix.

Alternate form of the condition for complete state controllability
Consider the state equation (3.1) $\dot{X}(t) = AX + BU$

If the eigen vectors of 'A' are distinct, then it is possible to find a transformation matrix 'P' such that

$$P^{-1}AP = D = \begin{bmatrix} \lambda_1 & 0 & 0 & \cdots & 0 \\ 0 & \lambda_2 & 0 & \cdots & 0 \\ 0 & & \lambda_3 & \cdots & 0 \\ & & & \vdots & \\ 0 & 0 & 0 & \cdots & \lambda_n \end{bmatrix}$$

Let us define $\quad X = PZ$

$$\Rightarrow \dot{X} = P\dot{Z} \tag{3.4}$$

Substituting equation (3.4) in equation (3.1), we get

$$P\dot{Z} = APZ + BU$$

$$\dot{Z} = P^{-1}APZ + P^{-1}BU \tag{3.5}$$

Let us define $P^{-1}B = F$

We can write equation (3.5) as

Now, the condition for complete state controllability is that if the eigen vectors are distinct, then the system is completely state controllable, if and only if no row of $P^{-1}B$ has all zero elements.

NOTE

To apply this condition, we have to keep the matrix $P^{-1}AP$ in diagonal form.

NOTE

If the eigen vectors of A are not distinct, then diagonalization is not possible. For example, if the eigen values of A are $\lambda_1, \lambda_1, \lambda_1, \lambda_4, \lambda_4, \lambda_6 \dots \lambda_n$ and has $(n-3)$ distinct eigen vectors, then we have to transform 'A' into Jordan canonical form.

$$J = \begin{bmatrix} \lambda_1 & 1 & 0 & & & & & 0 \\ 0 & \lambda_1 & 1 & & & & & \\ 0 & 0 & \lambda_1 & & & & & \\ & & & \lambda_4 & 1 & & & \\ & & & 0 & \lambda_4 & & & \\ & & & & & \lambda_6 & & \\ & & & & & & \ddots & \\ & & & & & & & \lambda_n \end{bmatrix}$$

The sequence sub-matrices are called Jordan blocks.

If we define a new state vector 'Z' by $X = SZ$ and substitute it in the state equation (3.1), we have

$$S\dot{Z} = ASZ + BU$$

$$\dot{Z} = S^{-1}ASZ + S^{-1}BU \tag{3.6}$$

$$\dot{Z} = JZ + S^{-1}BU$$

where $J = S^{-1}AS$.

The condition for complete state controllability of the system may be stated as follows: the system is completely state controllable if and only if

  i. No two Jordan blocks in 'J' of equation (3.6) are associated with same eigen values.

 ii. The elements of any row of $S^{-1}B$ that correspond to last row of each Jordan block are not all zero.

iii. The elements of any row of $S^{-1}B$ that correspond to distinct eigen values are not all zero.

Condition for Complete State Controllability in the s-Plane

The condition for complete state controllability can be stated in terms of transfer function. It can be proved that the necessary and sufficient condition for complete state controllability is that no cancellation occur in the transfer function or transfer matrix. If cancellation occurs, then the system cannot be controlled in the direction of the cancelled mode.

For example, if $\dfrac{X(s)}{Y(s)} = \dfrac{(s+3)}{(s+3)(s-1)}$

Clearly, cancellation of the factor $(s + 3)$ occurs in the numerator and denominator of the above transfer function. (One degree of freedom is lost.) Due to this cancellation, this system is not completely state controllable.

Output Controllability

Consider a system represented by state model as in equations (3.1) and (3.2)

$$\dot{X}(t) = AX + BU$$

$$Y(t) = CX + DU$$

$X - n \times 1; Y - q \times 1; U - p \times 1; A - n \times n; B - n \times p; C - q \times n; D - q \times p.$

The system described by above equations is said to be completely output controllable, if it is possible to construct an unconstrained control vector $u(t)$ that will transfer any given initial output $y(t_0)$ to any final output $y(t_1)$ in a finite time interval $t_0 \leq t \leq t_1$.

Thus, the system described by equations (3.1) and (3.2) is completely output controllable, if the matrix $\begin{bmatrix} CB & CAB & CA^2B & \cdots & CA^{n-1}B & D \end{bmatrix}$ is of the rank 'q'.

Uncontrollable System

An uncontrollable system has a subsystem that is physically disconnected from the input.

Stabilizability

For a partially controllable system, if the uncontrollable modes are stable and the unstable modes are controllable, the system is said to be stabilizable.

Illustration 1

$$\begin{bmatrix} \dot{x}_1 \\ \dot{x}_2 \end{bmatrix} = \begin{bmatrix} 1 & 0 \\ 0 & -1 \end{bmatrix} \begin{bmatrix} x_1 \\ x_2 \end{bmatrix} + \begin{bmatrix} 1 \\ 0 \end{bmatrix} u.$$

Show that the system is stabilizable.

Solution

Given

$$A = \begin{bmatrix} 1 & 0 \\ 0 & -1 \end{bmatrix}; \quad B = \begin{bmatrix} 1 \\ 0 \end{bmatrix}$$

$$AB = \begin{bmatrix} 1 & 0 \\ 0 & -1 \end{bmatrix} \begin{bmatrix} 1 \\ 0 \end{bmatrix} = \begin{bmatrix} 1+0 \\ 0+0 \end{bmatrix} = \begin{bmatrix} 1 \\ 0 \end{bmatrix}$$

$$\begin{bmatrix} B & AB \end{bmatrix} = \begin{bmatrix} 1 & 1 \\ 0 & 0 \end{bmatrix} = [0]$$

Here, the stable mode that corresponds to the eigen value of −1 is not controllable. The unstable mode that corresponds to eigen value of 1 is controllable. Such a system can be made stable by the use of suitable feedback. Thus, this system is stabilizable.

Observability

Let us consider the unforced system described by the following equations:

$$\dot{X}(t) = AX$$

$$Y(t) = CX$$

$x - n \times 1; y - q \times 1; u - p \times 1; A - n \times n; B - n \times p; C - q \times n; D - q \times p.$

The system is said to be completely observable, if every state $x(t_0)$ can be determined from the observation of $y(t)$ over a finite time interval $t_0 \leq t \leq t_1$.

NOTE: WHY OBSERVABILITY IS NEEDED?

The difficulty encountered with state feedback control is that, some of the state variables are not accessible for direct measurement, with the result that it becomes necessary to estimate the unmeasurable state variables in order to construct the control signals. Such estimates of state variables are possible, if and only if the system is observable.

In discussing observability conditions, we consider the unforced system, i.e., $\dot{X}(t) = AX$ and $Y(t) = CX$. The reason is as follows:

If the system is described by

$$\dot{X}(t) = AX + BU \quad \text{and}$$

$$Y(t) = CX + DU$$

Then, the solution is given by

$$X(t) = e^{At}x(0) + \int_0^t e^{A(t-\tau)} Bu(\tau) d\tau \tag{3.7}$$

$$X(t) = e^{At}x(0) + C\int_0^t e^{A(t-\tau)} Bu(\tau) d\tau + Du \tag{3.8}$$

Since, the matrices A, B, C and D are known and $u(t)$ is also known, the last two terms on the right-hand side of equation (3.8) are known quantities. Therefore, they can be subtracted from the observed value of $y(t)$.

Hence, for investigating necessary and sufficient condition for complete observability, it is sufficient to consider the unforced system represented by $\dot{X}(t) = AX$ and $Y(t) = CX$.

Complete Observability

We have $\dot{X}(t) = AX$ and $Y(t) = CX$ the solution is given by

$$\left.\begin{array}{l} \dot{X}(t) = e^{At}x(0) \quad \text{and} \\[2mm] Y(t) = Ce^{At}x(0) \end{array}\right\} \tag{3.9}$$

where

$$e^{At} = \sum_{k=0}^{n-1} \alpha_k(t)A^k$$

$$\left.\begin{array}{l} \therefore \quad Y(t) = \sum_{k=0}^{n-1} \alpha_k(t)CA^k x(0) \\[4mm] \text{Or} \quad Y(t) = \alpha_0(t)Cx(t_0) + \alpha_1(t)CAx(t_0) + \cdots + \alpha_{n-1}(t)CA^{n-1}x(t_0) \end{array}\right\} \tag{3.10}$$

For the system to be completely observable, if and only if the rank of matrix $\begin{bmatrix} C & AC & \cdots & A^{n-1}C \end{bmatrix}$ is of rank n. This matrix is called observability matrix.

Condition for Complete Observability in the s-Plane

The necessary and sufficient condition for complete observability is that no cancellations occur in the transfer function or transfer matrix. If cancellation occurs, the cancelled mode cannot be observed in the output.

Let the transfer function between $X_1(s)$ and $U(s)$ be

$$\frac{X_1(s)}{U(s)} = \frac{1}{(s+1)(s+2)(s+3)}$$

and the transfer function between $Y(s)$ and $X_1(s)$ be

$$\frac{Y(s)}{X_1(s)} = (s+1)(s+4)$$

Then the transfer function between $Y(s)$ and $U(s)$ is given by

$$\frac{Y(s)}{U(s)} = \frac{Y(s)\big/X_1(s)}{U(s)\big/X_1(s)} = \frac{(s+1)(s+4)}{(s+1)(s+2)(s+3)}$$

Clearly, the factors $(s+1)$ cancel each other. This means that there are non-zero initial states $x(0)$, which cannot be determined from the measurement, $y(t)$.

Alternate form of the condition to check complete observability
We have $\dot{X}(t) = AX$

$$Y(t) = CX$$

Suppose the transformation matrix 'P' transforms A into a diagonal matrix, i.e.,
$P^{-1}AP = D$ where D is a diagonal matrix

$$\text{Let us define}\quad X = PZ \Rightarrow \dot{X} = P\dot{Z}$$

$$AX = P\dot{Z}\quad \left[\because \dot{X} = AX\right]$$

$$APZ = P\dot{Z}\quad \left[\because X = PZ\right]$$

$$\dot{Z} = P^{-1}APZ = DZ$$

$$Y = CX = CPZ$$

$$\Rightarrow Y(t) = CPe^{\lambda t}z(0)$$

$$Y(t) = CP\begin{bmatrix} e^{\lambda_1 t} & & & 0 \\ & e^{\lambda_2 t} & & \\ & & \ddots & \\ 0 & & & e^{\lambda_n t} \end{bmatrix} \quad z(0) = CP\begin{bmatrix} e^{\lambda_1 t}z_1(0) \\ e^{\lambda_2 t}z_2(0) \\ \\ e^{\lambda_n t}z_n(0) \end{bmatrix} \quad (3.11)$$

The system is completely observable, if none of the columns of CP matrix consists of all zero elements. This is because, if the ith column of CP consists of all zero elements, then the state variables $z_i(0)$ will not appear in the output equation and therefore cannot be determined from observation of $y(t)$. Thus, $X(0)$ which is related to $x(t)$ by the non-singular matrix P cannot be determined.

NOTE

The above test is applicable only when the matrix $P^{-1}AP$ is in diagonal form.

If the matrix A cannot be transformed into a diagonal matrix, then by use of suitable transformation matrix S, we can transform A into a Jordan canonical form, i.e., $S^{-1}AS = J$, where J is in Jordan canonical form.

Let us define

$$X = SZ \qquad \text{(i)}$$

$$\dot{X} = S\dot{Z} \qquad \text{(ii)}$$

We have

$$\dot{X} = AX \qquad \text{(iii)}$$

$$Y = CX \qquad \text{(iv)}$$

From equations (ii), (iii) and (iv), we have

$$S\dot{Z} = AX \quad \left(\text{From equation (i)}\right)$$

$$S\dot{Z} = ASZ$$

$$\dot{Z} = S^{-1}ASZ$$

and

$$Y = CX = CSZ$$

$$y(t) = CSe^{Jt}z(0) \qquad \text{(3.12)}$$

The system is completely observable, if

i. No two Jordan blocks in 'J' are associated with same eigen values.

ii. No columns of CS that correspond to the first row of each Jordan block consist of zero elements.

iii. No columns of CS that correspond to distinct eigen values consist of zero elements.

Detectability

For a partially observable system, if the unobservable modes are stable and the observable modes are unstable, the system is said to be detectable. In fact, the concept of detectability is dual to the concept of stabilizability.

Kalman's Tests for Controllability and Observability

The concepts of controllability and observability were introduced by Kalman. They play important role in the design of control systems in state space. The solution to the problem may not exist, if the system considered is not controllable. Although most systems are controllable and observable, corresponding mathematical models may not possess the property of controllability and observability. In that case, it is necessary to know the conditions under which a system is controllable and observable. In fact, the condition of controllability and observability may govern the existence of a complete solution to the control system design problem.

NOTE

As per Kalman's controllability test, the system is said to be completely state controllable, if the rank of the matrix $\begin{bmatrix} B & AB & A^2B & \cdots & A^{n-1}B \end{bmatrix}$ is 'n' where 'n' is the order of the system [n is also equal to the number of state variables].

NOTE

As per Kalman's test, the system is said to be completely output controllable, if the rank of the matrix $\begin{bmatrix} CB & CAB & CA^2B & \cdots & CA^{n-1}BD \end{bmatrix}$ is 'q' where 'q' is the number of output vectors.

NOTE

As per Kalman's test, the system is said to be completely observable, if the rank of the matrix $\begin{bmatrix} C^T & A^TC^T & \cdots & (A^T)^{n-1}C^T \end{bmatrix}$ is 'n' where 'n' is the order of the system [n is also equal to the number of state variables].

Illustration 2

Given $\begin{bmatrix} \dot{x}_1 \\ \dot{x}_2 \end{bmatrix} = \begin{bmatrix} 7 & 3 \\ -2 & -5 \end{bmatrix}\begin{bmatrix} x_1 \\ x_2 \end{bmatrix} + \begin{bmatrix} 0 \\ 1 \end{bmatrix}u$ and $y = \begin{bmatrix} 1 & 0 \end{bmatrix}\begin{bmatrix} x_1 \\ x_2 \end{bmatrix}$

Comment whether the system is

    a. Completely state controllable
    b. Completely output controllable
    c. Completely observable.

Solution

To check for complete state controllability
    Let us find the controllability matrix and then its ranks. We have the controllability matrix given by $\begin{bmatrix} B & AB & A^2B & \cdots & A^{n-1}B \end{bmatrix}$

In our case, $n = 2$

$\therefore$ Controllability matrix $= \begin{bmatrix} B & AB \end{bmatrix}$

Given $A = \begin{bmatrix} 7 & 3 \\ -2 & -5 \end{bmatrix}$ and $B = \begin{bmatrix} 0 \\ 1 \end{bmatrix}$

$$\therefore AB = \begin{bmatrix} 7 & 3 \\ -2 & -5 \end{bmatrix}\begin{bmatrix} 0 \\ 1 \end{bmatrix} = \begin{bmatrix} 3 \\ -5 \end{bmatrix}$$

$\therefore$ Controllability matrix $= \begin{bmatrix} B & AB \end{bmatrix} = \begin{bmatrix} 0 & 3 \\ 1 & -5 \end{bmatrix}$ whose rank is 2. The number of state variables is also 2. Therefore, the system is completely state controllable.

To check for complete output controllability
    Let us find the controllability matrix and then its rank. We have the controllability matrix given by $\begin{bmatrix} CB & CAB & CA^2B & \cdots & CA^{n-1}BD \end{bmatrix}$

In our case $n = 2$

$\therefore$ Controllability matrix $= \begin{bmatrix} CB & CAB \end{bmatrix}$

Given $A = \begin{bmatrix} 7 & 3 \\ -2 & -5 \end{bmatrix}$; $B = \begin{bmatrix} 0 \\ 1 \end{bmatrix}$ and $C = \begin{bmatrix} 1 & 0 \end{bmatrix}$

$$\therefore CB = \begin{bmatrix} 1 & 0 \end{bmatrix}\begin{bmatrix} 0 \\ 1 \end{bmatrix} = \begin{bmatrix} 0 \end{bmatrix}$$

$$\therefore CAB = \begin{bmatrix} 1 & 0 \end{bmatrix}\begin{bmatrix} 7 & 3 \\ -2 & -5 \end{bmatrix}\begin{bmatrix} 0 \\ 1 \end{bmatrix} = \begin{bmatrix} 7 & 3 \end{bmatrix}\begin{bmatrix} 0 \\ 1 \end{bmatrix} = \begin{bmatrix} 3 \end{bmatrix}$$

∴Output controllability matrix $= \begin{bmatrix} CB & CAB \end{bmatrix} = \begin{bmatrix} 0 & 3 \end{bmatrix}$ whose rank is 1. The number of output vectors is also 1. Therefore, the system is completely output controllable.

To check for complete observability

Let us find the observability matrix $\begin{bmatrix} C^T & A^T C^T & \cdots & (A^T)^{n-1} C^T \end{bmatrix}$ and then its rank. In our case, $n = 2$

∴Observability matrix $= \begin{bmatrix} C^T & A^T C^T \end{bmatrix}$

Given $A = \begin{bmatrix} 7 & 3 \\ -2 & -5 \end{bmatrix}$ and $C = \begin{bmatrix} 1 & 0 \end{bmatrix}$

$$\therefore C^T = \begin{bmatrix} 1 \\ 0 \end{bmatrix}$$

$$\therefore A^T C^T = \begin{bmatrix} 7 & -2 \\ 3 & -5 \end{bmatrix} \begin{bmatrix} 1 \\ 0 \end{bmatrix} = \begin{bmatrix} 7 \\ 3 \end{bmatrix}$$

∴Observability matrix $= \begin{bmatrix} C^T & A^T C^T \end{bmatrix} = \begin{bmatrix} 1 & 7 \\ 0 & 3 \end{bmatrix}$ whose rank is 2. The number of state variables and order of the system are also 2. Therefore, the system is completely observable.

Illustration 3

A feedback system has the following state equations

$$\dot{x}_1 = 5x_1 - 2x_2 + u$$

$$\dot{x}_2 = x_1$$

$$y = x_1 - x_2 + u$$

Comment on the controllability.

Solution

The matrix form of state model can be written as follows:

$$\begin{bmatrix} \dot{x}_1 \\ \dot{x}_2 \end{bmatrix} = \begin{bmatrix} 5 & -2 \\ 1 & 0 \end{bmatrix} \begin{bmatrix} x_1 \\ x_2 \end{bmatrix} + \begin{bmatrix} 1 \\ 0 \end{bmatrix} u \quad \text{and} \quad y = \begin{bmatrix} 1 & -1 \end{bmatrix} \begin{bmatrix} x_1 \\ x_2 \end{bmatrix} + [1]u$$

The order of the system is 2, i.e., $n = 2$.

The controllability matrix is given by $\begin{bmatrix} B & AB \end{bmatrix}$

where $A = \begin{bmatrix} 5 & -2 \\ 1 & 0 \end{bmatrix}$; $B = \begin{bmatrix} 1 \\ 0 \end{bmatrix}$

$$AB = \begin{bmatrix} 5 & -2 \\ 1 & 0 \end{bmatrix} \begin{bmatrix} 1 \\ 0 \end{bmatrix} = \begin{bmatrix} 5 \\ 1 \end{bmatrix}$$

∴Controllability matrix $= \begin{bmatrix} B & AB \end{bmatrix} = \begin{bmatrix} 1 & 5 \\ 0 & 1 \end{bmatrix}$ whose rank is 2. The number of state variables is also 2. Therefore, the system is completely state controllable.

Illustration 4

A feedback system has the following state equations:

$$\dot{x}_1 = -5x_1 + 6x_2 + u$$

$$\dot{x}_2 = x_1 + u$$

$$y = x_1$$

Comment on the controllability.

Solution

The matrix form of state model can be written as follows:

$$\begin{bmatrix} \dot{x}_1 \\ \dot{x}_2 \end{bmatrix} = \begin{bmatrix} -5 & 6 \\ 1 & 0 \end{bmatrix} \begin{bmatrix} x_1 \\ x_2 \end{bmatrix} + \begin{bmatrix} 1 \\ 1 \end{bmatrix} u \quad \text{and} \quad y = \begin{bmatrix} 1 & 0 \end{bmatrix} \begin{bmatrix} x_1 \\ x_2 \end{bmatrix}$$

The order of the system is 2, i.e., $n = 2$.
The controllability matrix is given by $\begin{bmatrix} B & AB \end{bmatrix}$

where $A = \begin{bmatrix} -5 & 6 \\ 1 & 0 \end{bmatrix}$; $B = \begin{bmatrix} 1 \\ 1 \end{bmatrix}$

$$AB = \begin{bmatrix} -5 & 6 \\ 1 & 0 \end{bmatrix} \begin{bmatrix} 1 \\ 1 \end{bmatrix} = \begin{bmatrix} -5+6 \\ 1 \end{bmatrix} = \begin{bmatrix} 1 \\ 1 \end{bmatrix}$$

∴Controllability matrix $= \begin{bmatrix} B & AB \end{bmatrix} = \begin{bmatrix} 1 & 1 \\ 1 & 1 \end{bmatrix}$ whose rank is 1. However, the number of state variables is 2. Therefore, the system is not completely state controllable.

Illustration 5

The state model of a system is as given below. Comment on observability.

$$\begin{bmatrix} \dot{x}_1 \\ \dot{x}_2 \end{bmatrix} = \begin{bmatrix} 4 & -7 \\ 2 & 0 \end{bmatrix} \begin{bmatrix} x_1 \\ x_2 \end{bmatrix} + \begin{bmatrix} 1 \\ 1 \end{bmatrix} u \quad \text{and} \quad y = \begin{bmatrix} 1 & 0 \end{bmatrix} \begin{bmatrix} x_1 \\ x_2 \end{bmatrix}$$

Solution

The order of the system is 2, i.e., $n = 2$
  $\therefore$ Observability matrix $= \begin{bmatrix} C^T & A^T C^T \end{bmatrix}$

Given $A = \begin{bmatrix} 4 & -7 \\ 2 & 0 \end{bmatrix}$ and $C = \begin{bmatrix} 1 & 0 \end{bmatrix}$

$$\therefore C^T = \begin{bmatrix} 1 \\ 0 \end{bmatrix}$$

$$\therefore A^T C^T = \begin{bmatrix} 4 & 2 \\ -7 & 0 \end{bmatrix} \begin{bmatrix} 1 \\ 0 \end{bmatrix} = \begin{bmatrix} 4 \\ -7 \end{bmatrix}$$

$\therefore$ Observability matrix $= \begin{bmatrix} C^T & A^T C^T \end{bmatrix} = \begin{bmatrix} 1 & 4 \\ 0 & -7 \end{bmatrix}$ whose rank is 2. The number of state variables and order of the system are also 2. Therefore, the system is completely observable.

Illustration 6

The state model of a system is as given below. Comment on observability.

$$\begin{bmatrix} \dot{x}_1 \\ \dot{x}_2 \end{bmatrix} = \begin{bmatrix} 5 & 2 \\ 4 & 3 \end{bmatrix} \begin{bmatrix} x_1 \\ x_2 \end{bmatrix} + \begin{bmatrix} 1 \\ 0 \end{bmatrix} u \quad \text{and} \quad y = \begin{bmatrix} 1 & -1 \end{bmatrix} \begin{bmatrix} x_1 \\ x_2 \end{bmatrix}$$

Solution

The order of the system is 2, i.e., $n = 2$
  $\therefore$ Observability matrix $= \begin{bmatrix} C^T & A^T C^T \end{bmatrix}$

Given $A = \begin{bmatrix} 5 & 2 \\ 4 & 3 \end{bmatrix}$ and $C = \begin{bmatrix} 1 & -1 \end{bmatrix}$

$$\therefore C^T = \begin{bmatrix} 1 \\ -1 \end{bmatrix}$$

$$\therefore A^T C^T = \begin{bmatrix} 5 & 4 \\ 2 & 3 \end{bmatrix} \begin{bmatrix} 1 \\ -1 \end{bmatrix} = \begin{bmatrix} 5-4 \\ 2-3 \end{bmatrix} = \begin{bmatrix} 1 \\ -1 \end{bmatrix}$$

$\therefore$ Observability matrix $= \begin{bmatrix} C^T & A^T C^T \end{bmatrix} = \begin{bmatrix} 1 & 1 \\ -1 & -1 \end{bmatrix}$ whose rank is 2, the number of state variables and order of the system are also 2. Therefore, the system is completely observable.

State Space Representation in Canonical Forms

Canonical form of state variable representation is convenient for time domain analysis as each component in state variable equation is decoupled from all other components. There are many canonical forms available for representing the given transfer function in state space. They are classified as follows (Figure 3.1):

Consider a system represented by

$$\frac{Y(s)}{U(s)} = \frac{b_0 s^n + b_1 s^{n-1} + b_2 s^{n-2} + \cdots + b_{n-1} s + b_n}{s^n + a_1 s^{n-1} + a_2 s^{n-2} \cdots + a_{n-1} s + a_n} \tag{3.13}$$

where 'u' is the input and 'y' is the output.

Controllable Canonical Form

The controllable canonical form is useful in the control system design performed using pole placement technique. The system represented by equation (3.13) can be rewritten in state space controllable canonical form as given in equation (3.14)

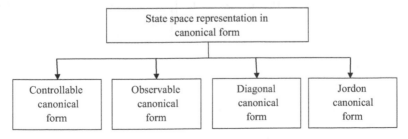

FIGURE 3.1
Classification of state space representation in canonical form.

$$
\begin{bmatrix} \dot{x}_1 \\ \dot{x}_2 \\ \vdots \\ \dot{x}_{n-1} \\ \dot{x}_n \end{bmatrix} = \overbrace{\begin{bmatrix} 0 & 1 & 0 & \cdots & 0 \\ 0 & 0 & 1 & \cdots & 0 \\ \vdots & \vdots & \vdots & & \vdots \\ 0 & 0 & 0 & \cdots & 1 \\ -a_n & -a_{n-1} & -a_{n-2} & \cdots & -a_1 \end{bmatrix}}^{A_c} \begin{bmatrix} x_1 \\ x_2 \\ \vdots \\ x_{n-1} \\ x_n \end{bmatrix} + \overbrace{\begin{bmatrix} 0 \\ 0 \\ \vdots \\ 0 \\ 1 \end{bmatrix}}^{B_c} u
$$

$$(3.14)$$

$$
y = \overbrace{\begin{bmatrix} (b_n - a_n b_0) & (b_{n-1} - a_{n-1} b_0) & \cdots & (b_1 - a_1 b_0) \end{bmatrix}}^{C_c} \begin{bmatrix} x_1 \\ x_2 \\ \vdots \\ x_n \end{bmatrix} + b_0 u \qquad (3.15)
$$

Observable Canonical Form

The system represented by equation (3.13) can be rewritten in state space observable canonical form as given in equation (3.16)

$$
\begin{bmatrix} \dot{x}_1 \\ \dot{x}_2 \\ \vdots \\ \dot{x}_{n-1} \\ \dot{x}_n \end{bmatrix} = \overbrace{\begin{bmatrix} 0 & 1 & 0 & \cdots & -a_n \\ 1 & 0 & 0 & \cdots & -a_{n-1} \\ \vdots & \vdots & \vdots & & \vdots \\ 0 & 0 & 0 & \cdots & \\ 0 & 0 & 0 & 1 & -a_1 \end{bmatrix}}^{A_o} \begin{bmatrix} x_1 \\ x_2 \\ \vdots \\ x_{n-1} \\ x_n \end{bmatrix} + \overbrace{\begin{bmatrix} (b_n - a_n b_0) \\ (b_{n-1} - a_{n-1} b_0) \\ \vdots \\ \vdots \\ (b_1 - a_1 b_0) \end{bmatrix}}^{B_o} u
$$

$$(3.16)$$

$$
y = \overbrace{\begin{bmatrix} 0 & 0 & \cdots & 1 \end{bmatrix}}^{C_o} \begin{bmatrix} x_1 \\ x_2 \\ \vdots \\ x_{n-1} \\ x_n \end{bmatrix} + b_0 u \qquad (3.17)
$$

NOTE

$$A_o = [A_c]^T ; \; B_o = [C_c]^T ; C_o = [B_c]^T$$

Diagonal Canonical Form

Let us consider a system transfer function with denominator polynomial having only distinct roots. In that case, the system can be represented by the equation (3.18)

$$\frac{Y(s)}{U(s)} = \frac{b_0 s^n + b_1 s^{n-1} + b_2 s^{n-2} + \cdots + b_{n-1} s + b_n}{(s+p_1)(s+p_2)\cdots(s+p_n)} \tag{3.18}$$

The equation (3.18) can be rewritten as

$$\frac{Y(s)}{U(s)} = b_0 + \frac{C_1}{(s+p_1)} + \frac{C_2}{(s+p_2)} + \cdots + \frac{C_n}{(s+p_n)} \tag{3.19}$$

The system represented by equation (3.19) can be rewritten in state space observable canonical form as given in equations (3.20) and (3.21)

$$
\begin{bmatrix} \dot{x}_1 \\ \dot{x}_2 \\ \vdots \\ \dot{x}_{n-1} \\ \dot{x}_n \end{bmatrix} =
\begin{bmatrix}
-p_1 & 0 & 0 & \cdots & 0 \\
0 & -p_1 & 0 & \cdots & 0 \\
0 & 0 & -p_1 & & 0 \\
\vdots & & & & \vdots \\
0 & 0 & 0 & \cdots & -p_1
\end{bmatrix}
\begin{bmatrix} x_1 \\ x_2 \\ \vdots \\ x_{n-1} \\ x_n \end{bmatrix} +
\begin{bmatrix} 1 \\ 1 \\ \vdots \\ 1 \\ 1 \end{bmatrix} u \tag{3.20}
$$

$$
y = \begin{bmatrix} C_1 & C_2 & \cdots & C_n \end{bmatrix}
\begin{bmatrix} x_1 \\ x_2 \\ \vdots \\ x_n \end{bmatrix} + b_0 u \tag{3.21}
$$

Jordan Canonical Form

Let us consider a system transfer function with denominator polynomial having multiple roots. In that case, the system can be represented by the equation (3.22)

$$\frac{Y(s)}{U(s)} = \frac{b_0 s^n + b_1 s^{n-1} + b_2 s^{n-2} + \cdots + b_{n-1} s + b_n}{(s+p_1)^3 (s+p_4)(s+p_5)\cdots(s+p_n)} \tag{3.22}$$

The equation (3.22) can be rewritten using partial fraction expansion as

$$\frac{Y(s)}{U(s)} = b_0 + \frac{C_1}{(s+p_1)^3} + \frac{C_2}{(s+p_1)^2} + \frac{C_3}{(s+p_1)} + \frac{C_4}{(s+p_4)} + \cdots + \frac{C_n}{(s+p_n)} \quad (3.23)$$

The system represented by equation (3.23) can be rewritten in state space Jordan canonical form as given in equations (3.24) and (3.25)

$$\begin{bmatrix} \dot{x}_1 \\ \dot{x}_2 \\ \dot{x}_3 \\ \dot{x}_4 \\ \vdots \\ \dot{x}_n \end{bmatrix} = \begin{bmatrix} -p_1 & 1 & 0 & 0 & \cdots & 0 \\ 0 & -p_1 & 1 & 0 & \cdots & 0 \\ 0 & 0 & -p_1 & 0 & \cdots & 0 \\ 0 & 0 & 0 & -p_4 & \cdots & 0 \\ \vdots & \vdots & \vdots & \vdots & & \vdots \\ & & & & \cdots & -p_n \end{bmatrix} \begin{bmatrix} x_1 \\ x_2 \\ x_3 \\ x_4 \\ \vdots \\ x_n \end{bmatrix} + \begin{bmatrix} 0 \\ 0 \\ 1 \\ 1 \\ \vdots \\ 1 \end{bmatrix} u$$

$$(3.24)$$

$$y = \begin{bmatrix} C_1 & C_2 & \cdots & C_n \end{bmatrix} \begin{bmatrix} x_1 \\ x_2 \\ \vdots \\ x_n \end{bmatrix} + b_0 u \quad (3.25)$$

In the subsequent section, the state space representations in different canonical forms are illustrated using numerical examples.

Illustration 7

Obtain the following canonical forms of the system represented by the transfer function

$$\frac{Y(s)}{U(s)} = \frac{(s+6)}{(s^2+5s+4)}$$

(i) Controllable canonical form, (ii) observable canonical form and (iii) diagonal canonical form.

Solution

Given transfer function is

$$\frac{Y(s)}{U(s)} = \frac{(s+6)}{(s^2+5s+4)} = \frac{0s^2+1s+6}{1s^2+5s+4} \quad \text{(i)}$$

We have the standard form of transfer function as

$$\frac{Y(s)}{U(s)} = \frac{b_0 s^n + b_1 s^{n-1} + b_2 s^{n-2} + \cdots + b_{n-1}s + b_n}{s^n + a_1 s^{n-1} + a_2 s^{n-2} \cdots + a_{n-1}s + a_n} \tag{ii}$$

Comparing (i) and (ii), we get
$b_0 = 0; b_1 = 1; b_2 = 6$
$a_1 = 5; a_2 = 4$
Also,

$$b_n - a_n b_0 = b_1 - a_1 b_0 \quad (\text{when } n = 1)$$

$$= 1 - 5(0) = 1$$

$$= b_2 - a_2 b_0 \quad (\text{when } n = 2)$$

$$= 6 - 4(0) = 6$$

i. Controllable Canonical Form

For the generalized controllable canonical form given in equations (3.14) and (3.15), we can write the controllable canonical form for the transfer function as

$$\begin{bmatrix} \dot{x}_1 \\ \dot{x}_2 \end{bmatrix} = \begin{bmatrix} 0 & 1 \\ -a_2 & -a_1 \end{bmatrix} \begin{bmatrix} x_1 \\ x_2 \end{bmatrix} + \begin{bmatrix} 0 \\ 1 \end{bmatrix} u$$

$$y = \begin{bmatrix} (b_2 - a_2 b_0) & (b_1 - a_1 b_0) \end{bmatrix} \begin{bmatrix} x_1 \\ x_2 \end{bmatrix} + \begin{bmatrix} b_0 \end{bmatrix} u$$

Thus, after substituting the values, the canonical form is given by

$$\begin{bmatrix} \dot{x}_1 \\ \dot{x}_2 \end{bmatrix} = \begin{bmatrix} 0 & 1 \\ -4 & -5 \end{bmatrix} \begin{bmatrix} x_1 \\ x_2 \end{bmatrix} + \begin{bmatrix} 0 \\ 1 \end{bmatrix} u; \quad y = \begin{bmatrix} 6 & 1 \end{bmatrix} \begin{bmatrix} x_1 \\ x_2 \end{bmatrix} + [0] u$$

ii. Observable Canonical Form

For the generalized observable canonical form given in equations (3.16) and (3.17), we can write the observable canonical form for the transfer function as

$$\begin{bmatrix} \dot{x}_1 \\ \dot{x}_2 \end{bmatrix} = \begin{bmatrix} 0 & -a_2 \\ 1 & -a_1 \end{bmatrix} \begin{bmatrix} x_1 \\ x_2 \end{bmatrix} + \begin{bmatrix} (b_2 - a_2 b_0) \\ (b_1 - a_1 b_0) \end{bmatrix} u$$

$$y = \begin{bmatrix} 0 & 1 \end{bmatrix} \begin{bmatrix} x_1 \\ x_2 \end{bmatrix} + \begin{bmatrix} b_0 \end{bmatrix} u$$

Thus, after substituting the values, the observable canonical form is given by

$$\begin{bmatrix} \dot{x}_1 \\ \dot{x}_2 \end{bmatrix} = \begin{bmatrix} 0 & -4 \\ 1 & -5 \end{bmatrix} \begin{bmatrix} x_1 \\ x_2 \end{bmatrix} + \begin{bmatrix} 6 \\ 1 \end{bmatrix} u$$

$$y = \begin{bmatrix} 0 & 1 \end{bmatrix} \begin{bmatrix} x_1 \\ x_2 \end{bmatrix} + \begin{bmatrix} 0 \end{bmatrix} u$$

iii. Diagonal Canonical Form

Given transfer function is

$$\frac{Y(s)}{U(s)} = \frac{(s+6)}{(s^2+5s+4)} = \frac{(s+6)}{(s+1)(s+4)} \tag{i}$$

For the generalized diagonal canonical form given by

$$\frac{Y(s)}{U(s)} = b_0 + \frac{C_1}{(s+p_1)^3} + \frac{C_2}{(s+p_1)^2} + \frac{C_3}{(s+p_1)} + \frac{C_4}{(s+p_4)} \cdots + \frac{C_n}{(s+p_n)} \tag{ii}$$

Equation (i) can be rewritten using partial fraction expansion as follows

$$\frac{(s+6)}{(s+1)(s+4)} = \frac{A}{(s+1)} + \frac{B}{(s+4)} = \frac{A(s+4)+(s+1)}{(s+1)(s+4)}$$

$$(s+6) = A(s+4) + B(s+1)$$

$$= As + 4A + Bs + B$$

$$(s+6) = s(A+B) + 4A + B$$

Equating 's' coefficients, we get

$$A + B = 1 \tag{iii}$$

Equating 'constant' coefficients, we get

$$4A + B = 6 \tag{iv}$$

Solving equations (iii) and (iv) for A and B, we get $A = 5/3$ and $B = -2/3$

$$\frac{(s+6)}{(s+1)(s+4)} = \frac{5/3}{(s+1)} + \frac{(-2/3)}{(s+4)} \tag{v}$$

Comparing equations (ii) and (iii), we get
$b_1 = 0; p_1 = 1; p_2 = 4$

$C_1 = 5/3; C_2 = -2/3$

The generalized diagonal canonical form is given by

$$\begin{bmatrix} \dot{x}_1 \\ \dot{x}_2 \end{bmatrix} = \begin{bmatrix} -p_1 & 0 \\ 0 & -p_2 \end{bmatrix} \begin{bmatrix} x_1 \\ x_2 \end{bmatrix} + \begin{bmatrix} 1 \\ 1 \end{bmatrix} u$$

$$y = \begin{bmatrix} C_1 & C_2 \end{bmatrix} \begin{bmatrix} x_1 \\ x_2 \end{bmatrix} + \begin{bmatrix} b_0 \end{bmatrix} u$$

Thus, after substituting the values, the diagonal canonical form of the given transfer is given by

$$\begin{bmatrix} \dot{x}_1 \\ \dot{x}_2 \end{bmatrix} = \begin{bmatrix} -1 & 0 \\ 1 & -4 \end{bmatrix} \begin{bmatrix} x_1 \\ x_2 \end{bmatrix} + \begin{bmatrix} 1 \\ 1 \end{bmatrix} u$$

$$y = \begin{bmatrix} 5/3 & -2/3 \end{bmatrix} \begin{bmatrix} x_1 \\ x_2 \end{bmatrix} + \begin{bmatrix} 0 \end{bmatrix} u$$

Illustration 8

For the system represented by the state model given below, obtain Jordan canonical form of state model.

$$\begin{bmatrix} \dot{x}_1 \\ \dot{x}_2 \\ \dot{x}_3 \end{bmatrix} = \begin{bmatrix} 4 & 1 & -2 \\ 1 & 0 & 2 \\ 1 & -1 & 3 \end{bmatrix} \begin{bmatrix} x_1 \\ x_2 \\ x_3 \end{bmatrix} + \begin{bmatrix} 0 \\ 2 \\ 0 \end{bmatrix} u \text{ and } y = \begin{bmatrix} 1 & 0 & 0 \end{bmatrix} \begin{bmatrix} x_1 \\ x_2 \\ x_3 \end{bmatrix}$$

Solution

For the given state model, we have $A = \begin{bmatrix} 4 & 1 & -2 \\ 1 & 0 & 2 \\ 1 & -1 & 3 \end{bmatrix}; B = \begin{bmatrix} 0 \\ 2 \\ 0 \end{bmatrix};$

$C = \begin{bmatrix} 1 & 0 & 0 \end{bmatrix}$

The various steps involved to obtain the Jordan canonical form of state model are as follows:

1. Determine the eigen values of A, i.e., $|\lambda I - A| = 0$
2. After checking that the eigen values are not distinct, find the elements of matrix M, given by

$$M = \begin{bmatrix} m_1 & m_2 & m_3 \end{bmatrix}$$

where $m_1 = \begin{bmatrix} C_{11} \\ C_{12} \\ C_{13} \end{bmatrix}$ where C_{11}, C_{12}, C_{13} are the cofactors of $[\lambda_1 I - A]$

along the first row.

$m_2 = \begin{bmatrix} C_{11} \\ C_{12} \\ C_{13} \end{bmatrix}$ where C_{11}, C_{12}, C_{13} are the cofactors of $[\lambda_2 I - A]$ along

the first row.

NOTE 1

If $\lambda_2 = \lambda_1$, then $m_2 = \begin{bmatrix} \dfrac{d}{d\lambda_1}C_{11} \\ \dfrac{d}{d\lambda_1}C_{12} \\ \dfrac{d}{d\lambda_1}C_{13} \end{bmatrix}$ where C_{11}, C_{12}, C_{13} are the cofactors of

$[\lambda_1 I - A]$ along the first row.

Similarly, $m_3 = \begin{bmatrix} C_{11} \\ C_{12} \\ C_{13} \end{bmatrix}$ where C_{11}, C_{12}, C_{13} are the cofactors of $[\lambda_3 I - A]$

along the first row.

NOTE 2

If $\lambda_3 = \lambda_2$, then $m_3 = \begin{bmatrix} \dfrac{d}{d\lambda_2}C_{11} \\ \dfrac{d}{d\lambda_2}C_{12} \\ \dfrac{d}{d\lambda_2}C_{13} \end{bmatrix}$ where C_{11}, C_{12}, C_{13} are the cofactors of

$[\lambda_2 I - A]$ along the first row.

NOTE 3

In case the cofactors along the first row give null solution (i.e., a row with all elements as zero), then m_1, m_2 and m_3 have to be formed with

cofactors along the second row. In that case, $m_1 = \begin{bmatrix} C_{21} \\ C_{22} \\ C_{23} \end{bmatrix}$

3. Determine the Jordan matrix, J given by $[J] = [M^{-1}][A][M]$
4. The Jordan canonical form of state model is given by

$$\begin{bmatrix} \dot{z}_1 \\ \dot{z}_2 \\ \dot{z}_3 \end{bmatrix} = [J] \begin{bmatrix} z_1 \\ z_2 \\ z_3 \end{bmatrix} + [\tilde{B}]u \text{ where } \tilde{B} = [M^{-1}][B] \text{ and } y = [\tilde{C}] \begin{bmatrix} z_1 \\ z_2 \\ z_3 \end{bmatrix}$$

where $\tilde{C} = [C][M]$

Step 1: To find the eigen values of the given $[A]$

$$[A] = \begin{bmatrix} 4 & 1 & -2 \\ 1 & 0 & 2 \\ 1 & -1 & 3 \end{bmatrix}$$

$$[\lambda I - A] = \lambda \begin{bmatrix} 1 & 0 & 0 \\ 0 & 1 & 0 \\ 0 & 0 & 1 \end{bmatrix} - \begin{bmatrix} 4 & 1 & -2 \\ 1 & 0 & 2 \\ 1 & -1 & 3 \end{bmatrix}$$

$$= \begin{bmatrix} (\lambda - 4) & -1 & -2 \\ -1 & \lambda & -2 \\ -1 & 1 & (\lambda - 3) \end{bmatrix}$$

$$|\lambda I - A| = (\lambda - 4)[\lambda(\lambda - 3) - (1)(-2)] + 1[(-1)(\lambda - 3) - (-1)(-2)]$$

$$+ 2[(-1)(1) - (-1)(\lambda)]$$

After simplification, we get

$$|\lambda I - A| = (\lambda - 1)(\lambda - 3)(\lambda - 3)$$

To find the eigen values, we have to solve the equation $|\lambda I - A| = 0$

$\therefore \lambda_1 = 1, \lambda_2 = 3, \lambda_3 = 3$.

NOTE

Since, one of the eigen values has multiplicity (in our case λ_2 and λ_3 are equal), the canonical form will be Jordan canonical form.

Step 2: To find the nodal matrix $[M]$, first we have to find m, where

$$m_1 = \begin{bmatrix} C_{11} \\ C_{12} \\ C_{13} \end{bmatrix} \text{ where } C_{11}, C_{12}, C_{13} \text{ are the cofactors of } [\lambda_1 I - A] \text{ along}$$

the first row.
We have

$$[\lambda_1 I - A] = \begin{bmatrix} 1 & 0 & 0 \\ 0 & 1 & 0 \\ 0 & 0 & 1 \end{bmatrix} - \begin{bmatrix} 4 & 1 & -2 \\ 1 & 0 & 2 \\ 1 & -1 & 3 \end{bmatrix} = \begin{bmatrix} -3 & -1 & 2 \\ -1 & 1 & -2 \\ -1 & 1 & -2 \end{bmatrix}$$

$$C_{11} = (-1)^{1+1} \begin{vmatrix} 1 & -2 \\ 1 & -2 \end{vmatrix} = 0$$

$$C_{12} = (-1)^{1+2} \begin{vmatrix} -1 & -2 \\ -1 & -2 \end{vmatrix} = 0$$

$$C_{13} = (-1)^{1+3} \begin{vmatrix} -1 & 1 \\ -1 & 1 \end{vmatrix} = 0$$

The cofactors of $[\lambda_1 I - A]$ along first row give null solution.
∴Let us take cofactors along the second row. In that case,

$$m_1 = \begin{bmatrix} C_{21} \\ C_{22} \\ C_{23} \end{bmatrix}$$

$$C_{21} = (-1)^{2+1} \begin{vmatrix} -1 & 2 \\ 1 & -2 \end{vmatrix} = 0$$

In our case, $C_{22} = (-1)^{2+2} \begin{vmatrix} -3 & 2 \\ -1 & -2 \end{vmatrix} = 8$

$$C_{23} = (-1)^{2+3} \begin{vmatrix} -3 & -1 \\ -1 & 1 \end{vmatrix} = 4$$

$$\therefore m_1 = \begin{bmatrix} 0 \\ 8 \\ 4 \end{bmatrix}$$

Similarly, we have to find the values of m_2 using the second eigen value.

$$m_2 = \begin{bmatrix} C_{11} \\ C_{12} \\ C_{13} \end{bmatrix} \text{ where } C_{11}, C_{12}, C_{13} \text{ are the cofactors of } [\lambda_2 I - A]$$

along the first row.
We have

$$[\lambda_2 I - A] = [3] \begin{bmatrix} 1 & 0 & 0 \\ 0 & 1 & 0 \\ 0 & 0 & 1 \end{bmatrix} - \begin{bmatrix} 4 & 1 & -2 \\ 1 & 0 & 2 \\ 1 & -1 & 3 \end{bmatrix} = \begin{bmatrix} -3 & -1 & 2 \\ -1 & 1 & -2 \\ -1 & 1 & -2 \end{bmatrix}$$

$$C_{11} = (-1)^{1+1} \begin{vmatrix} 3 & -2 \\ 1 & -2 \end{vmatrix} = 2$$

$$C_{12} = (-1)^{1+2} \begin{vmatrix} -1 & -2 \\ -1 & 0 \end{vmatrix} = 2$$

$$C_{13} = (-1)^{1+3} \begin{vmatrix} -1 & 3 \\ -1 & 1 \end{vmatrix} = 2$$

$$\therefore m_2 = \begin{bmatrix} 2 \\ 2 \\ 2 \end{bmatrix}$$

Since $\lambda_2 = \lambda_3$, the matrix m_3 can be written as

$$m_3 = \begin{bmatrix} \dfrac{d}{d\lambda_2} C_{11} \\[2ex] \dfrac{d}{d\lambda_2} C_{12} \\[2ex] \dfrac{d}{d\lambda_2} C_{13} \end{bmatrix}$$

We have

$$[\lambda_2 I - A] = [\lambda_2]\begin{bmatrix} 1 & 0 & 0 \\ 0 & 1 & 0 \\ 0 & 0 & 1 \end{bmatrix} - \begin{bmatrix} 4 & 1 & -2 \\ 1 & 0 & 2 \\ 1 & -1 & 3 \end{bmatrix} = \begin{bmatrix} \lambda_2 - 4 & -1 & 2 \\ -1 & \lambda_2 & -2 \\ -1 & 1 & \lambda_2 - 3 \end{bmatrix}$$

After simplification, we get

$$C_{11} = \lambda_2^2 - 3\lambda_2 + 2$$

$$C_{12} = \lambda_2 - 1$$

$$C_{13} = \lambda_2 - 1$$

$$\frac{d}{d\lambda_2}C_{11} = \frac{d}{d\lambda_2}(\lambda_2^2 - 3\lambda_2 + 2) = 2\lambda_2 - 3 = 3$$

$$\frac{d}{d\lambda_2}C_{12} = \frac{d}{d\lambda_2}(\lambda_2 - 1) = 1$$

$$\frac{d}{d\lambda_2}C_{13} = \frac{d}{d\lambda_2}(\lambda_2 - 1) = 1$$

$$\therefore m_2 = \begin{bmatrix} 3 \\ 1 \\ 1 \end{bmatrix}$$

$\therefore$Modal matrix $M = \begin{bmatrix} m_1 & m_2 & m_3 \end{bmatrix}$

$$M = \begin{bmatrix} 0 & 2 & 3 \\ 8 & 2 & 1 \\ 4 & 2 & 1 \end{bmatrix}$$

Step 3: To determine the Jordan matrix $[J]$
we have, $[J] = [M^{-1}][A][M]$
we have $[M]^{-1} = \dfrac{Adj(M)}{|M|}$

$$\text{Determinant of } [M] = |M| = \begin{vmatrix} 0 & 2 & 3 \\ 8 & 2 & 1 \\ 4 & 2 & 1 \end{vmatrix} = 16$$

Adjoint of $[M] = [\text{Cofactor of } [M]]^T$

$$= \begin{bmatrix} 0 & -4 & 8 \\ 4 & -12 & 8 \\ -4 & 24 & -16 \end{bmatrix}^T = \begin{bmatrix} 0 & 4 & -4 \\ -4 & -12 & 24 \\ 8 & 8 & -16 \end{bmatrix}$$

$$\therefore [M^{-1}] = \frac{1}{16} \begin{bmatrix} 0 & 4 & -4 \\ -4 & -12 & 24 \\ 8 & 8 & -16 \end{bmatrix} = \frac{1}{4} \begin{bmatrix} 0 & 1 & -1 \\ -1 & -3 & 6 \\ 2 & 2 & -4 \end{bmatrix}$$

$$[J] = [M^{-1}][A][M]$$

$$\therefore [J] = \frac{1}{4} \begin{bmatrix} 0 & 1 & -1 \\ -1 & -3 & 6 \\ 2 & 2 & -4 \end{bmatrix} \begin{bmatrix} 4 & 1 & -2 \\ 1 & 0 & 2 \\ 1 & -1 & 3 \end{bmatrix} \begin{bmatrix} 4 & 1 & -2 \\ 1 & 0 & 2 \\ 1 & -1 & 3 \end{bmatrix}$$

$$= \frac{1}{4} \begin{bmatrix} 0 & 1 & -1 \\ -1 & -3 & 6 \\ 2 & 2 & -4 \end{bmatrix} \begin{bmatrix} 0 & 2 & 3 \\ 8 & 2 & 1 \\ 4 & 2 & 1 \end{bmatrix}$$

$$[J] = \frac{1}{4} \begin{bmatrix} 4 & 0 & 0 \\ 0 & 12 & 4 \\ 0 & 0 & 12 \end{bmatrix} = \begin{bmatrix} 1 & 0 & 0 \\ 0 & 3 & 1 \\ 0 & 0 & 3 \end{bmatrix}$$

Step 4: To obtain the Jordan canonical form of state model, we have to find

$$\tilde{B} = [M^{-1}][B]$$

$$\tilde{B} = \frac{1}{4} \begin{bmatrix} 0 & 1 & -1 \\ -1 & -3 & 6 \\ 2 & 2 & -4 \end{bmatrix} \begin{bmatrix} 0 \\ 2 \\ 0 \end{bmatrix} = \frac{1}{4} \begin{bmatrix} 2 \\ -6 \\ 4 \end{bmatrix} = \begin{bmatrix} \frac{2}{4} \\ -\frac{6}{4} \\ \frac{4}{4} \end{bmatrix}$$

$$\tilde{B} = \begin{bmatrix} 0.5 \\ -1.5 \\ 1.0 \end{bmatrix}$$

$$\text{Also } \tilde{C} = [C][M] = \begin{bmatrix} 1 & 0 & 0 \end{bmatrix} \begin{bmatrix} 0 & 2 & 3 \\ 8 & 2 & 1 \\ 4 & 2 & 1 \end{bmatrix} = \begin{bmatrix} 0 & 2 & 3 \end{bmatrix}$$

The Jordan canonical form of state model is given by

$$\begin{bmatrix} \dot{z}_1 \\ \dot{z}_2 \\ \dot{z}_3 \end{bmatrix} = [J] \begin{bmatrix} z_1 \\ z_2 \\ z_3 \end{bmatrix} + [\tilde{B}] u; \quad y = [\tilde{C}] \begin{bmatrix} z_1 \\ z_2 \\ z_3 \end{bmatrix}$$

$$\begin{bmatrix} \dot{z}_1 \\ \dot{z}_2 \\ \dot{z}_3 \end{bmatrix} = \begin{bmatrix} 1 & 0 & 0 \\ 0 & 3 & 1 \\ 0 & 0 & 3 \end{bmatrix} \begin{bmatrix} z_1 \\ z_2 \\ z_3 \end{bmatrix} + \begin{bmatrix} 0.5 \\ -1.5 \\ 1.0 \end{bmatrix} u$$

$$Y = \begin{bmatrix} 0 & 2 & 3 \end{bmatrix} \begin{bmatrix} z_1 \\ z_2 \\ z_3 \end{bmatrix}$$

State Feedback Control (Pole Placement Technique)

In the design of feedback control system using conventional techniques, the design is carried out by manipulating the dominant closed loop poles using appropriate compensation, to meet desired performance specifications (damping ratio or undamped natural frequency). Though, this approach is useful, it has certain limitations. In this case, the design is based on dominant closed loop poles only. Moreover, the desired performance is achieved by using output alone as a feedback signal.

Whereas, in the control system design using state feedback, any inner parameter can be used as feedback. Moreover, the closed loop pole can be placed at any desired location by means of state feedback with the help of an appropriate state feedback gain matrix, and hence, this method is called pole placement technique. The condition to be satisfied by the system for arbitrary pole placement is that, the given system be completely state controllable. However, the limitation of this design technique is the assumption that all the state variables are measurable and are available for feedback (which may not be true in all cases). The state feedback gain matrix can be determined by three methods.

The state model of a system without state feedback is given by

$$\dot{X}(t) = AX + BU \tag{i}$$

$$Y(t) = CX + DU \tag{ii}$$

The block diagram of above system can be drawn as shown in Figure 3.2.

Now, let the same system be fed with state variable feedback. The block diagram of such a system can be drawn as shown in Figure 3.3.

The control signal 'u' is determined by an instantaneous state. Such a system is called state feedback control system.

$$\text{Thus, we have} \quad U = -K_f X \tag{3.26}$$

Substituting equation (3.26) in the state equation, we get

$$\dot{X}(t) = AX + B(-K_f X)$$
$$\dot{X}(t) = (A - BK_f)X \tag{3.27}$$

The solution of equation (3.27) is given by equation (3.28)

$$X(t) = e^{(A-BK_f)t}X(0) \tag{3.28}$$

where $X(0)$ is the initial state caused by external disturbances.

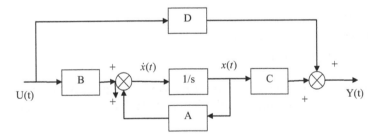

FIGURE 3.2
Block diagram of a system without feedback.

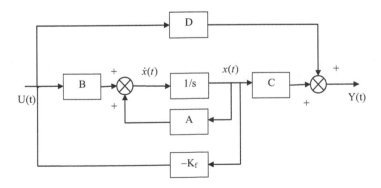

FIGURE 3.3
Block diagram of system with state feedback.

NOTE

The eigen values of matrix [A − BK] are called regulatory poles, and they influence the stability and transient response characteristics of the given system.

By choosing proper value of matrix K_f and by placing the regulatory poles on the left-hand side of s-plane, $x(t)$ can be made to approach zero as 't' approaches to infinity. Locating the regulatory pole at desired place is called pole placement technique.

Determination of State Feedback Gain Matrix (K_f)

The state feedback gain matrix (K_f) can be determined by the following three methods:

  i. Transformation matrix approach
 ii. Direct substitution method
iii. Ackermann's formula

In all the above three methods, the first step is to check whether the given system is completely controllable. This controllability check can be performed as follows:

1. Determine the controllability matrix

$$Q_c = \begin{bmatrix} B & AB & A^2B & \cdots & A^{n-1}B \end{bmatrix}$$

2. Determine the rank of Q_c
3. If the rank of Q_c is equal to the order of system 'n', then the given system is completely controllable.

Procedural steps for determining K_f using transformation matrix, T_f, approach.

Step 1: Verify the condition for complete controllability. If yes, then proceed with step 2.

Step 2: Find the characteristics polynomial $|\lambda I - A| = 0$ for the given original system.

$|\lambda I - A| = \lambda^n + a_1 \lambda^{n-1} + \cdots + a_{n-1}\lambda + a_n$, and hence, determine the values of $a_1, a_2, \ldots a_n$

Step 3: Determine the transformation matrix, T_f, which transforms the state equation into controllable canonical form

$$T_f = \begin{bmatrix} B & AB & A^2B & \cdots & A^{n-1}B \end{bmatrix} \begin{bmatrix} a_{n-1} & a_{n-2} & \cdots & a_1 & 1 \\ a_{n-2} & a_{n-3} & \cdots & 1 & 0 \\ \vdots & \vdots & & \vdots & \vdots \\ a_1 & 1 & & 0 & 0 \\ 0 & 0 & & 0 & \end{bmatrix}$$

NOTE

If the given system is already in controllable canonical form, then $T_f = I$

Step 4: Using the specified closed loop poles; $\mu_1, \mu_2, \mu_3, \ldots \mu_n$, write the desired characteristics polynomial as $(s - \mu_1)(s - \mu_2)\cdots(s - \mu_n) = \lambda^n + \alpha_1\lambda^{n-1} + \cdots + \alpha_{n-1}\lambda + \alpha_n$ and hence determine the values of $\alpha_1, \alpha_2, \ldots \alpha_n$.

Step 5: Determine the required state feedback gain matrix, K_f, using the following equation

$$K_f = \begin{bmatrix} \alpha_n - a_n & \alpha_{n-1} - a_{n-1} & \cdots & \alpha_{n-1} - a_{n-1} & \alpha_1 - a_1 \end{bmatrix}\begin{bmatrix} T_f^{-1} \end{bmatrix} \qquad (3.29)$$

Procedural steps for determining K_f using direct substitution method.

Step 1: Verify the condition for complete controllability. If yes, then proceed with step 2.

Step 2: Using the specified closed loop poles; $\mu_1, \mu_2, \mu_3, \ldots \mu_n$, write the desired characteristics polynomial as

$$(\lambda - \mu_1)(\lambda - \mu_2)\cdots(\lambda - \mu_n) = |\lambda I - (A - BK_f)|$$

$$\text{where} \quad K_f = \begin{bmatrix} K_{f1} & K_{f2} & K_{f3} & \cdots & K_{fn} \end{bmatrix} \qquad (3.30)$$

Step 3: Equate the coefficients of 'λ' on both the sides, and hence, find the values of $K_{f1}, K_{f2}, K_{f3}\ldots K_{fn}$.

> **NOTE**
>
> This method is suitable for lower-order systems ($n \leq 3$). For higher-order system, the calculations become tedious.

Procedural steps for determining K_f using Ackermann's formula

Step 1: Verify the condition for complete controllability. If yes, then proceed with step 2.

Step 2: Using the specified closed loop poles; $\mu_1, \mu_2, \mu_3, \ldots \mu_n$, write the desired characteristics polynomial as $(\lambda - \mu_1)(\lambda - \mu_2)\ldots(\lambda - \mu_n) = \lambda^n + \alpha_1 \lambda^{n-1} + \cdots + \alpha_{n-1}\lambda + \alpha_n$, and hence, determine the values of α_1, $\alpha_2, \ldots \alpha_n$.

Step 3: Determine the matrix $\Phi(A)$ using the coefficients of desired characteristic polynomial as

$$\Phi(A) = A^n + \alpha_1 A^{n-1} + \alpha_2 A^{n-2} + \cdots + \alpha_{n-1}A + \alpha_n I$$

Step 4: Compute the state feedback gain matrix, K_f using the Ackermann's formula as

$$K_f = \begin{bmatrix} 0 & 0 & \ldots & 0 & 1 \end{bmatrix} \begin{bmatrix} B & AB & A^2B & \ldots & A^{n-1}B \end{bmatrix}^{-1} \begin{bmatrix} \Phi(A) \end{bmatrix} \quad (3.31)$$

Illustration 9

Consider a system as given $\begin{bmatrix} \dot{x}_1 \\ \dot{x}_2 \\ \dot{x}_3 \end{bmatrix} = \begin{bmatrix} 0 & 1 & 0 \\ 0 & 0 & 1 \\ -1 & -2 & -3 \end{bmatrix} \begin{bmatrix} x_1 \\ x_2 \\ x_3 \end{bmatrix} + \begin{bmatrix} 0 \\ 0 \\ 1 \end{bmatrix} u.$

Design a feedback controller with a state feedback so that the closed loop poles are placed at 1, 2, 3.

Solution

Method 1

Step 1: Verify the condition for complete controllability

Given that $A = \begin{bmatrix} 0 & 1 & 0 \\ 0 & 0 & 1 \\ -1 & -2 & -3 \end{bmatrix}$ and $B = \begin{bmatrix} 0 \\ 0 \\ 1 \end{bmatrix}$

$$A^2 = A.A = \begin{bmatrix} 0 & 1 & 0 \\ 0 & 0 & 1 \\ -1 & -2 & -3 \end{bmatrix} \begin{bmatrix} 0 & 1 & 0 \\ 0 & 0 & 1 \\ -1 & -2 & -3 \end{bmatrix}$$

$$= \begin{bmatrix} 0 & 0 & 1 \\ -1 & -2 & -3 \\ 3 & 5 & 7 \end{bmatrix}$$

$$AB = \begin{bmatrix} 0 & 1 & 0 \\ 0 & 0 & 1 \\ -1 & -2 & -3 \end{bmatrix} \begin{bmatrix} 0 \\ 0 \\ 1 \end{bmatrix} = \begin{bmatrix} 0 \\ 1 \\ -3 \end{bmatrix}$$

$$A^2B = \begin{bmatrix} 0 & 0 & 1 \\ -1 & -2 & -3 \\ 3 & 5 & 7 \end{bmatrix} \begin{bmatrix} 0 \\ 0 \\ 1 \end{bmatrix} = \begin{bmatrix} 1 \\ -3 \\ 7 \end{bmatrix}$$

$$Q_c = \begin{bmatrix} B & AB & A^2B \end{bmatrix}$$

$$= \begin{bmatrix} 0 & 0 & 1 \\ 0 & 1 & -3 \\ 1 & -3 & 7 \end{bmatrix}$$

$$\text{Determinant of } Q_c = \Delta Q_c = \begin{vmatrix} 0 & 0 & 1 \\ 0 & 1 & -3 \\ 1 & -3 & 7 \end{vmatrix} = 1(0-1) = -1 \neq 0.$$

Therefore, the system is completely state controllable

Find Q_c^{-1}

$$Q_c^{-1} = \frac{[\text{Cofactor of } Q_c]^T}{\text{Determinant of } Q_c} = \frac{1}{-1} \begin{vmatrix} -2 & -3 & -1 \\ -3 & -1 & 0 \\ -1 & 0 & 0 \end{vmatrix}$$

$$\therefore Q_c^{-1} = \begin{vmatrix} 2 & 3 & 1 \\ 3 & 1 & 0 \\ 1 & 0 & 0 \end{vmatrix}$$

Step 2: Find the characteristics polynomial $|\lambda I - A| = 0$

The characteristic polynomial of original system is given by $|\lambda I - A| = 0$

$$[\lambda I - A] = \begin{bmatrix} \lambda & 0 & 0 \\ 0 & \lambda & 0 \\ 0 & 0 & \lambda \end{bmatrix} - \begin{bmatrix} 0 & 1 & 0 \\ 0 & 0 & 1 \\ -1 & -2 & -3 \end{bmatrix} = \begin{bmatrix} \lambda & -1 & 0 \\ 0 & \lambda & -1 \\ 1 & 2 & \lambda+3 \end{bmatrix}$$

$$= \lambda\left[\lambda(\lambda+3)+2\right]+1(1) = \lambda[\lambda^2+3\lambda+2]+1$$

$$= \lambda^3 + 3\lambda^2 + 2\lambda + 1$$

∴The characteristic polynomial of original system is $\lambda^3 + 3\lambda^2 + 2\lambda + 1 = 0$

Step 3: Determine the transformation matrix, T_f

$T_f = I$ (Since, the given system is already in controllable canonical form)

Step 4: Using the specified closed loop poles; $\mu_1, \mu_2, \mu_3, \ldots \mu_n$, write the desired characteristics polynomial.

The desired closed loop poles are $\mu_1 = 1, \mu_2 = 2$ and $\mu_3 = 3$

Hence, the desired characteristic polynomial is given by

$$(\lambda - m_1)(\lambda - m_2)(\lambda - m_3) = (\lambda - 1)(\lambda - 2)(\lambda - 3) = 0$$

$$(\lambda^2 - 2\lambda - 1\lambda + 2)(\lambda - 3) = 0$$

$$(\lambda^2 - 3\lambda + 2)(\lambda - 3) = 0$$

$$\lambda^3 - 3\lambda^2 - 3\lambda^2 + 9\lambda + 2\lambda - 6 = 0$$

$$\lambda^3 - 6\lambda^2 + 11\lambda - 6 = 0$$

Hence, the desired characteristic polynomial is $\lambda^3 - 6\lambda^2 + 11\lambda - 6 = 0$

Step 5: Determine the required state feedback gain matrix, K_f using the following equation

$$K_f = [\alpha_n - a_n \quad \alpha_{n-1} - a_{n-1} \quad \ldots \quad \alpha_{n-1} - a_{n-1} \quad \alpha_1 - a_1][T_f^{-1}]$$

$$= \left[(-6-1) \quad (11-2) \quad (-6-3)\right]$$

$$K_f = [-7 \quad 9 \quad -9]$$

$$K_f = \left[(-6-1) \quad (11-2) \quad (-6-3)\right]$$

$$K_f = [-7 \quad 9 \quad -9]$$

Method 2

Step 1: Check for controllability as in method 1.

Step 2: Using the specified closed loop poles; $\mu_1, \mu_2, \mu_3, \ldots \mu_n$, write the desired characteristics polynomial as $(\lambda - \mu_1)(\lambda - \mu_2)\ldots(\lambda - \mu_n) = |\lambda I - (A - BK)|$

The characteristic polynomial of the system is given by $|\lambda I - (A - BK)| = 0$

$$[\lambda I - A + BK] = \begin{bmatrix} \lambda & 0 & 0 \\ 0 & \lambda & 0 \\ 0 & 0 & \lambda \end{bmatrix} - \begin{bmatrix} 0 & 1 & 0 \\ 0 & 0 & 1 \\ -1 & -2 & -3 \end{bmatrix} + \begin{bmatrix} 0 \\ 0 \\ 1 \end{bmatrix} \begin{bmatrix} K_1 & K_2 & K_3 \end{bmatrix}$$

$$= \begin{bmatrix} \lambda & -1 & 0 \\ 0 & \lambda & -1 \\ 1 & 2 & \lambda+3 \end{bmatrix} + \begin{bmatrix} 0 & 0 & 0 \\ 0 & 0 & 0 \\ K_1 & K_2 & K_3 \end{bmatrix}$$

$$= \begin{bmatrix} \lambda & -1 & 0 \\ 0 & \lambda & -1 \\ 1+K_1 & 2+K_2 & \lambda+3+K_3 \end{bmatrix}$$

$$[\lambda I - A + BK] = \lambda\big[\lambda(\lambda+3+K_3)+(2+K_2)\big]+1\big[0+(1+K_1)\big]=0$$

$$\lambda\big[\lambda^2 + 3\lambda + K_3\lambda + 2 + K_2\big] + 1 + K_1 \quad = 0$$

$$\lambda^3 + 3\lambda^2 + K_3\lambda^2 + 2\lambda + K_2\lambda + 1 + K_1 \quad = 0$$

$$\lambda^3 + (3+K_3)\lambda^2 + (2+K_2)\lambda + 1 + K_1 \quad = 0$$

Step 3: Equate the coefficients of 'λ' and constant term on both the sides, and hence, find the vales of $K_{f1}, K_{f2}, K_{f3} \ldots K_{fn}$.

$1 + K1 = -6$	$2 + K2 = 11$	$3 + K3 = -6$
$K1 = -7$	$K2 = 9s$	$K3 = -9$

$$\Rightarrow K_f = \begin{bmatrix} K_1 & K_2 & K_3 \end{bmatrix} = [-7 \quad 9 \quad -9]$$

Method 3

Step 1: Verify the condition for complete controllability as in method 1.

Step 2: Using the specified closed loop poles; $\mu_1, \mu_2, \mu_3, \ldots \mu_n$, write the desired characteristics polynomial.

The desired closed loop poles are $\mu_1 = 1$, $\mu_2 = 2$ and $\mu_3 = 3$
Hence, the desired characteristic polynomial is given by

$$(\lambda - m_1)(\lambda - m_2)(\lambda - m_3) = (\lambda - 1)(\lambda - 2)(\lambda - 3) = 0$$

$$(\lambda^2 - 2\lambda - 1\lambda + 2)(\lambda - 3) = 0$$

$$(\lambda^2 - 3\lambda + 2)(\lambda - 3) = 0$$

$$\lambda^3 - 3\lambda^2 - 3\lambda^2 + 9\lambda + 2\lambda - 6 = 0$$

$$\lambda^3 - 6\lambda^2 + 11\lambda - 6 = 0$$

$$\Rightarrow a_1 = -6; \ a_2 = 11; \ a_3 = -6;$$

Step 3: Determine the matrix $\Phi(A)$ using the coefficients of desired characteristic polynomial as

$$\Phi(A) = A^3 + \alpha_1 A^2 + \alpha_2 A + \alpha_3 I$$

$$= \begin{bmatrix} -1 & -2 & -3 \\ 3 & 5 & 7 \\ -7 & -11 & -16 \end{bmatrix} + (-6) \begin{bmatrix} 0 & 0 & 1 \\ -1 & -2 & -3 \\ 3 & 5 & 7 \end{bmatrix}$$

$$+ 11 \begin{bmatrix} 0 & 1 & 0 \\ 0 & 0 & 1 \\ -1 & -2 & -3 \end{bmatrix} + (-6) \begin{bmatrix} 1 & 0 & 0 \\ 0 & 1 & 0 \\ 0 & 0 & 1 \end{bmatrix}$$

$$= \begin{bmatrix} -7 & 9 & -9 \\ 9 & 1 & 36 \\ -36 & -63 & -97 \end{bmatrix}$$

Step 4: Compute the state feedback gain matrix, K_f using the Ackermann's formula as

$$K_f = \begin{bmatrix} 0 & 0 & \dots & 0 & 1 \end{bmatrix} \begin{bmatrix} B & AB & A^2B & \dots & A^{n-1}B \end{bmatrix}^{-1} \begin{bmatrix} \Phi(A) \end{bmatrix}$$

$$= \begin{bmatrix} 0 & 0 & 1 \end{bmatrix} \times \begin{bmatrix} 2 & 3 & 7 \\ 3 & 1 & 0 \\ 1 & 0 & 0 \end{bmatrix} \times \begin{bmatrix} -7 & 9 & -9 \\ 9 & 1 & 36 \\ -36 & -63 & -97 \end{bmatrix}$$

$$K_f = [-7 \quad 9 \quad -9] \ \text{(After simplification)}$$

State Observers

In the state feedback control using pole placement technique, it is assumed that all the state variables are measurable and, hence, are available for feedback. However, in practice, all the state variables are not measurable and, hence, are not available for feedback. In such cases, the unmeasurable state variables need to be estimated. The estimation of unmeasurable state variable is called observation. The tool used for the purpose of observation or estimation is called 'observer'.

Full-Order State Observers

If all the state variables are observed by the state observer, then it is called full-order state observer.

Reduced-Order State Observers

An observer that estimates the states fewer than the 'n' state variables is called a reduced-order state observer.

Minimum-Order State Observers

If the order of the reduced order observer is the minimum possible order, then that reduced order observer is called minimum-order state observer.

Mathematical Model of an Observer

Let us consider a system with a state model represented by

$$\dot{X}(t) = AX + BU \qquad \text{(i)}$$

$$Y(t) = CX + DU \qquad \text{(ii)}$$

Assuming that the state variables are not available for feedback, a full-order state observer has to be designed to measure them.

Let $\hat{X}$ be the estimated state
$\hat{Y}$ be the estimated output
K_o be the state observer gain matrix

Thus, the block diagram of full-order state observer can be drawn as shown in Figure 3.4

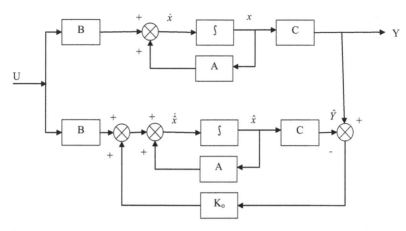

FIGURE 3.4
Block diagram of a system with full-order state observer.

Ideally, the state equation of observed system should be same as that of original system. The purpose of state observer design is to minimize the difference between system output (Y) and estimated output, $\hat{Y}$. The model equation of state observer will have an additional term that includes estimation error to compensate for inaccuracies in matrix $[A]$ and $[B]$

$$\therefore \dot{\hat{X}}(t) = A\hat{X} + BU + K_o(Y - \hat{Y})$$

But, $Y = CX$ and $\hat{Y} = C\hat{X}$

$$\therefore \dot{\hat{X}}(t) = A\hat{X} + BU + K_o(CX - C\hat{X})$$

$$\dot{\hat{X}}(t) = (A - K_o C)\hat{X} + BU + K_o CX$$

(3.32)

Subtracting equation (3.32) from the state equation, we get

$$\dot{X} - \dot{\hat{X}} = AX - (A - K_o C)\hat{X} - K_o CX$$

$$\dot{X} - \dot{\hat{X}} = (A - K_o C)X - (A - K_o CX)\hat{X}$$

(3.33)

$$\dot{X} - \dot{\hat{X}} = (A - K_o C)(X - \hat{X})$$

(3.34)

The difference between X and $\hat{X}$ is the error
$\therefore$ The equation (3.34) can be rewritten as

$$\dot{e} = (A - K_o C)e$$

Thus, the dynamic behaviour of error is determined by the eigen values of matrix $(A - K_oC)$. If the system is completely observable, then it can be proved that it is possible to choose matrix K_o such that $(A - K_oC)$ has arbitrarily desired eigen values.

Determination of State Observer Gain Matrix (K_o)

The state observer gain matrix can be determined by the following three methods:

   i. Transformation matrix approach
  ii. Direct substitution method
 iii. Ackermann's formula method

In all the above three methods, the first step is to check whether the given system is completely observable. This observability check can be performed as follows:

1. Determine the observability matrix

$$Q_o = \begin{bmatrix} C^T & A^T C^T & (A^T)^2 C^T & \cdots & (A^T)^{n-1} C^T \end{bmatrix}$$

2. Determine the rank of Q_o
3. If the rank of Q_o is equal to the order of system 'n' then, the given system is completely observable.

Procedural steps for determining K_o using transformation matrix T_o, approach.

   Step 1: Verify the condition for complete observability. If yes, then proceed with step 2

   Step 2: Find the characteristics polynomial $|\lambda I - A| = \lambda^n + a_1 \lambda^{n-1} + \cdots + a_{n-1}\lambda + a_n$, and hence, determine the values of $a_1, a_2, \ldots a_n$

   Step 3: Determine the transformation matrix, T_o.
      where

$$T_o = [WQ_oT]^{-1}$$

$$W = \begin{bmatrix} a_{n-1} & a_{n-2} & \cdots & a_1 & 1 \\ a_{n-2} & a_{n-3} & \cdots & 1 & 0 \\ \vdots & \vdots & & \vdots & \vdots \\ a_1 & 1 & & 0 & 0 \\ 1 & 0 & & 0 & 0 \end{bmatrix}$$

$$Q_o = \begin{bmatrix} C^T & A^T C^T & (A^T)^2 C^T & \cdots & (A^T)^{n-1} C^T \end{bmatrix}$$

NOTE

If the given system is already in observable canonical form, then $T_o = I$

Step 4: Using the specified closed loop poles; $\mu_1,\ \mu_2,\ \mu_3,\ \cdots\ \mu_n$, write the desired characteristics polynomial as $(s - \mu_1)(s - \mu_2)\ldots(s - \mu_n) = \lambda^n + \alpha_1 \lambda^{n-1} + \cdots + \alpha_{n-1} \lambda + \alpha_n$, and hence, determine the values of α_1, $\alpha_2,\ \ldots\ \alpha_n$

Step 5: Determine the required state observer gain matrix, K_o using the following equation

$$K_o = [T_o] \begin{bmatrix} \alpha_n - a_n \\ \alpha_{n-1} - a_{n-1} \\ \vdots \\ \alpha_1 - a_1 \end{bmatrix} \tag{3.35}$$

Procedural steps for determining K_o using direct substitution method.

Step 1: Verify the condition for complete observability. If yes, then proceed with step 2

Step 2: Using the specified closed loop poles; $\mu_1,\ \mu_2,\ \mu_3,\ \cdots\ \mu_n$, write the desired characteristics polynomial as

$$(\lambda - \mu_1)(\lambda - \mu_2)\ldots(\lambda - \mu_n) = |\lambda I - (A - B K_o)| \quad \text{where } K_0 = \begin{bmatrix} K_{o1} \\ K_{o2} \\ K_{o3} \\ \vdots \\ K_{on} \end{bmatrix} \tag{3.36}$$

Step 3: Equate the coefficients of 'λ' on both the sides, and hence, find the values of $K_{o1}, K_{o2}, \ldots K_{on}$.

NOTE

This method is suitable for lower-order systems ($n \leq 3$). For higher-order system, the calculations become tedious.

Procedural steps for determining K_o using Ackermann's formula

Step 1: Verify the condition for complete observability. If yes, then proceed with step 2

Step 2: Using the specified closed loop poles; $\mu_1, \mu_2, \mu_3, \ldots \mu_n$, write the desired characteristics polynomial as $(\lambda - \mu_1)(\lambda - \mu_2)\ldots(\lambda - \mu_n) = \lambda^n + \alpha_1 \lambda^{n-1} + \cdots + \alpha_{n-1}\lambda + \alpha_n$, and hence, determine the values of α_1, $\alpha_2, \ldots \alpha_n$

Step 3: Determine the matrix $\Phi(A)$ using the coefficients of desired characteristic polynomial as

$$\Phi(A) = A^n + \alpha_1 A^{n-1} + \alpha_2 A^{n-2} + \cdots + \alpha_{n-1}A + \alpha_n I$$

Step 4: Compute the state observer gain matrix, K_o using the Ackermann's formula as

$$K_o = \left[\Phi(A) \right] \begin{bmatrix} C \\ CA \\ \vdots \\ CA^{n-2} \\ CA^{n-1} \end{bmatrix}^{-1} \begin{bmatrix} 0 \\ 0 \\ \vdots \\ 0 \\ 1 \end{bmatrix} \tag{3.37}$$

Illustration 10

Consider a system as given $\begin{bmatrix} \dot{x}_1 \\ \dot{x}_2 \\ \dot{x}_3 \end{bmatrix} = \begin{bmatrix} 0 & 0 & 0 \\ 1 & 0 & -3 \\ 0 & 1 & -4 \end{bmatrix} \begin{bmatrix} x_1 \\ x_2 \\ x_3 \end{bmatrix} + \begin{bmatrix} 1 \\ 2 \\ 0 \end{bmatrix} u;$

$y = \begin{bmatrix} 0 & 0 & 1 \end{bmatrix} \begin{bmatrix} x_1 \\ x_2 \\ x_3 \end{bmatrix}$. Design a full-order state observer. The desired eigen values for the observer matrices are 1, 1, 2.

Method 1

Step 1: Verify the condition for complete observability.

Given that $A = \begin{bmatrix} 0 & 0 & 0 \\ 1 & 0 & -3 \\ 0 & 1 & -4 \end{bmatrix}$ and $C = \begin{bmatrix} 0 & 0 & 1 \end{bmatrix}$

$$(A^T)^2 = \begin{bmatrix} 0 & 1 & 0 \\ 0 & 0 & 1 \\ 0 & -3 & -4 \end{bmatrix} \begin{bmatrix} 0 & 1 & 0 \\ 0 & 0 & 1 \\ 0 & -3 & -4 \end{bmatrix} = \begin{bmatrix} 0 & 0 & 1 \\ 0 & -3 & -4 \\ 0 & 12 & 13 \end{bmatrix}$$

$$(A^T)^2 C^T = \begin{bmatrix} 0 & 0 & 1 \\ 0 & -3 & -4 \\ 0 & 12 & 13 \end{bmatrix} \begin{bmatrix} 0 \\ 0 \\ 1 \end{bmatrix} = \begin{bmatrix} 1 \\ -4 \\ 13 \end{bmatrix}$$

$$A^T C^T = \begin{bmatrix} 0 & 1 & 0 \\ 0 & 0 & 1 \\ 0 & -3 & -4 \end{bmatrix} \begin{bmatrix} 0 \\ 0 \\ 1 \end{bmatrix} = \begin{bmatrix} 0 \\ 1 \\ -4 \end{bmatrix}$$

$$Q_O = \begin{bmatrix} C^T & A^T C^T & (A^T)^2 C^T \end{bmatrix}$$

$$Q_O = \begin{bmatrix} 0 & 0 & 1 \\ 0 & 1 & -4 \\ 1 & -4 & 13 \end{bmatrix}$$

Determinant of $Q_O = \Delta Q_O = \begin{vmatrix} 0 & 0 & 1 \\ 0 & 1 & -4 \\ 1 & -4 & 13 \end{vmatrix} = 1(0-1) = -1 \neq 0$. The

system is completely observable.

Step 2: Find the characteristics polynomial $|\lambda I - A| = \lambda^n + a_1 \lambda^{n-1} + \cdots + a_{n-1}\lambda + a_n$ and hence determine the values of $a_1, a_2, \ldots, a_n$

The characteristic polynomial of original system is given by $|\lambda I - A| = 0$

$$[\lambda I - A] = \begin{bmatrix} \lambda & 0 & 0 \\ 0 & \lambda & 0 \\ 0 & 0 & \lambda \end{bmatrix} - \begin{bmatrix} 0 & 0 & 0 \\ 1 & 0 & -3 \\ 0 & 1 & -4 \end{bmatrix} = \begin{bmatrix} \lambda & 0 & 0 \\ -1 & \lambda & 3 \\ 0 & -1 & \lambda+4 \end{bmatrix}$$

$$= \lambda[\lambda(\lambda+4)+3] = 0$$

$$\lambda(\lambda^2 + 4\lambda + 3) = 0$$

$$\lambda^3 + 4\lambda^2 + 3\lambda = 0$$

The characteristic polynomial of original system is $\lambda^3 + 4\lambda^2 + 3\lambda = 0$
Here, $a_1 = 4$, $a_2 = 3$, $a_3 = 0$.

Step 3: Determine the transformation matrix, T_o

The given system is already in observable canonical form, then
$T_o = I$

Step 4: Using the specified closed loop poles; $\mu_1, \mu_2, \mu_3, \dots \mu_n$, write the desired characteristics polynomial

$$(\lambda - \mu_1)(\lambda - \mu_2)(\lambda - \mu_3) = (\lambda - 1)(\lambda - 1)(\lambda - 2) = 0$$

$$(\lambda^2 - 1\lambda - 1\lambda + 1)(\lambda - 2) = 0$$

$$(\lambda^2 - 2\lambda + 1)(\lambda - 2) = 0$$

$$\lambda^3 - 2\lambda^2 - 2\lambda^2 + 4\lambda + 1\lambda - 2 = 0$$

$$\lambda^3 - 4\lambda^2 + 5\lambda - 2 = 0$$

$$\Rightarrow \alpha_1 = -4, \ \alpha_2 = 5, \ \alpha_3 = -2$$

Step 5: Determine the required state observer gain matrix, K_o using the following equation

$$K_O = [T_o] \begin{bmatrix} \alpha_n - a_n \\ \alpha_{n-1} - a_{n-1} \\ \vdots \\ \alpha_1 - a_1 \end{bmatrix}$$

$$= \begin{bmatrix} (-2-0) \\ (5-3) \\ (-4-4) \end{bmatrix} = \begin{bmatrix} -2 \\ 2 \\ -8 \end{bmatrix}$$

Method 2

Step 1: Verify the condition for complete observability as in method 1

Step 2: Using the specified closed loop poles; $\mu_1, \mu_2, \mu_3, \dots \mu_n$, write the desired characteristics polynomial

$$(\lambda - \mu_1)(\lambda - \mu_2)(\lambda - \mu_3) = (\lambda - 1)(\lambda - 1)(\lambda - 2) = 0$$

$$(\lambda^2 - 1\lambda - 1\lambda + 1)(\lambda - 2) = 0$$

$$(\lambda^2 - 2\lambda + 1)(\lambda - 2) = 0$$

$$\lambda^3 - 2\lambda^2 - 2\lambda^2 + 4\lambda + 1\lambda - 2 = 0$$

$$\lambda^3 - 4\lambda^2 + 5\lambda - 2 = 0$$

$$\Rightarrow \alpha_1 = -4, \ \alpha_2 = 5, \ \alpha_3 = -2$$

The characteristic polynomial of original system is given by $|\lambda I - (A - K_o C)| = 0$

$$[\lambda I - A + K_o C] = \begin{bmatrix} \lambda & 0 & 0 \\ 0 & \lambda & 0 \\ 0 & 0 & \lambda \end{bmatrix} - \begin{bmatrix} 0 & 0 & 0 \\ 1 & 0 & -3 \\ 0 & 1 & -4 \end{bmatrix} + \begin{bmatrix} K_{o1} \\ K_{o2} \\ K_{o3} \end{bmatrix} \begin{bmatrix} 0 & 0 & 1 \end{bmatrix}$$

$$= \begin{bmatrix} \lambda & 0 & 0 \\ -1 & \lambda & 3 \\ 0 & -1 & \lambda + 4 \end{bmatrix} + \begin{bmatrix} 0 & 0 & K_{01} \\ 0 & 0 & K_{02} \\ 0 & 0 & K_{03} \end{bmatrix}$$

$$= \begin{bmatrix} \lambda & 0 & K_{01} \\ -1 & \lambda & 3 + K_{02} \\ 0 & -1 & \lambda + 4 + K_{03} \end{bmatrix}$$

$$= \lambda \big[\lambda(\lambda + 4 + K_{03}) + (3 + K_{02}) \big] + K_{01} = 0$$

$$\lambda \big[\lambda^2 + 4\lambda + K_{03}\lambda + 3 + K_{02} \big] + K_{01} = 0$$

$$\lambda^3 + 4\lambda^2 + K_{03}\lambda^2 + 3\lambda + K_{02}\lambda + K_{01} = 0$$

$$\lambda^3 + (4 + K_{03})\lambda^2 + (3 + K_{02})\lambda + K_1 = 0$$

Step 3: Equate the coefficients of 'λ' on both the sides, and hence, find the values of $K_{o1}, K_{o2}, \dots K_{on}$, we get

$$K_{01} = -2 \qquad \begin{vmatrix} 3 + K_{02} = 5 \\ K_{02} = 2 \end{vmatrix} \qquad \begin{vmatrix} 4 + K_{03} = -4 \\ K_{03} = -8 \end{vmatrix}$$

$$\therefore K_O = \begin{bmatrix} K_{o1} \\ K_{o2} \\ K_{o3} \end{bmatrix} = \begin{bmatrix} -2 \\ 2 \\ -8 \end{bmatrix}$$

Method 3

Step 1: Verify the condition for complete observability as in Method 1.

Step 2: Using the specified closed loop poles; $\mu_1, \mu_2, \mu_3, \dots \mu_n$, write the desired characteristics polynomial as $(\lambda - \mu_1)(\lambda - \mu_2)\dots(\lambda - \mu_n) = \lambda^n + \alpha_1\lambda^{n-1} + \dots + \alpha_{n-1}\lambda + \alpha_n$, and hence, determine the values of $\alpha_1, \alpha_2, \dots \alpha_n$.

$$(\lambda - \mu_1)(\lambda - \mu_2)(\lambda - \mu_3) = (\lambda - 1)(\lambda - 1)(\lambda - 2) = 0$$

$$(\lambda^2 - 1\lambda - 1\lambda + 1)(\lambda - 2) = 0$$

$$(\lambda^2 - 2\lambda + 1)(\lambda - 2) = 0$$

$$\lambda^3 - 2\lambda^2 - 2\lambda^2 + 4\lambda + 1\lambda - 2 = 0$$

$$\lambda^3 - 4\lambda^2 + 5\lambda - 2 = 0$$

$$\Rightarrow \alpha_1 = -4, \ \alpha_2 = 5, \ \alpha_3 = -2$$

Step 3: Determine the matrix $\Phi(A)$ using the coefficients of desired characteristic polynomial as

$$\Phi(A) = A^n + \alpha_1 A^{n-1} + \alpha_2 A^{n-2} + \cdots + \alpha_{n-1}A + \alpha_n I$$

$$= A^3 + \alpha_1 A^2 + \alpha_2 A + \alpha_3 I$$

$$= \begin{bmatrix} 0 & 0 & 0 \\ -3 & 12 & -39 \\ -4 & 13 & -40 \end{bmatrix} + (-4)\begin{bmatrix} 0 & 0 & 1 \\ 0 & -3 & 12 \\ 1 & -4 & 13 \end{bmatrix}$$

$$+ 5\begin{bmatrix} 0 & 0 & 0 \\ 1 & 0 & -3 \\ 0 & 1 & -4 \end{bmatrix} + (-2)\begin{bmatrix} 1 & 0 & 0 \\ 0 & 1 & 0 \\ 0 & 0 & 1 \end{bmatrix}$$

$$= \begin{bmatrix} -2 & 0 & -4 \\ 2 & 22 & -102 \\ -8 & 34 & -104 \end{bmatrix}$$

Step 4: Compute the state observer gain matrix, K_o using equation (3.37)

$$K_o = [\Phi(A)]\begin{bmatrix} C \\ CA \\ CA^2 \end{bmatrix}^{-1}\begin{bmatrix} 0 \\ 0 \\ 1 \end{bmatrix}$$

$$= \begin{bmatrix} -2 & 0 & -4 \\ 2 & 22 & -102 \\ -8 & 34 & -104 \end{bmatrix}\begin{bmatrix} 3 & 4 & 1 \\ 4 & 1 & 0 \\ 1 & 0 & 0 \end{bmatrix}^{-1}\begin{bmatrix} 0 \\ 0 \\ 1 \end{bmatrix}$$

$$\therefore K_o = \begin{bmatrix} -2 \\ 2 \\ -8 \end{bmatrix} \ \text{(After simplification)}$$

4

Non-Linear Systems and
Phase Plane Analysis

This chapter on 'Non-Linear Systems' discusses the topics, namely: Inherent characteristics of non-linear systems, types of nonlinearity (saturation, deadzone, friction, backlash, relay), linearization using describing function, derivation of describing functions for different non-linear elements (deadzone, saturation, deadzone and saturation, on–off controller with dead zone, backlash, relay with deadzone and hysteresis), phase plane definitions (phase plane, phase trajectory, phase portrait, singular point, behaviour of trajectories in the vicinity of singular point), limit cycle, phase plane analysis using analytical method, isocline method and delta method.

In practice, all the systems are non-linear in nature. The nonlinearity can be incidental or intentional. The characteristics of non-linear systems make it difficult to analyse. The non-linear element can be approximated as a linear element using describing function approach. The describing function analysis is based on Fourier series approach. Phase plane analysis is used for investigating system behaviour and hence to design system parameters to achieve a desired response.

Characteristics of Non-Linear Systems

- Non-linear equations, unlike linear ones, cannot in general be solved analytically.
- Powerful mathematical tools such as Laplace and Fourier transforms do not apply to non-linear systems.
- Non-linear systems do not obey principle of superposition and homogeneity.

  Thus, they respond quite differently to external inputs and initial conditions. In other words, the response of a non-linear system to a particular test signal is no guide to their behaviour to other inputs.

- The stability of non-linear system may depend on initial conditions.
- Non-linear systems frequently have more than one equilibrium point (where as linear system has unique equilibrium point).

- Non-linear system may exhibit limit cycles which are self-sustained oscillations of fixed amplitude and fixed period (without external excitation). These oscillations are called **limit cycles.**

- Non-linear system with a periodic input may exhibit a periodic output whose frequency is either sub-harmonic or a harmonic of the input frequency. For example, an input of frequency 10 Hz may result in an output of 5 Hz for sub-harmonic case or 30 Hz for a harmonic case.

- A non-linear system exhibits phenomena called frequency entrainment and asynchronous quenching.

- A non-linear system can also display a behaviour in its frequency response, in the form of hysteresis called 'jump resonance'.

Jump Resonance

Consider a mass–spring damper system

$$M\ddot{y} + B\dot{y} + K_1 y + K_2 y^3 = F \cos \omega t, \quad M > 0, B > 0, K_1 > 0, K_2 > 0.$$

where M – mass, B – damping coefficient, K_1 and K_2 – spring constants, F – force applied, y – response.

The response 'y' obtained by varying 'ω' (by keeping F constant) is shown in Figure 4.1. The response 'y' travels along P, Q and R. After the point R, any increase in the value of 'ω' results in a sudden jump to point 'S'. Further increase in 'ω' makes the curve travel along 'T'. On the other hand, if 'ω' is decreased, then the curve travels along T, S and U. Again, at the point 'U', any further decrease in the value of 'ω' will result in a sudden jump to point Q. The curve travels along 'QP' for further decrease in the value of 'ω'.

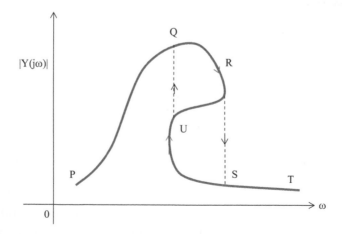

FIGURE 4.1
Frequency response of a system with jump resonance.

In fact, the curve never follows the path '*RU*'. The portion refers to a condition of unstable equilibrium, and this kind of behaviour of the system is called 'Jump resonance'.

Types of Nonlinearities

The nonlinearity can be broadly classified into two categories, namely: (1) Incidental nonlinearity, which is inherently present in the system, and (2) intentional nonlinearity, which is deliberately inserted in the systems. Further classification of nonlinearities is presented in Figure 4.2.

Saturation

Saturation nonlinearity is witnessed in servo motors, transistor amplifiers and also in actuators. As depicted in Figure 4.3, initially, the output increases linearly with the input. Later, even with the increase in the input, the output change does not show much increase. In fact, the output reaches to a maximum value and remains unaltered. This is experienced in the servo motor, where the output torque tends to get saturated due to the saturation magnetic material.

Deadzone

The motors, sensors, valve-controlled pneumatic and hydraulic actuators exhibit the non-linear phenomenon called deadzone. Here, the system

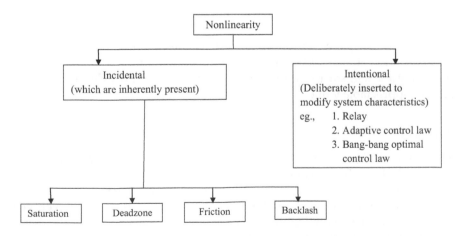

FIGURE 4.2
Types of nonlinearities.

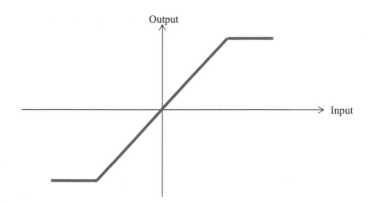

FIGURE 4.3
Saturation nonlinearity.

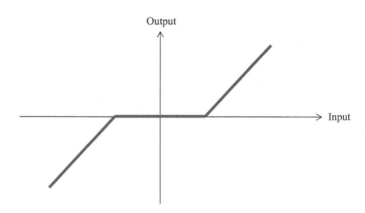

FIGURE 4.4
Deadzone nonlinearity.

responds to the given input only after it reaches certain value as shown in Figure 4.4. In the case of DC motor, due to static friction at the motor shaft, the armature will start rotating only after the torque provided has reached certain appreciable value.

Backlash

The mechanical components of control systems exhibit backlash nonlinearity. The backlash nonlinearity observed in the gear trains is shown in Figure 4.5. A small gap 'OU' is seen between the pair of mating gears. If the angle of rotation of driving gear is smaller than the gap 'OU', then the driven gear will not move. The path 'PQ' shown will be the path traversed by the driven gear after developing its contact with the driving gear. Similarly, during the reverse cycle, when the driving gear travels by a distance 2U, the driven

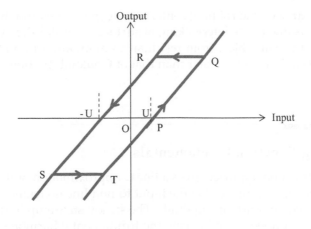

FIGURE 4.5
Backlash nonlinearity.

gear will not move which corresponds to the portion *QR*. Once the contact between the driven gear and the driving gear is re-established, the driven gear travels along the path *RS*. Thus, the driven gear travels along *TQRS* by following the periodic motion of driving gear.

Friction

Friction comes into existence when there is relative motion between contacting surfaces. The three important types of friction are viscous friction, Coulomb friction and stiction (Figure 4.6). The viscous friction is linear in nature and the frictional force is directly proportional to relative velocity of sliding surfaces. The Coulomb friction is a drag force which opposes

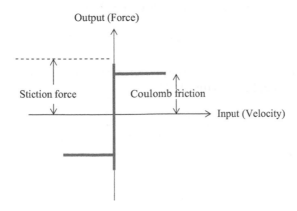

FIGURE 4.6
Ideal Coulomb friction and stiction.

motion (e.g., drag due to rubbing contact between brushes and the commutator). Stiction is due to the interlocking of surface irregularities. More force is required to move an object from rest than to maintain it in motion. Hence, force of stiction is always greater than that of Coulomb friction.

Describing Function Fundamentals

Describing function approach gives a linear approximation of the non-linear element. Here, it is assumed that the input to non-linear element is a sinusoidal input of known constant amplitude. The steady-state amplitude and phase relation is determined by comparing the fundamental harmonic component of non-linear element's output with sinusoidal input. Hence, this method can also be named as 'harmonic linearization' of a non-linear element.

Describing function analysis is based on 'Fourier series approach'. It predicts whether limit cycle oscillations exist or not and gives numerical estimates of amplitude and frequency of oscillation.

In order to understand the basic concept of describing function analysis, let us consider the block diagram of a non-linear system as shown in Figure 4.7, where $G_1(s)$ and $G_2(s)$ represent linear elements while the 'NL' represents non-linear element.

Let $x = X \sin \omega t$ be a sinusoidal input given to the non-linear system. The output 'y' of the non-linear element will be a non-sinusoidal periodic function, which may be expressed in terms of Fourier series

$$y = Y_o + A_1 \sin \omega t + B_1 \cos \omega t + A_2 \sin 2\omega t + B_2 \cos 2\omega t + \cdots \quad (4.1)$$

The non-linear characteristics are all odd symmetrical and odd half-wave symmetrical. Therefore, the mean value of Y_o is zero.

$$\therefore y = A_1 \sin \omega t + B_1 \cos \omega t + A_2 \sin 2\omega t + B_2 \cos 2\omega t + \cdots \quad (4.2)$$

In the absence of external input (i.e., when $r = 0$), the output of the non-linear element 'NL' is feedback to input.

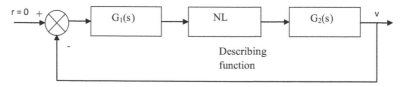

FIGURE 4.7
Non-linear system.

If $G_2(s)$ $G_1(s)$ has low pass characteristics, all the higher harmonics of 'y' are filtered out. In such a case, the input 'x' to the non-linear element 'NL' is mainly contributed by fundamental component (first component y_1) of y. Therefore, 'x' remains as sinusoidal.

Thus, when the second and third harmonics of y are neglected, then

$$y = y_1 = A_1 \sin \omega t + B_1 \cos \omega t$$
$$y = y_1 = Y_1 \sin(\omega t + \Phi_1) \tag{4.3}$$

$$Y_1 = \sqrt{A_1^2 + B_1^2} \quad \text{and} \quad \Phi_1 = \tan^{-1}\left[\frac{B_1}{A_1}\right] \tag{4.4}$$

y_1 is the amplitude of the fundamental harmonic component of output. Φ_1 is the phase shift of the fundamental harmonic component of the output with respect to input.

The coefficients of A_1 and B_1 of the Fourier series are given by

$$A_1 = \frac{2}{2\pi} \int_0^{2\pi} y \sin \omega t \, d\omega t \tag{4.5}$$

$$B_1 = \frac{2}{2\pi} \int_0^{2\pi} y \cos \omega t \, d\omega t \tag{4.6}$$

The amplitude ratio and phase shift can be combined such that

$$N(X) = \frac{Y_1(X)}{X} \lfloor \Phi_1(X) = \frac{B_1 + jA_1}{X} \tag{4.7}$$

where $N(X)$ is the describing function of nonlinearity.

Thus, the block diagram of non-linear system with non-linear element replaced by describing function is shown in Figure 4.8.

Derivation for describing function of certain nonlinearities [deadzone, saturation, deadzone and saturation, on–off controller with deadzone, backlash, relay with dead zone and hysteresis].

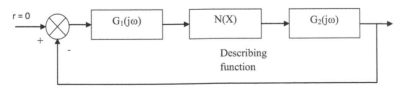

FIGURE 4.8
Non-linear system with nonlinearity replaced by describing function.

Describing Function of Deadzone

The input–output relationship of dead zone nonlinearity is shown in Figure 4.9.

When the input is less than $D/2$, the output is zero. It is linear when input is greater than $D/2$. When the deadzone nonlinearity is excited with a sinusoidal input signal $x = X \sin \omega t$, the output obtained is as shown in Figure 4.10

$$\text{The input} \quad x = X \sin \omega t \tag{i}$$

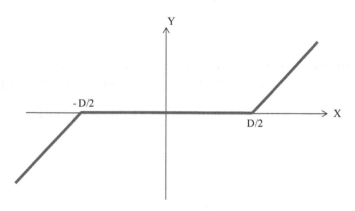

FIGURE 4.9
Input–output characteristics of deadzone nonlinearity.

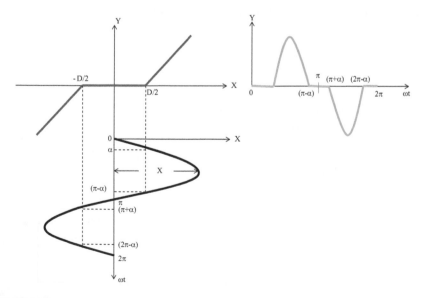

FIGURE 4.10
Sinusoidal response of deadzone nonlinearity.

where X is the maximum value of input
when $\omega t = \alpha$, $x = D/2$

$$\therefore D/2 = X \sin \alpha \Rightarrow \alpha = \sin^{-1}\left[\frac{D}{2X}\right] \qquad \text{(ii)}$$

The output y shown in Figure 4.10 can be divided into three regions in a period of 'π' and are represented by equation (iii).

$$y = \begin{cases} 0 & ; & 0 \le \omega t \le \alpha \\ K(x - D/2) & ; & \alpha \le \omega t \le (\pi - \alpha) \\ 0 & ; & (\pi - \alpha) \le \omega t \le \pi \end{cases} \qquad \text{(iii)}$$

The describing function is given by $N(X) = \dfrac{Y_1(X)}{X} \lfloor \Phi_1(X)$

where $Y_1 = \sqrt{A_1^2 + B_1^2}$ and $\Phi_1 = \tan^{-1}\left[\dfrac{B_1}{A_1}\right]$

The output y has half-wave and quarter-wave symmetries.

$$\therefore B_1 = 0 \text{ and } A_1 = \frac{2}{\pi/2} \int_0^{\pi/2} y \sin \omega t \, d(\omega t)$$

Since the output 'y' is zero in the range $0 \le \omega t \le \alpha$, the limits of integration in the above equation can be changed to α to $\pi/2$ instead of 0 to $\pi/2$.

$$\therefore A_1 = \frac{4}{\pi} \int_\alpha^{\pi/2} K\left(x - \frac{D}{2}\right) \sin \omega t \, d(\omega t) \qquad \text{(iv)}$$

Substitute $x = X \sin \omega t$ in equation (iv), we get

$$A_1 = \frac{4K}{\pi} \left[\int_\alpha^{\pi/2} \left(X \sin \omega t - \frac{D}{2}\right) \sin \omega t \, d(\omega t) \right]$$

$$= \frac{4K}{\pi} \left[\int_\alpha^{\pi/2} X \sin^2 \omega t \, d(\omega t) - \frac{D}{2} \int_\alpha^{\pi/2} \sin \omega t \, d(\omega t) \right]$$

$$= \frac{4K}{\pi} \left[\frac{X}{2} \int_\alpha^{\pi/2} (1 - \cos 2\omega t) \, d(\omega t) - \frac{D}{2} \int_\alpha^{\pi/2} \sin \omega t \, d(\omega t) \right]$$

$$= \frac{4K}{\pi} \left[\frac{X}{2} \left(\omega t - \frac{\sin 2\omega t}{2}\right)_\alpha^{\pi/2} - \frac{D}{2} [-\cos \omega t]_\alpha^{\pi/2} \right]$$

$$= \frac{4K}{\pi}\left[\frac{X}{2}\left(\frac{\pi}{2} - \frac{\sin \pi}{2} - \alpha + \frac{\sin 2\alpha}{2}\right) - \frac{D}{2}\left(-\cos\frac{\pi}{2} + \cos\alpha\right)\right]$$

$$= \frac{4K}{\pi}\left[\frac{X}{2}\left(\frac{\pi}{2} - \alpha + \frac{\sin 2\alpha}{2}\right) - \frac{D}{2}(\cos\alpha)\right]$$

But, $\dfrac{D}{2} = X \sin \alpha$ (From equation (ii))

$$\therefore A_1 = \frac{4K}{\pi}\left[\frac{X}{2}\left(\frac{\pi}{2} - \alpha + \frac{\sin 2\alpha}{2}\right) - X \sin \alpha \cos \alpha\right]$$

$$= \frac{4KX}{\pi}\left[\frac{\pi}{4} - \frac{\alpha}{2} + \frac{2\sin\alpha\cos\alpha}{4} - \sin\alpha\cos\alpha\right]$$

$$= \frac{4KX}{\pi}\left[\frac{\pi}{4} - \frac{\alpha}{2} - \frac{\sin\alpha\cos\alpha}{2}\right]$$

$$= KX\left[\frac{4}{\pi} \times \frac{\pi}{4} - \frac{4}{\pi} \times \frac{1}{2}(\alpha + \sin\alpha\cos\alpha)\right]$$

$$= KX\left[1 - \frac{2}{\pi}(\alpha + \sin\alpha\cos\alpha)\right]$$

$$\therefore Y_1 = \sqrt{A_1^2 + B_1^2} = Y_1 = \sqrt{A_1^2 + 0^2} = A_1 = KX\left[1 - \frac{2}{\pi}(\alpha + \sin\alpha\cos\alpha)\right]$$

$$\Phi_1 = \tan^{-1}\left[\frac{B_1}{A_1}\right] = \tan^{-1}(0) = 0$$

$\therefore$ The describing function $N(X) = \dfrac{Y_1(X)}{X}\underline{|\Phi_1(X)}$

$$N(X) = \frac{KX\left[1 - \frac{2}{\pi}(\alpha + \sin\alpha\cos\alpha)\right]}{X}\underline{|0^\circ}$$

$$N(X) = KX\left[1 - \frac{2}{\pi}(\alpha + \sin\alpha\cos\alpha)\right]\underline{|0^\circ}$$

Since $\alpha = \sin^{-1}\left[\dfrac{D}{2X}\right]$ and $\sin \alpha = \dfrac{D}{2X}$, we can rewrite the above equations as

$$N(X) = KX\left[1 - \frac{2}{\pi}\left(\sin^{-1}\left[\frac{D}{2X}\right] + \frac{D}{2X}\sqrt{1 - \left(\frac{D}{2X}\right)^2}\right)\right]\lfloor 0°$$

$$N(X) = 0 \quad \text{for } X < D/2$$

$$N(X) = K\left[1 - \frac{2}{\pi}(\alpha + \sin\alpha\cos\alpha)\right] \quad \text{for } X < D/2$$

(4.8)

Describing Function of Saturation Nonlinearity

The input/output relationship of saturation nonlinearity is shown in the Figure 4.11.

The input–output relation is linear for x varying from '0' to S. When the input x is greater than S, the output reaches a saturated value of KS. The saturation nonlinearity is excited with the sinusoidal signal input ($x = X \sin \omega t$). The output, thus obtained, is shown in Figure 4.12.

$$\text{The input} \quad x = X \sin \omega t \tag{i}$$

where X is the maximum value of input
When $\omega t = \beta, \quad x = S$
Thus, equation (i) can be rewritten as

$$S = X \sin \beta$$

$$\sin \beta = \frac{S}{X} \text{ or } \beta = \sin^{-1}\left[\frac{S}{X}\right] \tag{ii}$$

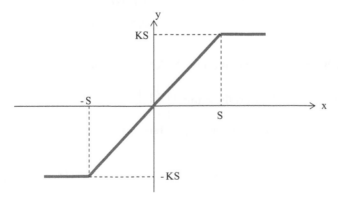

FIGURE 4.11
Input–output characteristics of saturation nonlinearity.

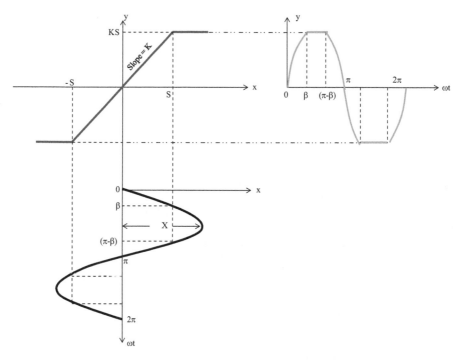

FIGURE 4.12
Sinusoidal response of a saturation nonlinearity.

In a period from '0' to 'π', the output 'y' of the saturation nonlinearity can be divided into three regions and is represented by equation (iii).

$$y = \begin{cases} Kx; & 0 \le \omega t \le \beta \\ KS; & \beta \le \omega t \le (\pi - \beta) \\ Kx; & (\pi - \beta) \le \omega t \le \pi \end{cases} \qquad \text{(iii)}$$

The describing function is given by $N(X) = \dfrac{Y_1(X)}{X} \lfloor \Phi_1(X)$

where $Y_1 = \sqrt{A_1^2 + B_1^2}$ and $\Phi_1 = \tan^{-1}\left[\dfrac{B_1}{A_1}\right]$

The output y has half-wave and quarter-wave symmetries.

$$\therefore B_1 = 0 \text{ and } A_1 = \frac{2}{\pi/2} \int_0^{\pi/2} y \sin \omega t \, d(\omega t)$$

The output 'y' is given by two different expressions in the period 0 to π/2. Hence, the above equation can be rewritten as follows:

$$\therefore A_1 = \frac{4}{\pi} \int_0^\beta y \sin \omega t\, d(\omega t) + \frac{4}{\pi} \int_\beta^{\pi/2} y \sin \omega t\, d(\omega t) \qquad \text{(iv)}$$

Substituting the values of 'y' from equation (iii) in equation (iv), we get

$$A_1 = \frac{4}{\pi} \int_0^\beta Kx \sin \omega t\, d(\omega t) + \frac{4}{\pi} \int_\beta^{\pi/2} KS \sin \omega t\, d(\omega t)$$

Substituting $x = X \sin \omega t$ in A_1, we get

$$A_1 = \frac{4K}{\pi} \int_0^\beta X \sin \omega t \sin \omega t\, d(\omega t) + \frac{4KS}{\pi} \int_\beta^{\pi/2} \sin \omega t\, d(\omega t)$$

$$= \frac{4KX}{\pi} \int_0^\beta \sin^2 \omega t\, d(\omega t) + \frac{4KS}{\pi} \int_\beta^{\pi/2} \sin \omega t\, d(\omega t)$$

$$= \frac{4KX}{\pi} \int_0^\beta \frac{(1 - \cos 2\omega t)}{2}\, d(\omega t) + \frac{4KS}{\pi} \int_\beta^{\pi/2} \sin \omega t\, d(\omega t)$$

$$= \frac{2KX}{\pi} \left[\omega t - \frac{\sin 2\omega t}{2} \right]_0^\beta + \frac{4KS}{\pi} \left[-\cos \omega t \right]_\beta^{\pi/2}$$

$$= \frac{2KX}{\pi} \left[\beta - \frac{\sin 2\beta}{2} \right] + \frac{4KS}{\pi} \left[-\cos \frac{\pi}{2} + \cos \beta \right]$$

$$= \frac{2KX}{\pi} \left[\beta - \frac{\sin 2\beta}{2} \right] + \frac{4KS}{\pi} \cos \beta$$

Substituting $S = X \sin \beta$ in A_1, we get

$$\therefore A_1 = \frac{2KX}{\pi} \left[\beta - \frac{\sin 2\beta}{2} \right] + \frac{4K}{\pi} X \sin \beta \cos \beta$$

$$= \frac{2KX}{\pi} \left[\beta - \frac{2 \sin \beta \cos \beta}{2} \right] + \frac{4KX}{\pi} \sin \beta \cos \beta$$

$$= \frac{2KX}{\pi} \left[\beta - \sin \beta \cos \beta + 2 \sin \beta \cos \beta \right]$$

$$= \frac{2KX}{\pi} \left[\beta + \sin \beta \cos \beta \right]$$

$$\therefore Y_1 = \sqrt{A_1^2 + B_1^2} = Y_1 = \sqrt{A_1^2 + 0^2} = A_1 = \frac{2KX}{\pi}\left[\beta + \sin\beta\cos\beta\right]$$

$$\Phi_1 = \tan^{-1}\left[\frac{B_1}{A_1}\right] = \tan^{-1}(0) = 0$$

$\therefore$The describing function $N(X) = \dfrac{Y_1(X)}{X}\underline{|\Phi_1(X)}$

$$N(X) = \frac{2K}{\pi}\left[\beta + \sin\beta\cos\beta\right]\underline{|0^\circ} \tag{4.9}$$

If $X < S$ then $\beta = \pi/2$ then $N(X) = K$

If $X > S$ then $N(X) = \frac{2K}{\pi}\left[\beta + \sin\beta\cos\beta\right]$

Since, $S = X\sin\beta$ and $\sin\beta = \dfrac{S}{X}$ and $\beta = \sin^{-1}\left[\dfrac{S}{X}\right]$, equation (4.9) can be rewritten as follows:

$$N(X) = \frac{2K}{\pi}\left[\sin^{-1}\left[\frac{S}{X}\right] + \frac{S}{X}\sqrt{1-\left(\frac{S}{X}\right)^2}\right] \quad \text{for } X > S \tag{4.10}$$

Describing Function of Deadzone and Saturation

The input/output relationship of nonlinearity with deadzone and saturation is shown in Figure 4. 13.

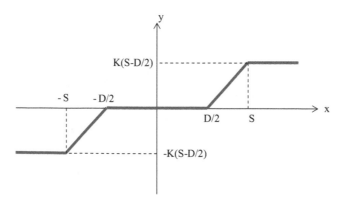

FIGURE 4.13
Input–output characteristics of deadzone and saturation nonlinearity.

The deadzone region is from $x = -D/2$ to $+D/2$, where output is zero. The input–output relation is linear for $x = \pm D/2$ to $\pm S$, and when the input $x > S$, the output reaches to a saturated value of $\pm K(S{-}D/2)$.

The output equation for the linear region can be obtained as follows:

$$\text{The general equation of a straight line is } Y = mx + C \qquad \text{(i)}$$

In the linear region, when $x = D/2$, $Y = 0$, substituting these values in equation (i), we get

$$0 = mD/2 + C \qquad \text{(ii)}$$

In the linear region, when $x = S$, $Y = K(S - D/2)$. Substituting these values of x and Y in equation (i), we get

$$K(S - D/2) = mS + C \qquad \text{(iii)}$$

Equation (iii) – equation (ii), we get

$$K(S - D/2) = mS + C - mD/2 - C$$
$$K(S - D/2) = m(S - D/2) \qquad \text{(iv)}$$
$$K = m$$

Substituting $K = m$ in equation (ii), we get

$$0 = KD/2 + C$$
$$C = -KD/2 \qquad \text{(v)}$$

From equations (i), (iv) and (v), we can write the output equation for linear region as follows:

$$Y = mx + C$$
$$= Kx - KD/2$$
$$Y = K(x - D/2)$$

Let us excite the deadzone and saturation nonlinearity by a sinusoidal input.

$$x = X \sin \omega t \qquad \text{(vi)}$$

where X is the amplitude of input.

The response of the nonlinearity for sinusoidal input is shown in Figure 4.14.

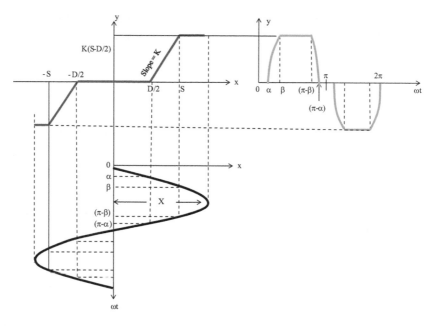

FIGURE 4.14
Sinusoidal response of a nonlinearity with deadzone and saturation.

In Figure 4.14, when $\omega t = \alpha$, $x = D/2$
    Hence, from equation (vi), we get

$$D/2 = X \sin \alpha \text{ or } \sin \alpha = \frac{D}{2X} \Rightarrow \alpha = \sin^{-1}\left[\frac{D}{2X}\right] \tag{vii}$$

Also, when $\omega t = \beta$, $x = S$
    Hence, from equation (vi), we get

$$S = X \sin \beta \text{ (or) } \sin \beta = \frac{S}{X} \Rightarrow \beta = \sin^{-1}\left[\frac{S}{X}\right] \tag{viii}$$

The output 'y' of the saturation nonlinearity can be divided into five regions in a period from of 'π', and the corresponding output equations are given in equation (ix).

$$y = \begin{cases} 0 & ; & 0 \le \omega t \le \alpha \\ K(x - D/2) & ; & \alpha \le \omega t \le \beta \\ K(S - D/2) & ; & \beta \le \omega t \le (\pi - \beta) \\ K(x - D/2) & ; & (\pi - \beta) \le \omega t \le (\pi - \alpha) \\ 0 & ; & (\pi - \alpha) \le \omega t \le \pi \end{cases} \tag{ix}$$

The describing function is given by $N(X) = \dfrac{Y_1(X)}{X} \lfloor \Phi_1(X)$

where $Y_1 = \sqrt{A_1^2 + B_1^2}$ and $\Phi_1 = \tan^{-1}\left[\dfrac{B_1}{A_1}\right]$

and $A_1 = \dfrac{2}{2\pi}\displaystyle\int_0^{2\pi} y \sin \omega t \, d(\omega t)$ and $B_1 = \dfrac{2}{2\pi}\displaystyle\int_0^{2\pi} y \cos \omega t \, d(\omega t)$

Here, the output y has half-wave and quarter-wave symmetries

$$\therefore B_1 = 0 \text{ and } A_1 = \dfrac{2}{\pi/2}\int_0^{\pi/2} y \sin \omega t \, d(\omega t) \qquad\qquad (x)$$

Substituting equations (ix) and (x), we get

$$A_1 = \dfrac{4}{\pi}\int_0^{\alpha} y \sin \omega t \, d(\omega t) + \dfrac{4}{\pi}\int_{\alpha}^{\beta} y \sin \omega t \, d(\omega t) + \dfrac{4}{\pi}\int_{\beta}^{\pi/2} y \sin \omega t \, d(\omega t)$$

$$A_1 = \dfrac{4}{\pi}\int_{\alpha}^{\beta} K(x - D/2)\sin \omega t \, d(\omega t) + \dfrac{4}{\pi} K(S - D/2)\sin \omega t \, d(\omega t) \qquad\qquad (xi)$$

Substituting $x = X \sin \omega t$ in equation (xi), we get

$$A_1 = \dfrac{4K}{\pi}\left[\int_{\alpha}^{\beta}\left(X\sin\omega t - \dfrac{D}{2}\right)\sin\omega t\, d(\omega t) + \int_{\beta}^{\pi/2}\left(S - \dfrac{D}{2}\right)\sin\omega t\, d(\omega t)\right]$$

$$= \dfrac{4K}{\pi}\left[\int_{\alpha}^{\beta} X\sin^2\omega t\, d(\omega t) - \int_{\alpha}^{\beta}\dfrac{D}{2}\sin\omega t\, d(\omega t) + \int_{\beta}^{\pi/2}\left(S - \dfrac{D}{2}\right)\sin\omega t\, d(\omega t)\right]$$

$$= \dfrac{4K}{\pi}\left[\dfrac{X}{2}\int_{\alpha}^{\beta}(1 - \cos 2\omega t)d(\omega t) - \dfrac{D}{2}\int_{\alpha}^{\beta}\sin\omega t\, d(\omega t) + \left(S - \dfrac{D}{2}\right)\int_{\beta}^{\pi/2}\sin\omega t\, d(\omega t)\right]$$

$$= \dfrac{4K}{\pi}\left[\dfrac{X}{2}\left(\omega t - \dfrac{\sin 2\omega t}{2}\right)_{\alpha}^{\beta} - \dfrac{D}{2}[-\cos\omega t]_{\alpha}^{\beta} + \left(S - \dfrac{D}{2}\right)[-\cos\omega t]_{\beta}^{\pi/2}\right]$$

$$= \dfrac{4K}{\pi}\left[\dfrac{X}{2}\left(\beta - \dfrac{\sin 2\beta}{2} - \alpha + \dfrac{\sin 2\alpha}{2}\right) - \dfrac{D}{2}(-\cos\beta + \cos\alpha)[\ ]\right.$$

$$\left. + \left(S - \dfrac{D}{2}\right)\left(-\cos\tfrac{\pi}{2} + \cos\beta\right)\right] \quad [\because \cos\pi/2 = 0]$$

$$A_1 = \frac{4K}{\pi}\left[\frac{X}{2}\left(\beta - \alpha - \frac{\sin 2\beta}{2} + \frac{\sin 2\alpha}{2}\right) + \frac{D}{2}\cos\beta - \frac{D}{2}\cos\alpha + S\cos\beta - \frac{D}{2}\cos\beta\right]$$

$$\text{(xii)}$$

But we know that, $\sin\alpha = \dfrac{D}{2X}$ or $X\sin\alpha = \dfrac{D}{2}$

$\sin\beta = \dfrac{S}{X}$ or $S = X\sin\beta$

On substituting for $D/2$ and S in the equation (xii), we get

$$A_1 = \frac{4K}{\pi}\left[\frac{X}{2}\left(\beta - \alpha - \frac{\sin 2\beta}{2} + \frac{\sin 2\alpha}{2}\right) - X\sin\alpha\cos\alpha + X\sin\beta\cos\beta\right]$$

$$= \frac{4KX}{\pi}\left[\frac{\beta}{2} - \frac{\alpha}{2} - \frac{\sin 2\beta}{4} + \frac{\sin 2\alpha}{4} - \frac{\sin 2\alpha}{2} + \frac{\sin 2\beta}{2}\right]$$

$$= \frac{4KX}{\pi}\left[\frac{\beta - \alpha}{2} + \frac{\sin 2\beta}{4} - \frac{\sin 2\alpha}{4}\right]$$

$$= \frac{2KX}{\pi}\left[(\beta - \alpha) + \frac{\sin 2\beta}{2} - \frac{\sin 2\alpha}{2}\right]$$

$$= \frac{KX}{\pi}\left[2(\beta - \alpha) + \sin 2\beta - \sin 2\alpha\right]$$

$$\therefore A_1 = \frac{KX}{\pi}\left[2(\beta - \alpha) + \sin 2\beta - \sin 2\alpha\right]$$

Thus, the describing function can be written as follows:

$$\therefore Y_1 = \sqrt{A_1^2 + B_1^2} = Y_1 = \sqrt{A_1^2 + 0^2} = A_1 = \frac{KX}{\pi}\left[2(\beta - \alpha) + \sin 2\beta - \sin 2\alpha\right]$$

$$\Phi_1 = \tan^{-1}\left[\frac{B_1}{A_1}\right] = \tan^{-1}(0) = 0$$

$\therefore$The describing function $N(X) = \dfrac{Y_1(X)}{X}\underline{|\Phi_1(X)}$

$$N(X) = \frac{KX}{\pi}\left[2(\beta - \alpha) + \sin 2\beta - \sin 2\alpha\right]\underline{|0^\circ}$$

If $X < D/2$ then $\alpha = \beta = \pi/2$ then $N(X) = 0$

If $D/2 < X < S$ then $\beta = \pi/2$ then $N(X) = K\left[1 - \frac{2}{\pi}(\alpha + \sin\alpha\cos\alpha)\right]$

If $X > S$ then $N(X) = \dfrac{K}{\pi}\left[2(\beta - \alpha) + \sin 2\beta - \sin 2\alpha\right]$ (4.11)

Describing Function of On–Off Controller with a Deadzone

The input/output relationship in the case of an on–off controller with a dead-zone is shown in Figure 4.15.

If x is less than 'Δ' (deadzone), the controller does not produce any output. However, when x is greater than Δ, the output reaches a value 'M'.

The on–off controller with a deadzone is excited with a sinusoidal input $x = X \sin \omega t$. The output thus obtained is shown in Figure 4.16.

$$\text{The input} \quad x = X \sin \omega t \tag{i}$$

where X is the maximum value of input.

Also, when $\omega t = \alpha$, $x = \Delta$

Thus, equation (i) can be written as, $\Delta = X \sin \alpha$ or $\alpha = \sin^{-1}\left[\dfrac{\Delta}{X}\right]$

In a period 0–2π, the output y can be described as follows:

$$y = \begin{cases} 0 & ; & 0 \le \omega t \le \alpha \\ M & ; & \alpha \le \omega t \le (\pi - \alpha) \\ 0 & ; & (\pi - \alpha) \le \omega t \le (\pi + \alpha) \\ -M & ; & (\pi + \alpha) \le \omega t \le (2\pi - \alpha) \\ 0 & ; & (2\pi - \alpha) \le \omega t \le 2\pi \end{cases} \tag{ii}$$

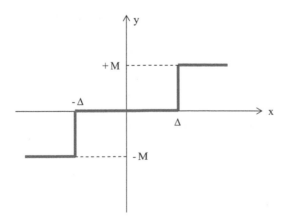

FIGURE 4.15
Input–output characteristics of on–off controller with deadzone.

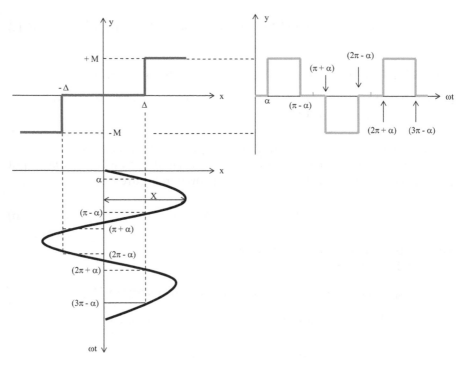

FIGURE 4.16
Sinusoidal response of on–off controller with deadzone.

The periodic function has odd symmetry

$$y(\omega t) = -y(-\omega t)$$

Therefore, the fundamental component of y is given by

$$y_1 = A_1 \sin \omega t$$

where $A_1 = \dfrac{1}{\pi} \displaystyle\int_0^{2\pi} y \sin \omega t \, d(\omega t)$

due to the symmetry of y, the coefficient A_1 can be calculated as follows:

$$A_1 = \frac{4}{\pi} \int_0^{\pi/2} y \sin \omega t \, d(\omega t) = \frac{4M}{\pi} \int_0^{\pi/2} y \sin \omega t \, d(\omega t) = \frac{4M}{\pi} \cos \alpha \tag{iii}$$

Since, B_1 (Fourier series cosine coefficient) is zero, the first harmonic component of 'y' is exactly in phase with $X \sin \omega t$. Thus, the describing function $N(X)$ of on–off controller with deadzone is given by

$$N(X) = \frac{Y_1(X)}{X} \lfloor \Phi_1(X)$$

where $Y_1 = \sqrt{A_1^2 + B_1^2}$ and $\Phi_1 = \tan^{-1}\left[\frac{B_1}{A_1}\right]$

$$N(X) = \frac{4M}{\pi X}\cos\alpha\lfloor\tan^{-1}[\alpha] = \frac{4M}{\pi X}\cos\alpha\lfloor 0^\circ$$

But we know that, $\alpha = \sin^{-1}\left[\frac{\Delta}{X}\right]$ or $\sin\alpha = \frac{\Delta}{X}; \cos\alpha = \sqrt{1 - \sin^2\alpha} = \sqrt{1 - \left(\frac{\Delta}{X}\right)^2}$

$$N(X) = \frac{4M}{\pi X}\sqrt{1 - \left(\frac{\Delta}{X}\right)^2}\lfloor 0^\circ; \ N(X) = 0; \ x < \Delta$$

$$N(X) = \frac{4M}{\pi X}\sqrt{1 - \left(\frac{\Delta}{X}\right)^2}; \ x \geq \Delta$$

(4.12)

Describing Function of Backlash Nonlinearity

The input–output relationship of backlash nonlinearity is shown in Figure 4.17.

Let the backlash nonlinearity be excited by a sinusoidal input.

$$x = X\sin\omega t \tag{i}$$

The response of the nonlinearity when excited with sinusoidal input is as shown in Figure 4.18.

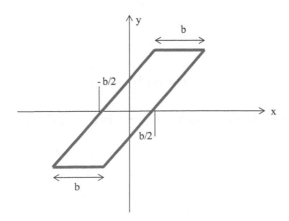

FIGURE 4.17
Input–output characteristics of backlash nonlinearity.

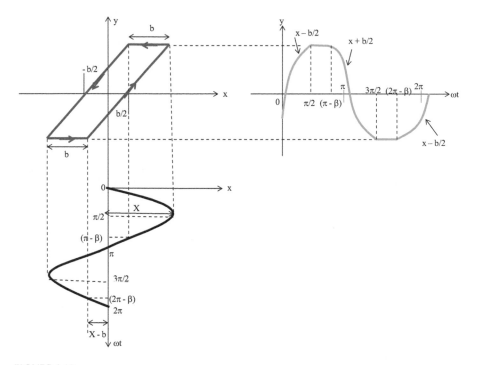

FIGURE 4.18
Sinusoidal response of backlash nonlinearity.

From Figure 4.18, it is clear that, when $\omega t = (\pi - \beta)$, $x = X - b$. Substituting these values in equation (i), we get

$$X - b = X \sin(\pi - \beta)$$

$$X - b = X \sin \beta$$

$$\sin \beta = \frac{X - b}{X} = 1 - \frac{b}{X}$$

$$\beta = \sin^{-1}\left[1 - \frac{b}{X}\right]$$

The output can be divided into five regions in a period of 2π.

$$Y = \begin{cases} x - b/2 & ; & 0 \le \omega t \le \pi/2 \\ X - b/2 & ; & \pi/2 \le \omega t \le (\pi - \beta) \\ x + b/2 & ; & (\pi - \beta) \le \omega t \le 3\pi/2 \\ -X + b/2 & ; & 3\pi/2 \le \omega t \le (2\pi - \beta) \\ x - b/2 & ; & (2\pi - \beta) \le \omega t \le 2\pi \end{cases} \qquad \text{(ii)}$$

The describing function $N(X) = \dfrac{Y_1(X)}{X}\lfloor\Phi_1(X)$

where $Y_1 = \sqrt{A_1^2 + B_1^2}$ and $\Phi_1 = \tan^{-1}\left[\dfrac{B_1}{A_1}\right]$

we have $A_1 = \dfrac{2}{\pi}\displaystyle\int_0^{\pi} y\sin\omega t\, d(\omega t)$

In the period '0' to 'π', the output y is given by three different equations. Hence, the expression for A_1 can be written as follows:

$$A_1 = \frac{2}{\pi}\int_0^{\pi/2}\left(x - \frac{b}{2}\right)\sin\omega t\, d(\omega t) + \frac{2}{\pi}\int_{\pi/2}^{\pi-\beta}\left(X - \frac{b}{2}\right)\sin\omega t\, d(\omega t)$$

$$+\frac{2}{\pi}\int_{\pi-\beta}^{\pi}\left(x + \frac{b}{2}\right)\sin\omega t\, d(\omega t)$$

(iii)

Substituting $x = X\sin\omega t$ in the equation (iii), we get

$$A_1 = \frac{2}{\pi}\int_0^{\pi/2}\left(X\sin\omega t - \frac{b}{2}\right)\sin\omega t\, d(\omega t) + \frac{2}{\pi}\int_{\pi/2}^{\pi-\beta}\left(X - \frac{b}{2}\right)\sin\omega t\, d(\omega t)$$

$$+\frac{2}{\pi}\int_{\pi-\beta}^{\pi}\left(X\sin\omega t + \frac{b}{2}\right)\sin\omega t\, d(\omega t)$$

$$= \frac{2X}{\pi}\int_0^{\pi/2}\sin^2\omega t\, d(\omega t) - \frac{b}{2}\int_0^{\pi/2}\sin^2\omega t\, d(\omega t) + \frac{2\left(X - \dfrac{b}{2}\right)}{\pi}\int_{\pi/2}^{\pi-\beta}\sin\omega t\, d(\omega t)$$

$$+\frac{2X}{\pi}\int_{\pi-\beta}^{\pi}\sin^2\omega t\, d(\omega t) + \frac{b}{\pi}\int_{\pi-\beta}^{\pi}\sin\omega t\, d(\omega t)$$

Substitute $\sin^2\omega t = \dfrac{1 - \cos\omega t}{2}$

$$A_1 = \frac{X}{\pi}\int_0^{\pi/2}(1 - \cos\omega t)\, d(\omega t) - \frac{b}{2}\int_0^{\pi/2}\sin\omega t\, d(\omega t) + \frac{2\left(X - \dfrac{b}{2}\right)}{\pi}\int_{\pi/2}^{\pi-\beta}\sin\omega t\, d(\omega t)$$

$$+\frac{X}{\pi}\int_{\pi-\beta}^{\pi}(1 - \cos\omega t)\, d(\omega t) + \frac{b}{\pi}\int_{\pi-\beta}^{\pi}\sin\omega t\, d(\omega t)$$

$$= \frac{X}{\pi}\left[\omega t - \frac{\sin 2\omega t}{2}\right]_0^{\pi/2} - \frac{b}{\pi}[-\cos\omega t]_0^{\pi/2} + \frac{2\left(X - \frac{b}{2}\right)}{\pi}[-\cos\omega t]_{\pi/2}^{\pi-\beta}$$

$$+ \frac{X}{\pi}\left[\omega t - \frac{\sin 2\omega t}{2}\right]_{\pi-\beta}^{\pi} + \frac{b}{\pi}[-\cos\omega t]_{\pi-\beta}^{\pi}$$

$$= \frac{X}{\pi}\left(\frac{\pi}{2}\right) - \frac{b}{\pi}(1) + \frac{2\left(X - \frac{b}{2}\right)}{\pi}(-\cos(\pi - \beta) + \cos\pi/2)$$

$$+ \frac{X}{\pi}\left[\pi - \frac{\sin 2\pi}{2} - (\pi - \beta) + \frac{\sin 2(\pi - \beta)}{2}\right] + \frac{b}{\pi}[-\cos\pi + \cos(\pi - \beta)]$$

$$= \frac{X}{2} - \frac{b}{\pi} + \frac{2}{\pi}\left(X - \frac{b}{2}\right)\cos\beta + \frac{X}{\pi}\left(\beta - \frac{\sin 2\beta}{2}\right) + \frac{b}{\pi}(1 - \cos\beta)$$

$$= \frac{X}{2} - \frac{b}{\pi} + \frac{2}{\pi}\left(X - \frac{b}{2}\right)\cos\beta + \frac{X\beta}{\pi} - \frac{X}{2\pi}\sin 2\beta + \frac{b}{\pi} - \frac{b}{\pi}\cos\beta$$

$$= \frac{X}{2} + \frac{X\beta}{\pi} + \frac{2}{\pi}\left(X - \frac{b}{2} - \frac{b}{2}\right)\cos\beta - \frac{X}{2\pi}\sin 2\beta$$

$$= \frac{X}{2} + \frac{X\beta}{\pi} + \frac{2X}{\pi}\left(1 - \frac{b}{X}\right)\cos\beta - \frac{X}{2\pi}\sin 2\beta$$

But, $\sin\beta = 1 - \dfrac{b}{X}$

$$\therefore A_1 = \frac{X}{2} + \frac{X\beta}{\pi} + \frac{2X}{\pi}\sin\beta\cos\beta - \frac{X}{2\pi}\sin 2\beta$$

$$= \frac{X}{2} + \frac{X\beta}{\pi} + \frac{X}{\pi}\sin 2\beta - \frac{X}{2\pi}\sin 2\beta$$

$$= \frac{X}{2} + \frac{X\beta}{\pi} + \frac{X}{2\pi}\sin 2\beta$$

$$= \frac{X}{\pi}\left[\frac{\pi}{2} + \beta + \frac{1}{2}\sin 2\beta\right]$$

$$B_1 = \frac{2}{\pi}\int_0^{\pi} y\cos\omega t \, d(\omega t)$$

In the period '0' to 'π', the output y is given by three different equations. Hence, the expression for B_1 can be written as follows:

$$B_1 = \frac{2}{\pi}\int_0^{\pi/2}\left(x - \frac{b}{2}\right)\cos\omega t\, d(\omega t) + \frac{2}{\pi}\int_{\pi/2}^{\pi-\beta}\left(X - \frac{b}{2}\right)\cos\omega t\, d(\omega t)$$

$$+\frac{2}{\pi}\int_{\pi-\beta}^{\pi}\left(x + \frac{b}{2}\right)\cos\omega t\, d(\omega t)$$

Substitute $x = X\sin\omega t$

$$B_1 = \frac{2}{\pi}\int_0^{\pi/2}\left(X\sin\omega t - \frac{b}{2}\right)\cos\omega t\, d(\omega t) + \frac{2}{\pi}\int_{\pi/2}^{\pi-\beta}\left(X - \frac{b}{2}\right)\cos\omega t\, d(\omega t)$$

$$+\frac{2}{\pi}\int_{\pi-\beta}^{\pi}\left(X\sin\omega t + \frac{b}{2}\right)\cos\omega t\, d(\omega t)$$

$$=\frac{X}{\pi}\int_0^{\pi/2}2\sin\omega t\cos\omega t\, d(\omega t) - \frac{b}{\pi}\int_0^{\pi/2}\cos\omega t\, d(\omega t) + \frac{2}{\pi}\left(X - \frac{b}{2}\right)\int_{\pi/2}^{\pi-\beta}\cos\omega t\, d(\omega t)$$

$$+\frac{X}{\pi}\int_{\pi-\beta}^{\pi}2\sin\omega t\cos\omega t\, d(\omega t) + \frac{b}{\pi}\int_{\pi-\beta}^{\pi}\cos\omega t\, d(\omega t)$$

$$=\frac{X}{\pi}\int_0^{\pi/2}\sin 2\omega t\, d(\omega t) - \frac{b}{\pi}\int_0^{\pi/2}\cos\omega t\, d(\omega t) + \frac{2}{\pi}\left(X - \frac{b}{2}\right)\int_{\pi/2}^{\pi-\beta}\cos\omega t\, d(\omega t)$$

$$+\frac{X}{\pi}\int_{\pi-\beta}^{\pi}\sin 2\omega t\, d(\omega t) + \frac{b}{\pi}\int_{\pi-\beta}^{\pi}\cos\omega t\, d(\omega t)$$

$$=\frac{X}{\pi}\left[-\frac{\cos 2\omega t}{2}\right]_0^{\pi/2} - \frac{b}{\pi}[\sin\omega t]_0^{\pi/2} + \frac{2}{\pi}\left(X - \frac{b}{2}\right)[\sin\omega t]_{\pi/2}^{\pi-\beta}$$

$$+\frac{X}{\pi}\left[-\frac{\cos 2\omega t}{2}\right]_{\pi-\beta}^{\pi} + \frac{b}{\pi}[\sin\omega t]_{\pi-\beta}^{\pi}$$

$$=\frac{X}{\pi}\left(\frac{-\cos\pi}{2} + \frac{\cos 0}{2}\right) - \frac{b}{\pi}\left(\sin\frac{\pi}{2} - 0\right) + \frac{2}{\pi}\left(X - \frac{b}{2}\right)(\sin(\pi - \beta) - \sin\pi/2)$$

$$+\frac{X}{\pi}\left[-\frac{\cos 2\pi}{2} + \frac{\cos 2(\pi - \beta)}{2}\right] + \frac{b}{\pi}[\sin\pi - \sin(\pi - \beta)]$$

$$= \frac{X}{\pi}\left(\frac{1}{2}+\frac{1}{2}\right) - \frac{b}{\pi}(1-0) + \frac{2}{\pi}\left(X - \frac{b}{2}\right)(\sin\beta - 1)$$

$$+ \frac{X}{\pi}\left(-\frac{1}{2}+\frac{\cos 2\beta}{2}\right) + \frac{b}{\pi}(0 - \sin\beta)$$

$$B_1 = \frac{X}{\pi} - \frac{b}{\pi} + \frac{2X}{\pi}\sin\beta - \frac{b}{\pi}\sin\beta - \frac{2X}{\pi} + \frac{b}{\pi} + \frac{X}{2\pi} + \frac{X}{2\pi}\cos 2\beta - \frac{b}{\pi}\sin\beta$$

$$= \frac{X}{\pi}\left(1 - 2 - \frac{1}{2}\right) + \frac{2X}{\pi}\sin\beta\left(1 - \frac{b}{X}\right) + \frac{X}{2\pi}\cos 2\beta$$

$$= \frac{-3X}{\pi} + \frac{2X}{\pi}\sin\beta\left(1 - \frac{b}{X}\right) + \frac{X}{2\pi}\cos 2\beta$$

Since, $\sin\beta = 1 - \dfrac{b}{X}$ and $\cos 2\beta = \left(1 - 2\sin^2\beta\right)$, we have

$$\therefore B_1 = \frac{-3X}{\pi} + \frac{2X}{\pi}\sin\beta\sin\beta + \frac{X}{2\pi}\left(1 - 2\sin^2\beta\right)$$

$$= \frac{-3X}{\pi} + \frac{2X}{\pi}\sin^2\beta + \frac{X}{2\pi} - \frac{X}{\pi}\sin^2\beta$$

$$= \frac{-X}{\pi} + \frac{X}{\pi}\sin^2\beta$$

$$= \frac{-X}{\pi} + \frac{X}{\pi}(1 - \cos^2\beta)$$

$$= \frac{-X}{\pi} + \frac{X}{\pi} - \frac{X}{\pi}\cos^2\beta$$

$$= -\frac{X}{\pi}\cos^2\beta$$

Thus, the describing function of backlash nonlinearity is given by

$$N(X) = \frac{Y_1(X)}{X}\lfloor\Phi_1(X)$$

where $Y_1 = \sqrt{A_1^2 + B_1^2} = \sqrt{\left[\frac{X}{\pi}\left(\frac{\pi}{2} + \beta + \frac{1}{2}\sin 2\beta\right)\right]^2 + \left(-\frac{X}{\pi}\cos^2\beta\right)^2}$

$$\therefore Y_1 = \frac{X}{\pi}\sqrt{\left[\left(\frac{\pi}{2} + \beta + \frac{1}{2}\sin 2\beta\right)^2 + \cos^4\beta\right]}$$

$$\Phi_1 = \tan^{-1}\left[\frac{B_1}{A_1}\right] = \tan^{-1}\left[\frac{-\dfrac{X}{\pi}\cos^2\beta}{\left[\dfrac{X}{\pi}\left(\dfrac{\pi}{2}+\beta+\dfrac{1}{2}\sin 2\beta\right)\right]}\right]$$

$$\therefore N(x) = \frac{1}{\pi}\sqrt{\left(\frac{\pi}{2}+\beta+\frac{1}{2}\sin 2\beta\right)^2 + \cos^4\beta} \ \angle\Phi_1 \qquad (4.13)$$

Describing Function of Relay with Deadzone and Hysteresis

The input/output characteristic of a relay with a dead zone and hysteresis nonlinearity is shown in Figure 4.19.

In the presence of deadzone, the relay will respond only after a definite value of input. Due to the presence of hysteresis, the output follows different paths for changes in the values of input. When the input x is increased from zero, the output follows the path PQRS, and when the input is decreased from a maximum value, the output follows the path SRTP. For increasing values of input, the output is zero when $x < D/2$ and output is M, when $x > D/2$. Similarly, for decreasing values of input, the output is M when $x > (D/2 - H)$ and output is zero, when $x < (D/2 - H)$.

The response of the relay with deadzone and hysteresis nonlinearity, when excited with a sinusoidal input, is shown in Figure 4.20.

$$\text{The input} \quad x = X\sin\omega t \qquad (i)$$

where X is the maximum value of input

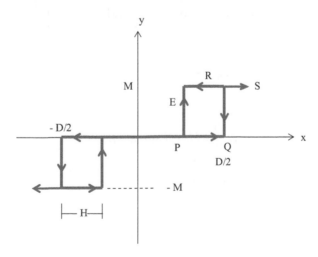

FIGURE 4.19
Input–output characteristics of relay with deadzone and hysteresis nonlinearity.

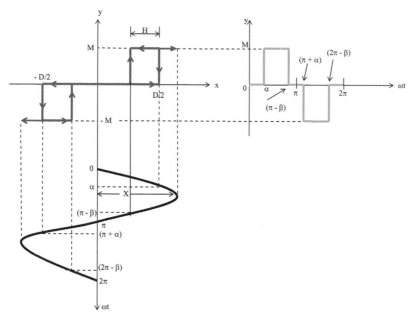

FIGURE 4.20
Sinusoidal response of relay with deadzone and hysteresis nonlinearity.

In Figure 4.20, when $\omega t = \alpha, \quad x = D/2$

$$\therefore D/2 = X \sin \alpha \Rightarrow \sin \alpha = \frac{D}{2X} \text{ and } \alpha = \sin^{-1}\left[\frac{D}{2X}\right] \qquad \text{(ii)}$$

Also, in Figure 4.20, when $\omega t = \pi - \beta, \ x = \dfrac{D}{2} - H$

$$\therefore \frac{D}{2} - H = X \sin(\pi - \beta)$$

$$\Rightarrow \sin(\pi - \beta) = \sin \beta = \left(\frac{D}{2} - H\right)\frac{1}{X} \qquad \text{(iii)}$$

$$\Rightarrow \beta = \sin^{-1}\left[\frac{1}{X}\left(\frac{D}{2} - H\right)\right]$$

In a period $0-2\pi$, the output y can be described as follows:

$$y = \begin{cases} 0 & ; & 0 \le \omega t \le \alpha \\ M & ; & \alpha \le \omega t \le (\pi - \alpha) \\ 0 & ; & (\pi - \alpha) \le \omega t \le (\pi + \alpha) \\ -M & ; & (\pi + \alpha) \le \omega t \le (2\pi - \alpha) \\ 0 & ; & (2\pi - \alpha) \le \omega t \le 2\pi \end{cases} \tag{iv}$$

The describing function is given by

$$N(X) = \frac{Y_1(X)}{X} \lfloor \Phi_1(X)$$

where $Y_1 = \sqrt{A_1^2 + B_1^2}$ and $\Phi_1 = \tan^{-1}\left[\frac{B_1}{A_1}\right]$

$$A_1 = \frac{2}{\pi} \int_0^\pi y \sin \omega t \, d(\omega t)$$

$$= \frac{2}{\pi}\left[\int_0^\alpha y \sin \omega t \, d(\omega t) + \int_\alpha^{\pi-\beta} y \sin \omega t \, d(\omega t) + \int_{\pi-\beta}^\pi y \sin \omega t \, d(\omega t)\right]$$

$$\therefore A_1 = 0 + \frac{2}{\pi}\int_\alpha^{\pi-\beta} M \sin \omega t \, d(\omega t) + 0$$

$$= \frac{2M}{\pi}[-\cos \omega t]_\alpha^{\pi-\beta} = \frac{2M}{\pi}[-\cos(\pi - \beta) + \cos \alpha]$$

$$\therefore A_1 = \frac{2M}{\pi}[\cos \beta + \cos \alpha] \tag{v}$$

From equation (ii), $\sin \alpha = \dfrac{D}{2X}$ and $\cos \alpha = \sqrt{1 - \left(\dfrac{D}{2X}\right)^2}$

From equation (iii), $\sin \beta = \left(\dfrac{D}{2} - H\right)\dfrac{1}{X} = \dfrac{D}{2X} - \dfrac{H}{X}$

$$\Rightarrow \cos \beta = \sqrt{1 - \left(\frac{D}{2X} - \frac{H}{X}\right)^2}$$

Substituting α and β in equation (v), we get

$$A_1 = \frac{2M}{\pi}\left[\sqrt{1-\left(\frac{D}{2X}\right)^2}+\sqrt{1-\left(\frac{D}{2X}-\frac{H}{X}\right)^2}\right]$$

Now,

$$B_1 = \frac{2}{\pi}\int_0^\pi y\cos\omega t\, d(\omega t)$$

$$= \frac{2}{\pi}\left[\int_0^\alpha y\cos\omega t\, d(\omega t)+\int_\alpha^{\pi-\beta} y\cos\omega t\, d(\omega t)+\int_{\pi-\beta}^\pi y\cos\omega t\, d(\omega t)\right]$$

$$= 0+\frac{2}{\pi}\int_\alpha^{\pi-\beta} M\cos\omega t\, d(\omega t)+0 = \frac{2M}{\pi}\left[\sin\omega t\right]_\alpha^{\pi-\beta}$$

$$= \frac{2M}{\pi}\left[\sin(\pi-\beta)-\sin\alpha\right]=\frac{2M}{\pi}\left[\sin\beta-\sin\alpha\right]$$

Substituting for $\sin\beta$ and $\sin\alpha$ in B_1, we get

$$B_1 = \frac{2M}{\pi}\left[\frac{D}{2X}-\frac{H}{X}-\frac{D}{2X}\right]=-\frac{2M}{\pi}\left[\frac{H}{X}\right]$$

$$\therefore Y_1 = \sqrt{A_1^2+B_1^2} = \sqrt{\frac{4M^2}{\pi^2}\left\{\sqrt{1-\left(\frac{D}{2X}\right)^2}+\sqrt{1-\left(\frac{D}{2X}-\frac{H}{X}\right)^2}\right\}+\left[\frac{4M^2}{\pi^2}\left(\frac{H^2}{X^2}\right)\right]}$$

$$\text{(vi)}$$

$$\Phi_1 = \tan^{-1}\left[\frac{B_1}{A_1}\right]=\tan^{-1}\left[\frac{-\dfrac{2M}{\pi}\left[\dfrac{H}{X}\right]}{\dfrac{2M}{\pi}\left[\sqrt{1-\left(\dfrac{D}{2X}\right)^2}+\sqrt{1-\left(\dfrac{D}{2X}-\dfrac{H}{X}\right)^2}\right]}\right] \quad \text{(vii)}$$

From equations (vi) and (vii), we can obtain the describing functions of the following three cases: (1) Ideal relay, (2) relay with deadzone and (3) relay with hysteresis.

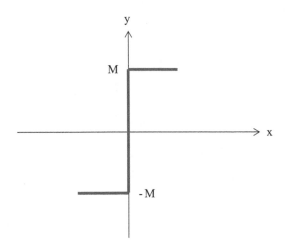

FIGURE 4.21
Input-output characteristic of ideal relay.

1. Ideal relay (Figure 4.21)

   In the case of ideal relay, $D = H = 0$

   Substituting these values in equations (vi) and (vii), we get

   $Y_1 = \dfrac{4M}{\pi}$ and $\Phi_1 = 0$

   $$\therefore N(X) = \frac{Y_1(X)}{X}\lfloor\Phi_1(X) = \frac{4M}{\pi X}\lfloor 0° = \frac{4M}{\pi X} \tag{4.14}$$

2. Relay with deadzone (Figure 4.22)

   In the case of relay with deadzone, $H = 0$

   Substituting $H = 0$ in equations (vi) and (vii), we get

   $$\therefore Y_1 = \sqrt{\frac{4M^2}{\pi^2}\left\{2\sqrt{1-\left(\frac{D}{2X}\right)^2}\right\}} = \frac{4M}{\pi}\sqrt{1-\left(\frac{D}{2X}\right)^2}$$

   $$\Phi_1 = 0$$

   $$\therefore N(X) = \frac{Y_1(X)}{X}\lfloor\Phi_1(X) = \begin{cases} 0 & ; \quad x < D/2 \\[2ex] \dfrac{4M}{\pi}\sqrt{1-\left(\dfrac{D}{2X}\right)^2} & ; \quad x > D/2 \end{cases} \tag{4.15}$$

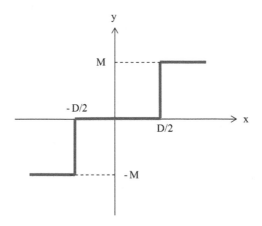

FIGURE 4.22
Input–output characteristics of relay with deadzone.

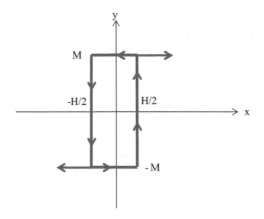

FIGURE 4.23
Input–output characteristics of relay with hysteresis.

3. Relay with hysteresis (Figure 4.23)
    In the case of relay with hysteresis, $D = H$
    Substituting this value in equations (vi) and (vii), we get

$$\therefore Y_1 = \sqrt{\frac{4M^2}{\pi^2}\left\{\sqrt{1-\left(\frac{D}{2X}\right)^2}+\sqrt{1-\left(\frac{D}{2X}-\frac{H}{X}\right)^2}\right\}^2 + \left[\frac{4M^2}{\pi^2}\left(\frac{H^2}{X^2}\right)\right]}$$

$$Y_1 = \frac{2M}{\pi}\sqrt{\left[4\left(1-\frac{H^2}{4X^2}\right)+\left(\frac{H^2}{X^2}\right)\right]} = \frac{2M}{\pi}\sqrt{4-\frac{H^2}{X^2}+\frac{H^2}{X^2}} = \frac{4M}{\pi} \qquad \text{(viii)}$$

Substituting $D = H$ in equation (vii), we get

$$\Phi_1 = \tan^{-1}\left[\frac{\dfrac{2M}{\pi}\left[\dfrac{-H}{X}\right]}{\dfrac{2M}{\pi}\left[\sqrt{1-\left(\dfrac{H}{2X}\right)^2} + \sqrt{1-\left(-\dfrac{H}{2X}\right)^2}\right]}\right]$$

$$= \tan^{-1}\left[\frac{\dfrac{-H}{X}}{2\sqrt{1-\dfrac{H^2}{4X^2}}}\right]$$

$$= -\tan^{-1}\left[\frac{\dfrac{H}{2X}}{\sqrt{1-\dfrac{H^2}{4X^2}}}\right]$$

$$-\Phi_1 = \tan^{-1}\left[\frac{\dfrac{H}{2X}}{\sqrt{1-\dfrac{H^2}{4X^2}}}\right] \quad \text{i.e.,} \quad \tan(-\Phi_1) = \left[\frac{\dfrac{H}{2X}}{\sqrt{1-\dfrac{H^2}{4X^2}}}\right] \qquad \text{(ix)}$$

From equation (ix), we can construct a right-angled triangle as shown in Figure 4.24. From the triangle, we have

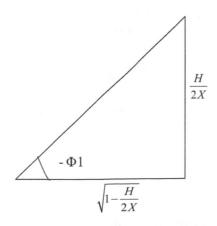

$$\frac{H}{2X}$$

$$-\Phi 1$$

$$\sqrt{1-\frac{H}{2X}}$$

FIGURE 4.24

$$\sin(-\Phi_1) = \frac{H}{2X}$$

$$-\Phi_1 = \sin^{-1}\left[\frac{H}{2X}\right] \tag{x}$$

$$\Phi_1 = -\sin^{-1}\left[\frac{H}{2X}\right]$$

From equations (ix) and (x), the describing function of relay with hysteresis can be written as follows:

$$\therefore N(X) = \frac{Y_1(X)}{X}\underline{|\Phi_1(X)} = \begin{cases} 0 & ; \quad x < H/2 \\ \frac{4M}{\pi X}\underline{\left[\left(-\sin^{-1}\frac{H}{2X}\right)\right.} & ; \quad x > H/2 \end{cases} \tag{4.16}$$

For quick reference, the describing functions for various non-linear elements are tabulated in Table 4.1.

Phase Plane Analysis

The phase plane concept was introduced in the 19th century by the mathematician Henri Poincare. It is a graphical method of investigating system behaviour and to design system parameters to achieve a desired response of second-order systems.

In order to understand the basic concepts of phase plane analysis, let us consider the free motion of a second-order non-linear system given in equation (i)

$$\ddot{y} + g(y,\dot{y})\dot{y} + h(y,\dot{y})y = 0 \tag{i}$$

By using state variables $x_1 = y$ and $x_2 = \dot{y}$, equation (i) can be rewritten in the form of canonical set of state equations as given in equation (ii)

$$\dot{x}_1 = \frac{dx_1}{dt} = x_2 \tag{ii}$$

$$\dot{x}_2 = \frac{dx_2}{dt} = \ddot{y} = -g(x_1,x_2)x_2 - h(x_1,x_2)x_1 \tag{iii}$$

Dividing equation (iii) by (ii), we get

$$\frac{dx_2}{dx_1} = -\frac{[g(x_1,x_2)x_2 - h(x_1,x_2)x_1]}{x_2} \tag{iv}$$

where x_1 is an independent variable and x_2 is a dependent variable.

TABLE 4.1

Describing Functions of Non-Linear Elements (Input, $x = \sin \omega t$)

Sl. No	Nonlinearity with Pictorial Representation	Describing Function
1.	 Deadzone	$N(X) = KX\left[1 - \dfrac{2}{\pi}\left(\sin^{-1}\left[\dfrac{D}{2X}\right] + \dfrac{D}{2X}\sqrt{1 - \left(\dfrac{D}{2X}\right)^2}\right)\right]\Big\|_{\underline{0^\circ}}$ $N(X) = 0 \quad \text{for } X < D/2$ $N(X) = K\left[1 - \dfrac{2}{\pi}(\alpha + \sin\alpha\cos\alpha)\right] \quad \text{for } X > D/2$
2.	 Saturation	$N(X) = \dfrac{2K}{\pi}\left[\sin^{-1}\left[\dfrac{S}{X}\right] + \dfrac{S}{X}\sqrt{1 - \left(\dfrac{S}{X}\right)^2}\right] \quad \text{for } X > S$

(Continued)

TABLE 4.1 (*Continued*)

Describing Functions of Non-Linear Elements (Input, $x = \sin\omega t$)

Sl. No	Nonlinearity with Pictorial Representation	Describing Function
3.	Deadzone and saturation	$N(X) = \dfrac{KX}{\pi}\left[2(\beta - \alpha) + \sin 2\beta - \sin 2\alpha\right]\underline{\vert 0^\circ}$ If $X < D/2$ then $\alpha = \beta = \pi/2$ then $N(X) = 0$ If $D/2 < X < S$ then $\beta = \pi/2$ then $N(X) = K\left[1 - \dfrac{2}{\pi}(\alpha + \sin\alpha\cos\alpha)\right]$ If $X > S$ then $N(X) = \dfrac{K}{\pi}\left[2(\beta - \alpha) + \sin 2\beta - \sin 2\alpha\right]$
4.	On-off controller with deadzone	$N(X) = \dfrac{4M}{\pi X}\sqrt{1 - \left(\dfrac{\Delta}{X}\right)^2}\;\underline{\vert 0^\circ};\; N(X) = 0;\; x < \Delta$ $N(X) = \dfrac{4M}{\pi X}\sqrt{1 - \left(\dfrac{\Delta}{X}\right)^2}\;;\; x \geq \Delta$

(*Continued*)

TABLE 4.1 (*Continued*)

Describing Functions of Non-Linear Elements (Input, $x = \sin\omega t$)

Sl. No	Nonlinearity with Pictorial Representation	Describing Function
5.	Backlash	$\therefore N(x) = \frac{1}{\pi}\sqrt{\left(\frac{\pi}{2} + \beta + \frac{1}{2}\sin 2\beta\right)^2 + \cos^4\beta}\;\underline{\lfloor \Phi_1}$
6.	Ideal relay	$N(X) = \frac{Y_1(X)}{X}\underline{\lfloor \Phi_1(X)} = \frac{4M}{\pi X}\underline{\lfloor 0^\circ} = \frac{4M}{\pi X}$

(*Continued*)

TABLE 4.1 (Continued)

Describing Functions of Non-Linear Elements (Input, $x = \sin\omega t$)

Sl. No	Nonlinearity with Pictorial Representation	Describing Function
7.	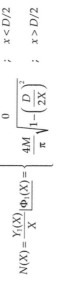 Relay with deadzone	$$N(X) = \frac{Y_1(X)}{X}\lfloor \Phi_1(X) = \begin{cases} 0 & ; \quad x < D/2 \\ \dfrac{4M}{\pi}\sqrt{1-\left(\dfrac{D}{2X}\right)^2} & ; \quad x > D/2 \end{cases}$$
8.	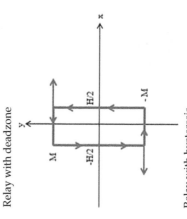 Relay with hysteresis	$$N(X) = \frac{Y_1(X)}{X}\lfloor \Phi_1(X) = \begin{cases} 0 & ; \quad x < H/2 \\ \dfrac{4M}{\pi X}\left\lfloor\left(-\sin^{-1}\dfrac{H}{2X}\right)\right. & ; \quad x > H/2 \end{cases}$$

Phase Plane

The state of a system at any moment can be represented by a point with coordinates $(y, \dot{y})$ or (x_1, x_2) in a system of rectangular coordinates. Such a coordinate plane is called a 'phase plane'.

Phase Trajectory

In a phase plane, the solution of equation (iv) may be represented as a single curve, for a given set of initial conditions $\left(x_1(0), x_2(0)\right)$. Here, the coordinates are x_1 and x_2. As time t varies from '0' to '∞', the curve traced out by the state point $\left(x_1(t), x_2(t)\right)$ forms the 'phase trajectory'.

Phase Portrait

The family of phase plane trajectories corresponding to various initial conditions forms the phase portrait of the system.

Phase Trajectory of a Linear Second-Order Servo System

Consider a linear second-order servo system represented by the equation (v),

$$\ddot{y} + 2\xi\dot{y} + y = 0$$

$$y(0) = y^0, \quad \dot{y}(0) = 0; \quad 0 < \xi < 1 \tag{v}$$

By using state variables $x_1 = y$ and $x_2 = \dot{y}$, the system model can be obtained as given in equation (vi)

$$\dot{x}_1 = x_2$$

$$\dot{x}_2 = -2\xi x_2 - x_1$$

$$x_1(0) = y^0 \tag{vi}$$

$$x_2(0) = 0$$

The output response and its corresponding phase trajectory of the linear second-order system are as shown in Figures 4.25 and 4.26, respectively.

The origin of phase $\left(x_1(0), x_2(0)\right)$ forms the equilibrium point of the system. At the equilibrium point, the derivatives $\dot{x}_1$ and $\dot{x}_2$ are zero.

The phase trajectory of the given linear second-order system is shown in Figure 26. From the phase trajectory, we can infer the transient behaviour of the given system.

The phase trajectory starts from point 'P' and returns to next position, i.e., to the origin by exhibiting a damped oscillatory behaviour.

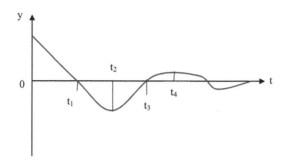

FIGURE 4.25
Output response of linear second-order system.

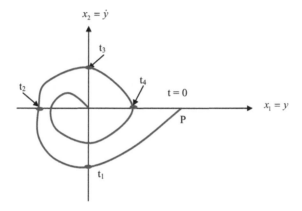

FIGURE 4.26
Phase trajectory of linear second-order system.

NOTE

1. In the case of time-invariant systems, the phase plane is covered with trajectories having only one curve passing through each point of the plane.

2. If the system parameters vary with time then, two or more trajectories may pass through single point in phase plane. In such cases, phase portrait becomes complex, and hence, the phase plane analysis is restricted to second-order system with constant parameters and constant or zero input.

Advantages of Phase Plane Analysis

- The phase plane analysis is a graphical analysis and the solution of phase trajectories can be represented by curves in a plane.
- It provides easy visualization of the system, qualitatively.
- The behaviour of the non-linear system can be studied for different initial conditions, without analytically solving the non-linear equations.
- It applies for both smooth and hard nonlinearities.

Disadvantages of Phase Plane Analysis

- It is restricted mainly to second-order system.
- The graphical study of higher-order system is computationally and geometrically complex.

Singular Point

A singular point deals with the important information about properties of system, namely: stability. Singular points are only points where several trajectories pass or approach them.

Consider a second-order linear system described by equation (vii).

$$\ddot{y} + 2\xi\omega_n\dot{y} + \omega_n^2 y = 0 \tag{vii}$$

If the characteristic roots of equation (vii) are assumed to be λ_1 and λ_2, then $(s - \lambda_1)(s - \lambda_2) = 0$

The corresponding canonical state model is given by equation (viii)

$$\dot{x}_1 = x_2; \quad \dot{x}_2 = -2\xi\omega_n\dot{y} - \omega_n^2 y$$
$$\Rightarrow \dot{x}_2 = -2\xi\omega_n x_2 - \omega_n^2 x_1 \tag{viii}$$

where $x_1 = y$, $x_2 = \dot{y} = \dot{x}_1$ and $\dot{x}_2 = \ddot{y}$

The differential equation of trajectories is given by equation (ix)

$$\frac{dx_2}{dx_1} = \frac{-2\xi\omega_n x_2 - \omega_n^2 x_1}{x_2} \tag{ix}$$

When $\dot{x}_1 = 0$ and $\dot{x}_2 = 0$, the slope $\dfrac{dx_2}{dx_1}$ becomes indeterminate, which corresponds to the name 'singular'.

NOTE

- Since, at singular points on the phase plane, $\dot{x}_1 = \dot{x}_2 = 0$, these points correspond to equilibrium states of the given non-linear system.
- For linear system, there is only one singular point, whereas for a non-linear system, there may be many singular points.
- Non-linear systems often have multiple equilibrium states and limit cycles.

Behaviour of Trajectories in the Vicinity of Singular Point

Depending on the values of λ_1 and λ_2, we can identify six types of singular points, namely: (1) Stable focus, (2) unstable focus, (3) centre, (4) stable node, (5) unstable node and (6) saddle point.

Stable Focus Singular Point

Consider a second-order system whose characteristic roots λ_1 and λ_2 are given by $\lambda_1 = -\alpha + j\beta$; $\lambda_2 = -\alpha - j\beta$; where $\alpha > 0$ and $\beta > 0$. In other words, the system considered is a stable system with complex roots. The phase portrait of such a system can be constructed with $x_1 = y$, $x_2 = \dot{y}$ as shown in Figure 4.27. The phase trajectories appear like a logarithmic, entering into the singular point. Such a singular point is called 'stable focus'.

Unstable Focus Singular Point

Consider a second-order system whose characteristic roots λ_1 and λ_2 are given by $\lambda_1 = \alpha + j\beta$; $\lambda_2 = \alpha - j\beta$ where $\alpha > 0$ and $\beta > 0$. In other words, the system considered is an unstable system with complex roots. The phase portraits of

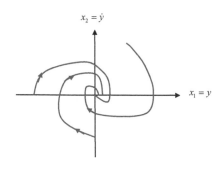

FIGURE 4.27
Stable focus.

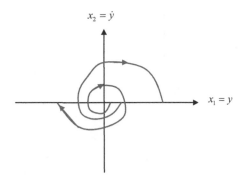

FIGURE 4.28
Unstable focus.

such a system are shown in Figure 4.28. The phase trajectories appear like a logarithmic spiral coming out of the singular point. Such a singular point is called 'unstable focus'.

'Vortex' or 'Centre' Singular Point

Consider a second-order system whose characteristic roots λ_1 and λ_2 are given by $\lambda_1 = j\beta$; $\lambda_2 = -j\beta$ where $\alpha = 0$ and $\beta > 0$. In other words, the system under consideration is a marginally stable system with complex roots (real part, $\alpha = 0$). The phase portrait of such a system is shown in Figure 4.29. The phase trajectories appear like concentric elliptical curves. Such a singular point is called 'vortex' or 'centre'.

Stable Node Singular Point

Consider a second-order system whose characteristic roots λ_1 and λ_2 are the two real and distinct roots in the left half of s-plane. In other words, the system under consideration is a stable system with real roots. The phase portrait

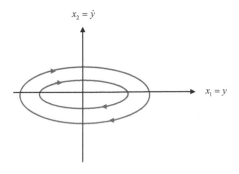

FIGURE 4.29
Centre.

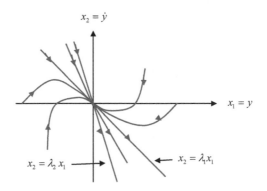

FIGURE 4.30
Stable node.

of such a system is shown in Figure 4.30, which has two straight line trajectories defined by the equations

$x_2(t) = \lambda_1 x_1(t)$ and $x_2(t) = \lambda_2 x_1(t)$. As t increases, the trajectories become tangential to the straight line $x_2(t) = \lambda_1 x_1(t)$, at the origin. Such a singular point is called 'stable node'.

Unstable Node Singular Point

Consider a second-order system whose characteristic roots λ_1 and λ_2 are the two real and distinct roots in the right half of s-plane. In other words, the system under consideration is an unstable system with positive real roots. The phase portrait of such a system is shown in Figure 4.31. The phase trajectories move out f singular point and travel towards infinity. Such a singular point is called 'unstable node'.

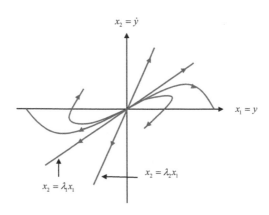

FIGURE 4.31
Unstable node.

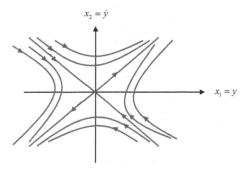

FIGURE 4.32
Saddle point.

Unstable System with One Negative Real Root and One Positive Real Root

Consider a second-order system with one negative real root and one positive real root. The phase portrait of such a system is shown in Figure 4.32. The straight line phase trajectory due to negative real root moves towards the singular point, whereas the straight line trajectory due to positive real root moves away from the singular point. Such a singular point is called 'saddle point'.

NOTE

Phase plane analysis of piecewise linear system (important class of non-linear systems) can be carried out using analytical approach. In graphical approach, there are two methods, namely: Isocline method and delta method.

Phase Plane Analysis (Analytical Approach – Illustration)

Analytical method can be approached in two ways:

a. Integration method: Here the slope equation $\dfrac{dx_2}{dx_1}$ is integrated to relate the state variables.

b. Time domain method: Here x_1 and x_2 are solved with respect to time 't'.

Illustration 1

Construct phase plane trajectories for the system shown in Figure 4.33, with an 'ideal relay' as the non-linear element, using analytical approach of phase plane analysis. (Take $r = 5$, $M = 1$, $y(0) = 2$, $\dot{y}(0) = 1$)

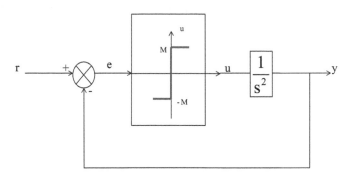

FIGURE 4.33
Non-linear system with ideal relay.

Solution

Given $r = 5$, $M = 1$, $y(0) = 2$, $\dot{y}(0) = 1$

We have to find out the slope equation, $\dfrac{dx_2}{dx_1}$

The differential equation describing the dynamics of the above system is written as follows:

$$\frac{y}{u} = \frac{1}{s^2} \Rightarrow y = \frac{u}{s^2} \Rightarrow \ddot{y} = u$$

Choosing the state vectors x_1 and x_2 as $x_1 = y$ and $x_2 = \dot{y}$
After differentiating, we get

$$\dot{x}_1 = \dot{y} = x_2$$

$$\dot{x}_2 = \ddot{y} = u$$

For an ideal relay, the output 'u' is as given below
$u = M \, \text{sgn}(e)$
But $e = r - y = r - x_1$
$\therefore u = M \, \text{sgn}(r - x_1)$
i.e., $\dot{x}_2 = M \, \text{sgn}(r - x_1)$

$$\frac{dx_2}{dx_1} = \frac{dx_2/dt}{dx_1/dt} = \frac{M \, \text{sgn}(r - x_1)}{x_2}$$

Separate the variables,
$x_2 \, dx_2 = M \, \text{sgn}(r - x_1) \, dx_1$
Integrating on both sides, we get

$$\int_{x_2(0)}^{x_2} x_2 \, dx_2 = \int_{x_1(0)}^{x_1} M.sgn(r - x_1) \, dx_1$$

After simplification, we get

$$x_2^2 = 2Mx_1 - 2Mx_1(0) + x_2^2(0) \quad (\text{for } x_1 < r)$$

$$x_2^2 = -2Mx_1 + 2Mx_1(0) + x_2^2(0) \quad (\text{for } x_1 > r)$$

Now, we know that $y(0) = 2$, $\dot{y}(0) = 1$ implies $x_1(0) = 2$, $x_2(0) = 1$
Substitute initial conditions in the equation

$$x_2^2 = 2Mx_1 - 2Mx_1(0) + x_2^2(0) \quad (\text{for } x_1 < r)$$

$$x_2^2 = 2(1)x_1 - 2(1)(2) + 1$$

$$x_2^2 = 2(x_1 - 2) + 1$$

$$x_2^2 = 2x_1 - 4 + 1 = 2x_1 - 3$$

$$\Rightarrow x_2 = \pm\sqrt{2x_1 - 3} \quad (\text{for } x_1 < r)$$

$$x_2^2 = -2Mx_1 + 2Mx_1(0) + x_2^2(0) \quad (\text{for } x_1 > r)$$

$$x_2^2 = -2(1)x_1 + 2(1)(2) + 1$$

$$x_2^2 = -2(x_1 - 2) + 1$$

$$x_2^2 = -2x_1 + 4 + 1 = -2x_1 + 5$$

$$\Rightarrow x_2 = \pm\sqrt{(-2x_1 + 5)} \quad (x_1 > r)$$

The values of x_2 computed for different values of x_1 (for both the conditions namely $x_1 < r$ and $x_1 > r$ are tabulated in Table 4.2.

TABLE 4.2

Values of x_1 & x_2 for $x_1 > r$ and $x_1 < r$

$x_1 > r$			$x_1 < r$		
x_1	x_2		x_1	x_2	
2	1	−1	1.5	0	0
2.1	0.89	−0.89	1.6	0.44	−0.44
2.2	0.77	−0.77	1.7	0.63	−0.63
2.3	0.63	−0.63	1.8	0.77	−0.77
2.4	0.44	−0.44	1.9	0.89	−0.89
2.5	0	0	2	1	−1

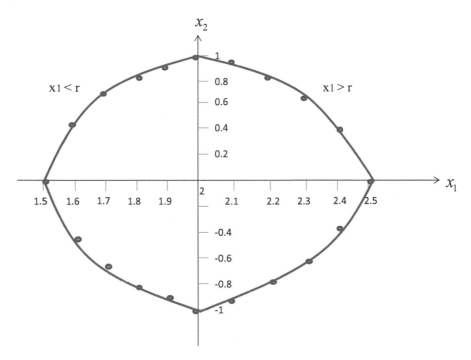

FIGURE 4.34
Phase plane trajectory for $x_1(0) = 2$, $x_2(0) = 1$.

Thus, the phase plane trajectory can be drawn as shown in Figure 4.34.
For different initial conditions, i.e., for different $x_1(0)$ and $x_2(0)$ values, we get different phase trajectories on the phase plane.

Illustration 2

Construct phase plane trajectories for the system shown in Figure 4.35, with an 'ideal relay' as the non-linear element, using analytical approach of phase plane analysis. (Take $r = 5$, $M = 1$, $y(0) = 2$, $\dot{y}(0) = 0$)

Solution

Given $r = 5$, $M = 1$, $y(0) = 2$, $\dot{y}(0) = 0$

We have to find out the slope equation, $\dfrac{dx_2}{dx_1}$

The differential equation describing the dynamics of the above system is given as follows:

$$\frac{y}{u} = \frac{1}{s^2} \Rightarrow y = \frac{u}{s^2} \Rightarrow \ddot{y} = u$$

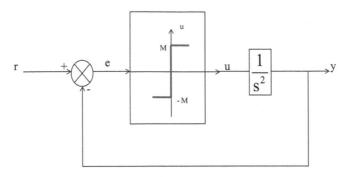

FIGURE 4.35
Non-linear system with ideal relay.

Choosing the state vectors x_1 and x_2 as $x_1 = y$ and $x_2 = \dot{y}$
  After differentiating, we get

$$\dot{x}_1 = \dot{y} = x_2$$

$$\dot{x}_2 = \ddot{y} = u$$

For an ideal relay, the output 'u' is as given below
  $u = M \, \text{sgn}(e)$
  But $e = r - y = r - x_1$
  $\therefore u = M \, \text{sgn}(r - x_1)$
  i.e., $\dot{x}_2 = M \, \text{sgn}(r - x_1)$

$$\frac{dx_2}{dx_1} = \frac{dx_2/dt}{dx_1/dt} = \frac{M \, \text{sgn}(r - x_1)}{x_2}$$

Separate the variables,
  $x_2 \, dx_2 = M.\text{sgn}(r - x_1) \, dx_1$
  Integrating on both sides, we get

$$\int_{x_2(0)}^{x_2} x_2 \, dx_2 = \int_{x_1(0)}^{x_1} M.\text{sgn}(r - x_1) \, dx_1$$

After simplification, we get

$$x_2^2 = 2Mx_1 - 2Mx_1(0) + x_2^2(0) \quad (\text{for } x_1 < r)$$

$$x_2^2 = -2Mx_1 + 2Mx_1(0) + x_2^2(0) \quad (\text{for } x_1 > r)$$

When $x_1 < 5$

Now when $M = 1$, $r = 5$ and $x_1 < 5$; $\dfrac{dx_2}{dx_1} = M \operatorname{sgn}(r - x_1)$

$$\frac{dx_2}{dx_1} = 1 \Rightarrow dx_2 = dt \Rightarrow \int dx_2 = \int dt$$

(i)

$$x_2 = t + C_1$$

But $x_2 = \dot{x}_1$

$$x_1(t) = \int x_2(t)\,dt$$

$$= \int (t + C_1)\,dt$$

(ii)

$$x_1(t) = \frac{t^2}{2} + C_1 t + C_2$$

We have $y(0) = x_1(0) = 2$, $\dot{y}(0) = x_2(0) = 0$
Substituting the above value in equation (i)
we get $C_1 = 0$
Substituting the above value in equation (ii)
we get $C_2 = 2$
Substituting $C_1 = 0$ and $C_2 = 2$ in equations (i) and (ii)
we get

$$x_2(t) = t$$

(iii)

$$x_1(t) = \frac{t^2}{2} + 2$$

(iv)

When $x_1 > 5$
Similarly, when $M = 1$, $r = 5$ and $x_1 > 5$;

$$\frac{dx_2}{dx_1} = -1 \Rightarrow dx_2 = -dt \Rightarrow \int dx_2 = \int -dt$$

$$x_2(t) = -t + C_3$$

(v)

But $x_2(t) = \dot{x}_1(t)$

$$x_1(t) = \int x_2(t)\,dt$$

$$= \int (-t + C_3)\,dt$$

$$x_1(t) = \frac{-t^2}{2} + C_3 t + C_4 \qquad \text{(vi)}$$

When $x_1 = 5$

$$5 = \frac{t^2}{2} + 2$$

$$\Rightarrow t = 2.449$$

Substituting t in equations (v) and (vi)

$$\Rightarrow x_2(t) = -t + C_3$$

$$t = -t + C_3$$

$$2.449 = -2.449 + C_3$$

$$C_3 = 4.898$$

$$\Rightarrow x_1(t) = \frac{-t^2}{2} + C_3 t + C_4$$

$$5 = \frac{-(2.449)^2}{2} + (4.898)(2.449) + C_4$$

$$5 = -2.9988 + 11.9952 + C$$

$$C_4 = -3.9964$$

When $x_1 \geq 5$
Substituting these values in equations (v) and (vi), we get

$$\therefore x_2(t) = -t + 4.898 \qquad \text{(vii)}$$

$$x_1(t) = \frac{-t^2}{2} + 4.898t - 3.9964 \qquad \text{(viii)}$$

The values of x_1 and x_2 computed for different values of t are tabulated in Table 4.3.
Thus, the phase plane trajectory can be drawn as follows:
For different initial conditions, i.e., for different $x_1(0)$ and $x_2(0)$ values, we get different phase trajectories (Figure 4.36).

TABLE 4.3

Values of $x_1(t)$ & $x_2(t)$ for zone-I and zone-II for different values of t

t	Zone	$x_1(t) = \dfrac{+t^2}{2} + 2$ (Equation iv)	$x_2(t) = +t$ (Equation iii)
0	I	2	0
1	I	2.5	1
2	I	4	2
2.449	I	5	2.449
		$x_1(t) = \dfrac{-t^2}{2} + 4.898t - 3.9964$ (Equation viii)	$x_2(t) = -t - 4.898$ (Equation vii)
4	II	7.5956	0.898
5	II	7.9936	−0.102
6	II	7.3916	−1.102
7.347	II	5	−2.449

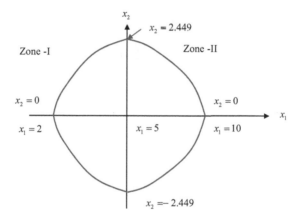

FIGURE 4.36
Phase plane trajectory for $x_1(0) = x_2(0) = 0$.

Isocline Method [Illustration]

Consider a time-invariant system described by the state equation, $\dot{x} = f(x, u)$. If the input vector 'u' is constant, then $\dot{X} = F(X)$. Such a system represented by the above equation is called autonomous system. In the component form, the above equation can be written as two-scalar first-order differential equations, $\dot{x}_1 = f_1(x_1, x_2)$ and $\dot{x}_2 = f_2(x_1, x_2)$.

Dividing we get

$$\frac{dx_2}{dx_1} = \frac{f_2(x_1, x_2)}{f_1(x_1, x_2)} = S \tag{i}$$

where 'S' is the slope of the phase trajectory. For a specific trajectory,

$$f_2(x_1, x_2) = S_1 f_1(x_1, x_2) \tag{ii}$$

Equation (ii) defines locus of all such points in the phase plane at which the slope of phase trajectory is S_1. Such a locus is called an isocline.

Procedure for Construction of Phase Trajectories Using Isocline Method

Figure 4.37a shows isoclines for different values of slopes 'S_1', 'S_2' and 'S_3' etc., where S_1, S_2 and S_3 are the slopes corresponding to isoclines 1, 2, 3 etc.
   Thus, we have $\alpha_1 = \tan^{-1}(S_1); \alpha_2 = \tan^{-1}(S_2); \alpha_3 = \tan^{-1}(S_3)$

1. Locate P be the point on isoclines-1 corresponding to a set of given initial conditions. The phase trajectory will leave the point P and reach the point Q on isocline-2.
2. At point P, draw two lines at angles α_1 and α_2 which intersect isocline-2 at A and B, respectively.
3. Mark the point 'Qa', midway between A and B.
4. The phase trajectory will cross isocline-2 at Q.
5. At point Q, draw two lines at angles α_2 and α_3 which intersect isocline-3 at C and D, respectively.
6. Mark the point 'R', midway between C and D.
7. The phase trajectory will cross isocline-3 at R. This procedure is repeated for other isoclines.
8. Now join the points P, Q, R to get the requirements phase trajectory.

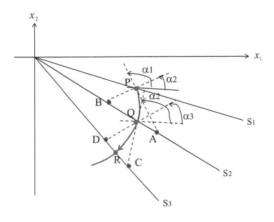

FIGURE 4.37A
Isoclines for different values of slopes.

Illustration 3

Construct a phase plane trajectory by isocline method for a non-linear system represented by the differential equation $\ddot{y} + \dot{y} + |y| = 0$

Solution

Given $\ddot{y} + \dot{y} + |y| = 0$

Let us take the state variables as x_1 and x_2

$\therefore x_1 = y, x_2 = \dot{y}$

Substitute the state variables in the given equation

$$\dot{x}_2 + x_2 + |x_1| = 0$$

$$\dot{x}_1 = x_2$$

$$\dot{x}_2 = -|x_1| - x_2$$

Constructing the slope equation,

$$\frac{dx_2}{dx_1} = \frac{-|x_1| - x_2}{x_2}$$

$$S = \frac{-|x_1| - x_2}{x_2}$$

$$x_2 = \frac{-|x_1|}{S+1}$$

The values of x_1 and x_2 computed for different values of s are tabulated in Table 4.4.

TABLE 4.4

Tangent Table

S	$x_1 > 0$ $x_2 = \dfrac{-x_1}{S+1}$	$x_1 < 0$ $x_2 = \dfrac{x_1}{S+1}$	$\tan^{-1}(S)$
0	−1	1	0
1	−0.5	0.5	45°
2	−0.33	0.33	63°.43
3	−0.25	0.25	71°.56
4	−0.2	0.2	75°.96
−1	0.5	−0.5	−45°
−2	0.33	−0.33	−63°.43
−3	0.25	−0.25	−71°.56
−4	0.2	−0.2	−75°.96

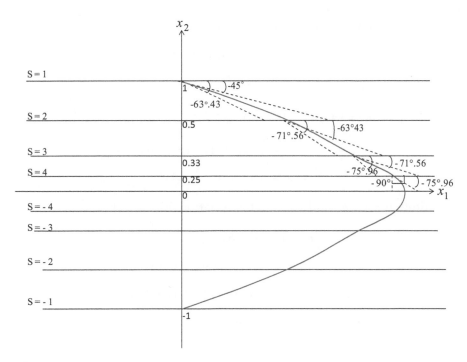

FIGURE 4.37B
Construction of phase trajectory using isocline method.

Using the values obtained in Table 4.4, now construct the phase trajectory, as shown in Figure 4.37b.

Construction of Phase Trajectories Using Delta Method (Graphical)

The delta method can be used to construct a single trajectory for system with describing differential equations of the form given in equation (i)

$$\ddot{x} + f(x, \dot{x}, t) = 0 \tag{i}$$

Equation (i) can be rewritten in the form given in equation (ii)

$$\ddot{x} + K^2[x + \delta(x, \dot{x}, t)] = 0 \tag{ii}$$

Let $K = \omega_n$, the undamped natural frequency of the system when $\delta = 0$

Then, the equation (ii) can be rewritten as follows:

$$\ddot{x} + \omega_n^2 [x + \delta(x, \dot{x}, t)] = 0$$

$$\ddot{x} + \omega_n^2 (x + \delta) = 0$$

(iii)

Choosing the state variables as $x_1 = x$; $x_2 = \dfrac{\dot{x}_1}{\omega_n}$

$$\dot{x}_1 = x_2 \omega_n \quad \text{(or)} \quad \dot{x} = x_2 \omega_n$$

$$\dot{x}_2 \Rightarrow \ddot{x}_1 = \dot{x}_2 \omega_n \quad \text{or} \quad \ddot{x} = \dot{x}_2 \omega_n$$

Substituting $\ddot{x} = \dot{x}_2 \omega_n$ and $x = x_1$ in the equation (iii), we get

$$\dot{x}_2 \omega_n + \omega_n^2 (x_1 + \delta) = 0$$

$$\dot{x}_2 = -\omega_n (x_1 + \delta)$$

Therefore, the state equations can be rewritten as follows:

$$\dot{x}_1 = x_2 \omega_n$$

$$\dot{x}_2 = -\omega_n (x_1 + \delta)$$

Thus, the slope of the trajectory can be expressed as follows:

$$\frac{dx_2}{dx_1} = \frac{-(x_1 + \delta)}{x_2}$$

(iv)

Using the above slope equation (iv), a short segment of trajectory can be drawn from the knowledge of 'δ' at any point A in the trajectory. Let 'A' be a point on the phase trajectory with coordinates (x_1, x_2) as shown in Figure 4.38. The various steps involved to construct phase trajectory using delta method are as given below.

1. Convert the equation of given non-linear system into standard form.
2. Calculate 'δ' by substituting values of coordinates of 'A' in equation (iv).
3. Mark point 'C' such that $OC = \delta$.
4. Through point A, draw the perpendicular lines AC and DE. (Now the line DE will have a slope of $\dfrac{-(x_1 + \delta)}{x_2}$, which is also the slope of the trajectory at A.)
5. Draw a small circular arc AF, such that AF arc becomes a section of phase trajectory with slope $\dfrac{-(x_1 + \delta)}{x_2}$.

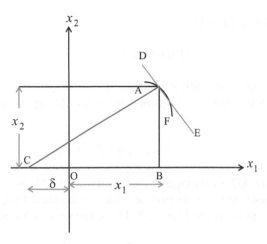

FIGURE 4.38
Geometry of δ method.

6. Using the coordinates 'F', a new value of 'δ' can be calculated and the above procedure is repeated to obtain more points, etc., to complete the phase trajectory.

Illustration 4

Construct a phase plane trajectory by delta method for a non-linear system represented by the differential equation $\ddot{x} + |\dot{x}|\dot{x} + x = 0$ with initial conditions as $x(0) = 1.0$ and $\dot{x}(0) = 0$.

Solution

Given that the nonlinear system is represented by a differential equation

$$\ddot{x} + |\dot{x}|\dot{x} + x = 0 \tag{i}$$

Convert the above equation into standard form as given in equation (ii)

$$\ddot{x} + \omega_n^2(x + \delta) = 0 \tag{ii}$$

Thus, equation (i) can be rewritten as

$$\ddot{x} + \left[x + |\dot{x}|\dot{x}\right] = 0 \tag{iii}$$

Comparing (ii) and (iii), we get

$$\delta = |\dot{x}|\dot{x} = |x_2|x_2 = |x_2|x_2 \tag{iv}$$

Given $x(0) = 1.0 \Rightarrow \dot{x}_1(0) = 1.0$

$$\dot{x}(0) = 0 \Rightarrow x_2(0) = 0$$

Thus, the coordinates of point A are $(1.0, 0)$

The value of δ at point A is obtained by substituting the coordinates of A in equation (iii).

$$\therefore \delta = |0|0 = 0$$

Now draw an arc AD with origin 'O' as centre and 'OA' as radius.

The arc segment AD is a section of phase trajectory. From the graph the value of x_2 corresponding to D is -0.2. Thus, the new value of δ is given by

$$\delta = |x_2|x_2 = |-0.2|(-0.2) = -0.04$$

Similarly, points for E to Z can be identified by finding new values of δ which were found to be -0.09, -0.16, -0.25, -0.36, -0.49, -0.49, -0.36, -0.25, -0.16, -0.09, -0.04, -0.01, 0.01, 0.04, 0.09, 0.04, 0.01, -0.01, -0.04, -0.01, 0.01 and -0.008, respectively.

The complete trajectory thus obtained is as shown in Figure 4.39.

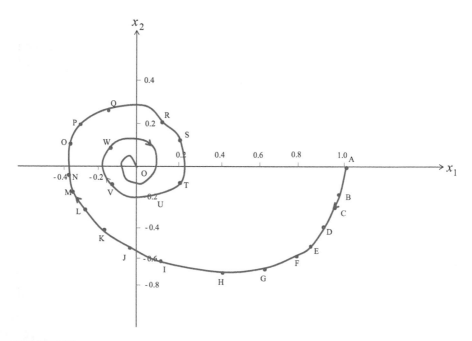

FIGURE 4.39
Phase trajectory by delta method.

Limit Cycle [For Non-Linear Systems]

The limit cycles are the oscillations of fixed amplitude and period exhibited by non-linear systems. The limit cycle is acceptable, only when its amplitude is within tolerance limits. Otherwise it is considered as undesirable. On the phase plane, a limit cycle is defined as an isolated closed orbit. For limit cycle, the trajectory should be closed, indicating the periodic nature of motion and be isolated, indicating the limiting nature of the cycle with neighbouring trajectories converging to or diverging from it.

Let us consider the standard form of Vander Pol's differential equation representing the non-linear system.

$$\frac{d^2x}{dt^2} \pm \mu(1-x^2)\frac{dx}{dt} + x = 0 \tag{i}$$

which describes physical situations in many non-linear systems.

By comparing this with the standard form of differential equation with damping factor 'ξ' as given in equation (ii),

$$\frac{d^2x}{dt^2} + 2\xi\frac{dx}{dt} + x = 0 \tag{ii}$$

We get the damping factor as $\frac{\pm\mu}{2}(1-x^2)$ which is a function of x. Now, we have two cases, one for stable limit cycle and the other for unstable limit cycle.

1. Stable limit cycle (when $\xi = \frac{-\mu}{2}(1-x^2)$):

   If $|x| \gg 1$, then the damping factor will have positive value. Hence, the system acts like an overdamped system. When 'x' decreases, the damping factor also decreases and the system state finally enters a limit cycle as shown by the outer trajectory in Figure 4.40.

   On the other hand, if $|x| \ll 1$, the damping factor will have negative value where the amplitude of x increases as the limit cycle enters

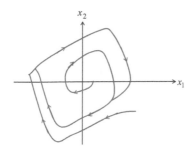

FIGURE 4.40
Stable limit cycle.

the inner trajectory. The limit cycle shown in Figure 4.39 is stable one, trajectories in the neighbourhood converge towards the limit cycle, the limit cycle is referred as stable limit cycle.

2. Unstable limit cycles (when $\xi = \dfrac{+\mu}{2}(1 - x^2)$):

   In this case, the phase trajectories in the neighbourhood diverge from the limit cycle, as shown in Figure 4.41. Such a limit cycle is called unstable limit cycle.

NOTE

1. The limit cycle in general is an undesirable characteristic of control system.
2. It is acceptable only when its amplitude is within the specified limits.

Limit Cycle [For Linear Systems]

In linear systems, when oscillations occur, the resulting trajectories will be closed curves as shown in Figure 4.42.

NOTE

1. The amplitude of oscillations changes with size of initial conditions.
2. Variations in system parameters may wipe out the oscillations.

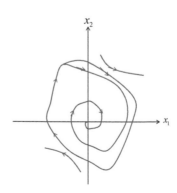

FIGURE 4.41
Unstable limit cycle.

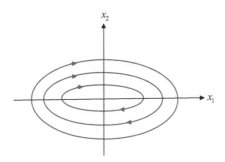

FIGURE 4.42
Phase portrait showing the limit cycle behaviour in linear system.

Example 1: Stable limit cycle

Consider a non-linear system represented by the following equations:

$$\dot{x}_1 = x_2 - x_1(x_1^2 + x_2^2 - 1)$$

$$\dot{x}_2 = -x_1 - x_2(x_1^2 + x_2^2 - 1)$$

with the polar coordinates as $(x_1 = r \cos(\theta), x_2 = r \sin(\theta))$

$$\dot{r} = -r(r^2 - 1)$$

Then,

$$\dot{\theta} = -1$$

If the trajectories start on the unit circle $(x_1^2(0) + x_2^2(0) = r^2 = 1)$, then $\dot{r} = 0$.
Hence, the trajectory will circle the origin of the phase plane with a period of $\dfrac{1}{2\pi}$.

When $r < 1$, then $\dot{r} > 0$. The trajectories converge to the unit circle from inside.

When $r > 1$, then $\dot{r} < 0$. The trajectories converge to the unit circle from outside.

Such a limit cycle is referred as stable limit cycle.

Example 2: Unstable limit cycle

Consider a non-linear system represented by the following equations:

$$\dot{x}_1 = x_2 + x_1(x_1^2 + x_2^2 - 1)$$

$$\dot{x}_2 = -x_1 + x_2(x_1^2 + x_2^2 - 1)$$

with the polar coordinates as $x_1 = r \cos(\theta), x_2 = r \sin(\theta)$

$$\dot{r} = r(r^2 - 1)$$

Then,

$$\dot{\theta} = -1$$

If the trajectories start on the unit circle $(x_1^2(0) + x_2^2(0) = r^2 = 1)$, then $\dot{r} = 0$. Hence, the trajectory will circle the origin of the phase plane with a period of $\dfrac{1}{2\pi}$.

When $r < 1$, then $\dot{r} > 0$. The trajectories diverge to the unit circle from inside. When $r > 1$, then $\dot{r} < 0$. The trajectories diverge to the unit circle from outside. Such a limit cycle is referred as unstable limit cycle.

Example 3: Semi-stable limit cycle

Consider a non-linear system represented by the following equations

$$\dot{x}_1 = x_2 - x_1(x_1^2 + x_2^2 - 1)$$

$$\dot{x}_2 = -x_1 - x_2(x_1^2 + x_2^2 - 1)$$

with the polar coordinates as $x_1 = r\cos(\theta)$, $x_2 = r\sin(\theta)$

$$\dot{r} = -r(r^2 - 1)$$

Then,

$$\dot{\theta} = -1$$

If the trajectories start on the unit circle $(x_1^2(0) + x_2^2(0) = r^2 = 1)$, then $\dot{r} = 0$. Hence, the trajectory will circle the origin of the phase plane with a period of $\dfrac{1}{2\pi}$.

When $r < 1$, then $\dot{r} < 0$. The trajectories diverge to the unit circle from inside.

When $r > 1$, then $\dot{r} < 0$. The trajectories converge to the unit circle from outside.

Such a limit cycle is referred as semi-stable limit cycle.

5

Stability of Non-Linear Systems

This chapter deals with concept of stability, conditions for stability, Bounded Input Bounded Output (BIBO) and asymptotic stability, equilibrium points, stability of non-linear systems, describing function approach for stability analysis of non-linear systems, Lyapunov stability criteria, Krasovskii's method, variable gradient method and Popov's stability criteria.

Concept of Stability

All the systems experience transient behaviour for a brief period before reaching the steady state. The stability studies enable us to know the ability of the given system to reach its steady state after experiencing the transient stage. In other words, in a stable system, the output does not result in large changes, even when there is a change in input or system parameter or initial condition. Based on the system behaviour, the concept of stability can be classified as shown in Figure 5.1

Stability Conditions for a Linear Time-Invariant System

Condition 1: For a bounded input excitation, the output of the system should be bounded.

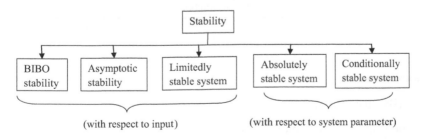

FIGURE 5.1
Classification of stability.

Condition 2: In the absence of input excitation, the output should tend to zero (or equilibrium state) irrespective of the initial conditions.

Thus, with respect to input excitation, we can have three definitions for the stability.

BIBO stability:	When the bounded input excitation results in a bounded output, then such a stability concept is called 'BIBO stability'.

NOTE

If the bounded input excitation produces an unbounded output, then such a system is called unstable system.

Asymptotic stability:	In the absence of input excitation, if the output tends to zero (or equilibrium state), irrespective of the initial conditions, then such a stability is called 'asymptotic stability'.
Limitedly stable system:	In the absence of input excitation, if the output does not tend to zero (but is bounded), then such a system is called 'limitedly stable system'.

We can give two more definitions for stability, by considering the behaviour of the system with respect to changes in system parameter.

Absolutely stable system:	A system is said to be absolutely stable with respect to a parameter, if the system is stable for all values of that parameter.
Conditionally stable system:	A system is said to be conditionally stable with respect to a parameter, if the system is stable for certain bounded ranges of that parameter.

Equilibrium Point

It is the point in phase plane where the derivatives of all state variables are zero. The system when placed at equilibrium point, continues to remain there, as long as it is not disturbed.

Stability Analysis of Non-Linear System Using Describing Function Method

The describing function approach can be employed to perform stability analysis of non-linear systems. Consider a non-linear system with unity feedback as shown in Figure 5.2a, where 'N' represents the non-linear element. The non-linear element can be linearized by replacing it with a describing function, $K_N (X,\omega)$ or K_N, as shown in Figure 5.2b.

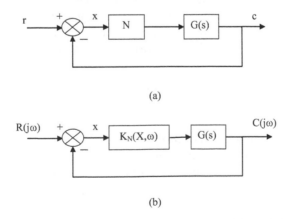

(a)

(b)

FIGURE 5.2
(a) Non-linear system and (b) non-linear system with non-linear element replaced by describing function.

Assumptions

1. $G(s)$ is assumed to have no poles in the right half of s-plane. (In other words, the linear part of the system is stable.)
2. The input to the non-linear element is pure sinusoidal, i.e., $x = X \sin \omega t$.
3. To ensure the condition stated in (2), the input (r) given to the system should be zero.

The closed loop transfer function of the linearized system shown in Figure 5.2b can be given by equation (5.1).

$$\frac{C(j\omega)}{R(j\omega)} = \frac{K_N G(j\omega)}{1 + K_N G(j\omega)} \tag{5.1}$$

Thus, the characteristic equation of the linearized system can be written as follows:

$$1 + K_N G(j\omega) \tag{5.2}$$

The Nyquist stability criterion can also be applied to perform the stability analysis of non-linear systems. According to Nyquist stability criterion, the system represented in Figure 5.2b will exhibit sustained oscillations when

$$K_N G(j\omega) = -1 \tag{5.3}$$

The condition given in equation (5.3) implies that the sustained oscillations (limit cycle) will occur, when the $K_N\ G(j\omega)$ plot drawn in complex plane passes through the critical point, $(-1 + j0)$. The amplitude and frequency of limit cycles can be obtained by rewriting the equation (5.3) as follows:

$$G(j\omega) = \frac{-1}{K_N} \tag{5.4}$$

Accordingly, the intersection of the plot $G(j\omega)$ with the critical locus $\dfrac{-1}{K_N}$ will give the amplitude and frequency of limit cycles.

The stability analysis can be carried out by inspecting the relative position of polar plots of $G(j\omega)$ and $\dfrac{-1}{K_N}$ with ω varying from 0 to ∞ in the complex plane, as shown in Figure 5.3a–c. By observing the plots drawn in Figure 5.3, the following conclusions can be made, with respect to stability of non-linear systems.

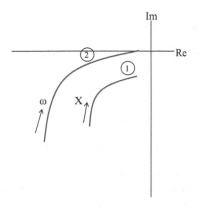

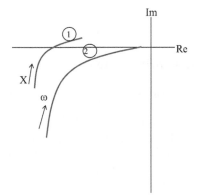

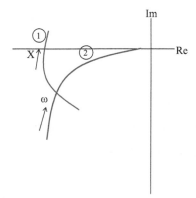

FIGURE 5.3

Stability analysis of non-linear system using polar plots of $G(j\omega)$ and $-\dfrac{1}{K_N}$. (a) Enclosed, unstable, (b) non-enclosure, stable and (c) intersecting; ① locus of $-\dfrac{1}{K_N}$ and ② locus of $G(j\omega)$.

1. If the $\dfrac{-1}{K_N}$ locus lies to the right of an observer traversing the $G(j\omega)$ locus, in the direction of increasing values of ω as shown in Figure 5.3a, then the system is said to be unstable.

2. If the $\dfrac{-1}{K_N}$ locus lies to the left of $G(j\omega)$ locus, as shown in Figure 5.3b, then the system is said to be stable.

3. On the other hand, if the $\dfrac{-1}{K_N}$ locus intersects $G(j\omega)$ locus, as shown in Figure 5.3c, then the region on the right is said to be unstable region and the region on left is said to be stable region, while travelling through $G(j\omega)$ plot, in the direction of increasing ω.

Stable and Unstable Limit Cycles

While drawing the locus of $\dfrac{-1}{K_N(X,\omega)}$ and locus of $G(j\omega)$ in the complex plane, at times they intersect at one point (Figure 5.4a) or at more than one point (Figure 5.4b and c). Every intersection point corresponds to possibility of periodic oscillation resulting a limit cycle. These limit cycles can be either stable or unstable depending on the vales of x.

Case 1: Stable Limit Cycle

If $\dfrac{-1}{K_N(X,\omega)}$ locus travels in unstable region and intersects the $G(j\omega)$ locus at point P_2 (Figure 5.4b) or at point P_5 (Figure 5.4c), to enter into stable region, then the limit cycles corresponding to P_2 and P_5 are called stable limit cycles.

Case 2: Unstable Limit Cycle

If $\dfrac{-1}{K_N(X,\omega)}$ locus travels in stable region and intersects the $G(j\omega)$ locus at point P_1 (Figure 5.4a) or at point P_3 (Figure 5.4b) or at point P_4 (Figure 5.4c), to enter into unstable region, then the limit cycles corresponding to P_1, P_3 and P_4 are called unstable limit cycles.

Procedure for Investigating Stability of a Given Non-Linear System

Step 1: Split the given linear part of the system, $G(j\omega)$ into real part (G_R) and imaginary part (G_I).

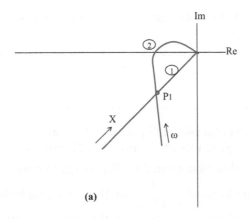

(a)

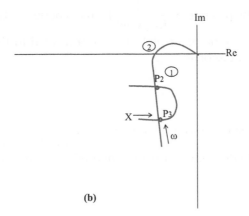

(b)

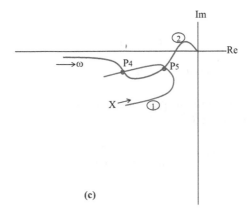

(c)

FIGURE 5.4

(a–c) Stability of limit cycles. ① Locus of $-\dfrac{1}{K_N(X,\omega)}$ and ② locus of $G(j\omega)$, P_2 and P_5-stable limit cycles, P_1, P_3 and P_4-unstable limit cycles.

Step 2: Find the values of G_R and G_I for different values of 'ω' and tabulate the results.

ω	ω_1	ω_2	ω_3	ω_4	ω_5	ω_6
G_R						
G_I						

Step 3: Draw the polar plot of $G(j\omega)$ using the values of 'ω' (as tabulated in step 2) on a graph sheet, as shown in Figure 5.9.

Step 4: Find the describing function (K_N) of given non-linear element.

Step 5: Draw the locus of $\dfrac{-1}{K_N(X,\omega)}$ on the same graph sheet.

Step 6: Comment on the stability of the given non-linear system by observing the position of $\dfrac{-1}{K_N(X,\omega)}$ locus (whether to the left or right) with respect to the polar plot of $G(j\omega)$ traversed in the direction of increasing values of ω.

Illustration 1

Comment on the stability of the system $G(s) = \dfrac{K}{s(2s+1)(0.5s+1)}$ with on–off nonlinearity using describing function method. Assume the relay amplitude as unity.

Solution

Given that

$$G(s) = \frac{K}{s(2s+1)(0.5s+1)} \text{ and}$$

The non-linear element is on–off relay.

Step 1: Split the given linear part of the system, $G(j\omega)$ into real part (G_R) and imaginary part (G_I).

$$G(j\omega) = \frac{K}{j\omega(2j\omega+1)(0.5j\omega+1)}$$

when $K = 1$

$$|G(j\omega)| = \frac{1}{\omega\sqrt{1+4\omega^2}\sqrt{1+0.25\omega^2}}$$

$$= \angle G(j\omega) = -90° - \tan^{-1}2\omega - \tan^{-1}0.5\omega$$

Also,

$$G(j\omega) = G_R + G_I = \frac{-2.5\omega^2}{(-2.5\omega^2)^2 + (\omega - \omega^3)^2} - j\frac{(\omega - \omega^3)}{(-2.5\omega^2)^2 + (\omega - \omega^3)^2}$$

$$G_R = \frac{-2.5\omega^2}{(-2.5\omega^2)^2 + (\omega - \omega^3)^2}$$

$$G_I = \frac{-(\omega - \omega^3)}{(-2.5\omega^2)^2 + (\omega - \omega^3)^2}$$

Step 2: Find the values of G_R and G_I for different values of 'ω' and tabulate the results.

ω	0.4	0.8	1	1.2	1.6	2
G_R	−1.46	−0.60	−0.4	−0.27	−0.13	−0.07
G_I	−1.23	−0.10	0	0.03	0.05	0.04

Step 3: Draw the polar plot of $G(j\omega)$ using the values of 'ω' (as tabulated in step 2) on a graph sheet. This is shown in Figure 5.5.

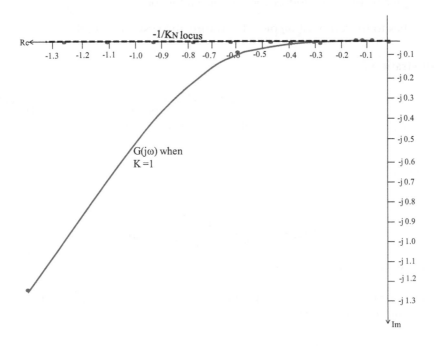

FIGURE 5.5
Polar plot.

Step 4: Find the describing function (K_N) of given non-linear element. From the describing function of on–off relay, the phase and amplitude of the nonlinearity can be given as follows.

$$K_N = \frac{4M}{X\pi} \angle 0° = \frac{4}{X\pi} \angle 0° \quad (\because M = 1)$$

The $-1/K_N$ locus can be drawn using the following table

X	0	0.2	0.4	0.6	0.8	1	1.2	1.4	1.6	2
K_N	∞	6.36	3.18	2.12	1.59	1.27	1.06	0.9	0.19	0.6
$\|-1/K_N\|$	0	0.15	0.31	0.47	0.62	0.78	0.94	1.11	1.26	1.66
$\angle K_N$	0	0	0	0	0	0	0	0	0	0
$\angle(-1/K_N)$	−180	−180	−180	−180	−180	−180	−180	−180	−180	−180
Re	0	−0.15	−0.31	−0.47	−0.62	−0.78	−0.94	−1.11	−1.26	−1.66
Im	0	0	0	0	0	0	0	0	0	0

Step 5: Draw the locus of $\dfrac{-1}{K_N(X,\omega)}$ on the same graph sheet.

Step 6: Comment on the stability of the given non-linear system.

From Figure 5.5, it is clear that, as the observer travels along the $G(j\omega)$ plot in the direction of increasing values of 'ω', the $\dfrac{-1}{K_N(X,\omega)}$ locus lies to the left of $G(j\omega)$ plot. Therefore, the system is stable.

Illustration 2

Comment on the stability of the system $G(s) = \dfrac{8}{(s+1)(0.5s+1)(0.2s+1)}$ with on–off nonlinearity using describing function method. Assume the relay amplitude as unity.

Solution

Given that

$$G(s) = \frac{8}{(s+1)(0.5s+1)(0.2s+1)} \text{ and}$$

The non-linear element is on–off relay.

Step 1: Split the given linear part of the system, $G(j\omega)$ into real part (G_R) and imaginary part (G_I).

$$G(j\omega) = \frac{8}{(j\omega+1)(0.5j\omega+1)(0.2j\omega+1)}$$

$$G(j\omega) = G_R + G_I$$

$$= \frac{8(1-0.8\omega^2)}{(1-0.8\omega^2)^2 + (1.7\omega - 0.1\omega^3)^2} - j\frac{8(1.7\omega - 0.1\omega^3)}{(1-0.8\omega^2)^2 + (1.7\omega - 0.1\omega^3)^2}$$

$$G_R = \frac{8(1-0.8\omega^2)}{(1-0.8\omega^2)^2 + (1.7\omega - 0.1\omega^3)^2}$$

$$G_I = -\frac{8(1.7\omega - 0.1\omega^3)}{(1-0.8\omega^2)^2 + (1.7\omega - 0.1\omega^3)^2}$$

Step 2: Find the values of G_R and G_I for different values of 'ω' and tabulate the results.

ω	0	0.1	0.5	0.9	1.2	1.5	2	2.5	3	10	20
G_R	8	7.83	4.77	1.25	−0.34	−1.15	−2.22	−1.37	−1.12	−0.40	−0.00
G_I	0	−1.37	−4.99	−5.18	−4.25	−3.19	−1.79	−0.92	−0.43	0.05	0.00

Step 3: Draw the polar plot of $G(j\omega)$ using the values of 'ω' (as tabulated in step 2) on a graph sheet. This is shown in Figure 5.6.

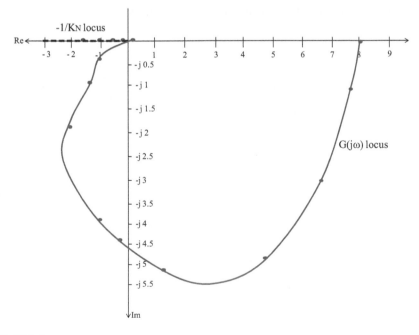

FIGURE 5.6
Polar plot.

Step 4: Find the describing function (K_N) of given non-linear element.
From the describing function of on–off relay, the phase and amplitude of the nonlinearity can be given as follows:

$$K_N = \frac{4M}{X\pi}\angle 0° = \frac{4}{X\pi}\angle 0° \quad (\because M = 1)$$

The $-1/K_N$ locus can be drawn using the following table

X	0	0.2	0.4	0.6	0.8	1	1.2	1.4	1.6	2
K_N	∞	6.36	3.18	2.12	1.59	1.27	1.06	0.9	0.19	0.6
$|-1/K_N|$	0	0.15	0.31	0.47	0.62	0.78	0.94	1.11	1.26	1.66
$\angle K_N$	0	0	0	0	0	0	0	0	0	0
$\angle(-1/K_N)$	−180	−180	−180	−180	−180	−180	−180	−180	−180	−180
Re	0	−0.15	−0.31	−0.47	−0.62	−0.78	−0.94	−1.11	−1.26	−1.66
Im	0	0	0	0	0	0	0	0	0	0

Step 5: Draw the locus of $\dfrac{-1}{K_N(X,\omega)}$ on the same graph sheet.

Step 6: Comment on the stability of the given non-linear system.
From Figure 5.6, it is clear that, as the observer travels along the $G(j\omega)$ plot in the direction of increasing values of 'ω', the $\dfrac{-1}{K_N(X,\omega)}$ locus lies to the left of $G(j\omega)$ plot. Therefore, the system is stable.

Illustration 3

Comment on the stability of the system $G(s) = \dfrac{K}{s(0.2s+1)(5s+1)}$ with saturation nonlinearity using describing function method. Find the maximum value of K for which system remains stable.

Solution

Given that

$$G(s) = \frac{K}{s(0.2s+1)(5s+1)} \quad \text{and}$$

The non-linear element is saturation.

Step 1: Split the given linear part of the system, $G(j\omega)$ into real part (G_R) and imaginary part (G_I).

$$G(j\omega) = \frac{K}{j\omega(0.2j\omega+1)(5j\omega+1)}$$

when $K = 1$

$$|G(j\omega)| = \frac{1}{\omega\sqrt{1 + 0.04\omega^2}\sqrt{1 + 25\omega^2}}$$

$$\angle G(j\omega) = -90° - \tan^{-1}0.2\omega - \tan^{-1}5\omega$$

Also,

$$G(j\omega) = G_R + G_I = \frac{-5.2\omega^2}{(-5.2\omega^2)^2 + (\omega - \omega^3)^2} - j\frac{(\omega - \omega^3)}{(-5.2\omega^2)^2 + (\omega - \omega^3)^2}$$

$$G_R = \frac{-5.2\omega^2}{(-5.2\omega^2)^2 + (\omega - \omega^3)^2}$$

$$G_I = -\frac{(\omega - \omega^3)}{(-5.2\omega^2)^2 + (\omega - \omega^3)^2}$$

Step 2: Find the values of G_R and G_I for different values of 'ω' and tabulate the results.

For $K = 1$

Ω	0.4	0.8	1.2	1.6	2.0	2.4
G_R	−1	−0.29	−0.13	−0.07	−0.04	−0.02
G_I	−0.46	−0.02	0.009	0.013	0.012	0.011

For $K = 2$

ω	0.4	0.8	1.2	1.6	2.0	2.4
G_R	−2	−0.58	−0.26	−0.14	−0.08	−0.05
G_I	−0.93	−0.05	0.01	0.02	0.02	0.02

For $K = 6.5$

ω	0.4	0.8	1.2	1.6	2.0	2.4
G_R	−6.5	−1.88	−0.85	−0.45	−0.26	−0.18
G_I	−3.03	−0.16	0.06	0.08	0.08	0.07

Step 3: Draw the polar plot of $G(j\omega)$ using the values of 'ω' (as tabulated in step 2) on a graph sheet. This is shown in Figure 5.7

Step 4: Find the describing function (K_N) of given non-linear element. From the describing function of saturation, the phase and amplitude of the nonlinearity is written.

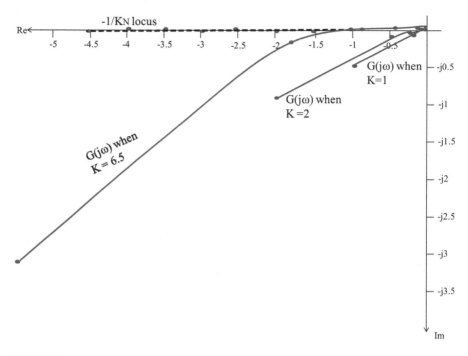

FIGURE 5.7
Polar plot.

Step 5: Draw the locus of $\dfrac{-1}{K_N(X,\omega)}$ on the same graph sheet.

To find the maximum value of K for stability

w.k.t

$$\left|G(j\omega)\right| = 1 \text{ and } \angle G(j\omega) = -180°$$

Let ω_s be the frequency of $G(j\omega) = -1$

Now at $\omega_s,$

$$\angle G(j\omega) = -90 - \tan^{-1} 0.2\omega - \tan^{-1} 5\omega = -180°$$

$$\tan^{-1} 0.2\omega + \tan^{-1} 5\omega = 90°$$

Taking tan on both sides

$$\tan(\tan^{-1} 0.2\omega + \tan^{-1} 5\omega) = \tan 90°$$

$$\frac{\tan(\tan^{-1} 0.2\omega) + \tan(\tan^{-1} 5\omega)}{1 - \tan(\tan^{-1} 0.2\omega)\tan(\tan^{-1} 5\omega)} = \tan 90°$$

$$\frac{0.2\omega + 5\omega}{1 - 0.2\omega * 5\omega} = \infty$$

$$1 - \omega^2 = 0$$

$$\omega = 1 \text{ rad/s}$$

Substituting ω in

$$|G(j\omega)| \quad = \frac{K}{\omega\sqrt{1 + 0.04\omega^2}\sqrt{1 + 25\omega^2}} = 1$$

$$1 = \frac{K}{1\sqrt{1 + 0.04.1^2}\sqrt{1 + 25.1^2}}$$

$$K = 6.308$$

Step 6: Comment on the stability of the given non-linear system.
     Thus, the system $G(s)$ will remain stable when $K \leq 6.3$ and becomes unstable for greater values of K. This can be shown in the polar plot drawn for the values of $K = 1$, $K = 2$ and $K = 6.5$.

Illustration 4

Comment on the stability of the system $G(s) = \dfrac{K}{s(s+1)(0.8s+1)}$ with backlash nonlinearity using describing function method. Find the maximum value of K for the system to remain stable. Also determine the frequency of oscillation of the limit cycle.

Solution

Given that

$$G(s) = \frac{K}{s(s+1)(0.8s+1)} \text{ and}$$

The non-linear element is backlash.

     Step 1: Split the given linear part of the system, $G(j\omega)$ into real part (G_R) and imaginary part (G_I).

$$G(j\omega) = \frac{K}{j\omega(j\omega + 1)(0.8j\omega + 1)}$$

when $K = 1$

$$|G(j\omega)| = \frac{1}{\omega\sqrt{1+\omega^2}\sqrt{1+0.64\omega^2}}$$

$$\angle G(j\omega) = -90° - \tan^{-1}\omega - \tan^{-1}0.8\omega$$

Also,

$$G(j\omega) = G_R + G_I = \frac{-1.8\omega^2}{(-1.8\omega^2)^2 + (\omega - 0.8\omega^3)^2} - j\frac{(\omega - 0.8\omega^3)}{(-1.8\omega^2)^2 + (\omega - 0.8\omega^3)^2}$$

$$G_R = \frac{-1.8\omega^2}{(-1.8\omega^2)^2 + (\omega - 0.8\omega^3)^2}$$

$$G_I = -\frac{(\omega - 0.8\omega^3)}{(-1.8\omega^2)^2 + (\omega - 0.8\omega^3)^2}$$

Step 2: Find the values of G_R and G_I for different values of 'ω' and tabulate the results.

For $K = 1$

ω	0.2	0.4	0.6	0.8	1	1.2	1.4
G_R	−1.68	−1.40	−1.07	−0.77	−0.54	−0.38	−0.26
G_I	−4.5	−1.70	−0.70	−0.26	−0.06	0.02	0.06

For $K = 1.5$

ω	0.2	0.4	0.6	0.8	1	1.2	1.4
G_R	−2.52	−2.1	−1.61	−1.16	−0.82	−0.57	−0.39
G_I	−6.75	−2.55	−1.05	−0.39	−0.09	0.03	0.09

For $K = 2.5$

ω	0.2	0.4	0.6	0.8	1	1.2	1.4
G_R	−4.2	−3.5	−2.68	−1.94	−1.37	−0.95	−0.65
G_I	−11.2	−4.25	−1.75	−0.65	−0.15	0.05	0.15

Step 3: Draw the polar plot of $G(j\omega)$ using the values of 'ω' (as tabulated in step 2) on a graph sheet. This is shown in Figure 5.8

Step 4: Find the describing function (K_N) of given non-linear element.

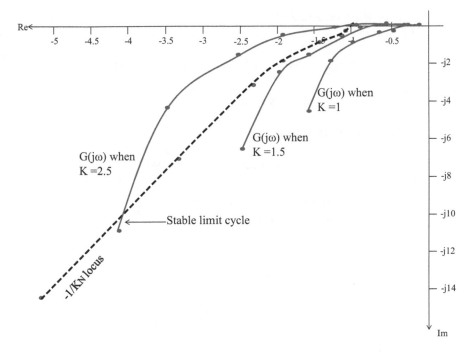

FIGURE 5.8
Polar plot.

From the describing function of backlash, the phase and amplitude of the nonlinearity can be written as.

b/X	0	0.2	0.4	1	1.4	1.6	1.8	1.9	2
K_N	1	0.95	0.88	0.59	0.36	0.24	0.12	0.06	0
$\lvert -1/K_N \rvert$	1	1.05	1.13	1.69	2.72	4.03	8.0	15.63	∞
$\angle K_N$	0	−6.7	−13.4	−32.5	−46.6	−55.2	−66	−69.8	−90
$\angle(-1/K_N)$	−180	−173	−166	−148	−133	−125	−114	−110	−90
Re	−1.0	−1.04	−1.1	−1.4	−1.9	−2.3	−3.3	−5.3	0
Im	0	−0.1	−0.3	−0.9	−2.0	−3.3	−7.3	−14.7	∞

Step 5: Draw the locus of $\dfrac{-1}{K_N(X,\omega)}$ on the same graph sheet.

To find the maximum value of K for stability

We know that

$$\lvert G(j\omega) \rvert = 1 \text{ and } \angle G(j\omega) = -180°$$

Let ω_s be the frequency of $G(j\omega) = -1$
Now at ω_s

$$\angle G(j\omega) = -90 - \tan^{-1} 0.2\omega - \tan^{-1} 5\omega = -180°$$

$$\tan^{-1} \omega + \tan^{-1} 0.8\omega = 90°$$

Taking tan on both sides

$$\tan(\tan^{-1} \omega + \tan^{-1} 0.8\omega) = \tan 90°$$

$$\frac{\tan(\tan^{-1} \omega) + \tan(\tan^{-1} 0.8\omega)}{1 - \tan(\tan^{-1} \omega)\tan(\tan^{-1} 0.8\omega)} = \tan 90°$$

$$\frac{\omega + 0.8\omega}{1 - \omega * 0.8\omega} = \infty$$

$$1 - 0.8\omega^2 = 0$$

$$\omega = 1.11 \text{ rad/s}$$

Substituting ω in

$$|G(j\omega)| = \frac{K}{\omega\sqrt{1+\omega^2}\sqrt{1+0.64\omega^2}} = 1$$

$$1 = \frac{K}{1\sqrt{1+1.11^2}\sqrt{1+0.64*1.11^2}}$$

$$K = 2.19$$

Step 6: Comment on the stability of the given non-linear system.
    Thus, the system $G(s)$ will remain stable when $K \le 2.19$ and becomes unstable for greater values of K. This can be shown in the polar plot drawn for both the values of $K = 1$, $K = 1.5$ and $K = 2.5$.
    To find the frequency of oscillation of limit cycle

$$|G(j\omega)| = 1 \text{ and } \angle G(j\omega) = -180°$$

$$\angle G(j\omega) = -90 - \tan^{-1} \omega - \tan^{-1} 0.8\omega = -180°$$

$$\tan^{-1} \omega + \tan^{-1} 0.8\omega = 90°$$

Taking tan on both sides

$$\tan(\tan^{-1}\omega + \tan^{-1}0.8\omega) = \tan 90°$$

$$\frac{\tan(\tan^{-1}\omega) + \tan(\tan^{-1}0.8\omega)}{1 - \tan(\tan^{-1}\omega)\tan(\tan^{-1}0.8\omega)} = \tan 90°$$

$$\frac{\omega + 0.8\omega}{1 - \omega * 0.8\omega} = \infty$$

$$1 - 0.8\omega^2 = 0$$

$$\omega = 1.11 \text{ rad/s}$$

Thus, the frequency of oscillation is $\omega = 1.11$ rad/s.

Illustration 5

Comment on the stability of the system $G(s) = \dfrac{7}{s(s+1)}$ with backlash nonlinearity using describing function method.

Solution

Given that

$$G(s) = \frac{7}{s(s+1)} \text{ and}$$

The non-linear element is backlash.

Step 1: Split the given linear part of the system, $G(j\omega)$ into real part (G_R) and imaginary part (G_I).

$$G(j\omega) = \frac{7}{j\omega(j\omega + 1)}$$

$$|G(j\omega)| = \frac{7}{\omega\sqrt{1 + \omega^2}}$$

$$\angle G(j\omega) = -90° - \tan^{-1}\omega$$

$$G(j\omega) = G_R + G_I = \frac{-7\omega^2}{(-\omega^2)^2 + (\omega)^2} - j\frac{7\omega}{(-\omega^2)^2 + (\omega)^2}$$

$$G_R = \frac{-7\omega^2}{(-\omega^2)^2 + (\omega)^2}$$

$$G_I = -\frac{7\omega}{(-\omega^2)^2 + (\omega)^2}$$

Step 2: Find the values of G_R and G_I for different values of 'ω' and tabulate the results.

ω	0.5	0.8	1	1.2	1.5	2	5
G_R	−5.6	−4.26	−3.5	−2.86	−2.15	−0.4	−0.07
G_I	−11.2	−5.33	−3.5	−2.39	−1.43	−0.2	−0.01

Step 3: Draw the polar plot of $G(j\omega)$ using the values of 'ω' (as tabulated in step 2) on a graph sheet, as shown in Figure 5.9.

Step 4: Find the describing function (K_N) of given non-linear element. From the describing function of backlash, the phase and amplitude of the nonlinearity can be written as.

b/X	0	0.2	0.4	1	1.4	1.6	1.8	1.9	2
K_N	1	0.95	0.88	0.59	0.36	0.24	0.12	0.06	0
$|-1/K_N|$	1	1.05	1.13	1.69	2.72	4.03	8.0	15.63	∞
$\angle K_N$	0	−6.7	−13.4	−32.5	−46.6	−55.2	−66	−69.8	−90
$\angle(-1/K_N)$	−180	−173	−166	−148	−133	−125	−114	−110	−90
Re	−1.0	−1.04	−1.1	−1.4	−1.9	−2.3	−3.3	−5.3	0
Im	0	−0.1	−0.3	−0.9	−2.0	−3.3	−7.3	−14.7	∞

Step 5: Draw the locus of $\dfrac{-1}{K_N(X,\omega)}$ on the same graph sheet.

To find the maximum value of K for stability
We know that

$$|G(j\omega)| = 1 \text{ and } \angle G(j\omega) = -180°$$

Let ω_s be the frequency of $G(j\omega) = -1$
Now at ω_s,

$$\angle G(j\omega) = -90 - \tan^{-1}\omega - \tan^{-1}0.8\omega = -180°$$

$$\tan^{-1}\omega + \tan^{-1}0.8\omega = 90°$$

Taking tan on both sides

$$\tan(\tan^{-1}\omega + \tan^{-1}0.8\omega) = \tan 90°$$

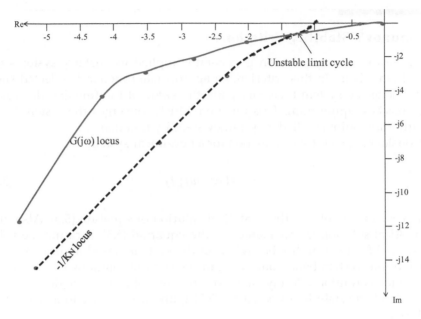

FIGURE 5.9
Polar plot.

$$\frac{\tan(\tan^{-1}\omega)+\tan(\tan^{-1}0.8\omega)}{1-\tan(\tan^{-1}\omega)\tan(\tan^{-1}0.8\omega)}=\tan 90°$$

$$\frac{\omega+0.8\omega}{1-\omega*0.8\omega}=\infty$$

$$1-0.8\omega^2=0$$

$$\omega=1.11 \text{ rad/s}$$

Substituting ω in

$$|G(j\omega)|=\frac{K}{\omega\sqrt{1+\omega^2}\sqrt{1+0.64\omega^2}}=1$$

$$1=\frac{K}{1\sqrt{1+1.11^2}\sqrt{1+0.64*1.11^2}}$$

$$K=2.19$$

Step 6: Comment on the stability of the given non-linear system.
      Thus, the system $G(s)$ is unstable and exhibits an unstable limit cycle.

Lyapunov's Stability Criterion

Lyapunov's stability criterion is generally applied to unstable systems for stabilizing them. In this criterion, an appropriate function is selected such that, along the system trajectory path, the value of the function decreases and reaches equilibrium. This function which converges the system to its equilibrium point is called a Lyapunov stability function.

Consider a general state equation for a non-linear system

$$\dot{x} = f(x(t), u(t), t) \tag{5.5}$$

It is very rare to obtain the analytical solution of equation (5.5). Also, if a numerical solution is considered for the equation (5.5), it cannot be fully answered for the stability behaviour of the system. Based on prior knowledge of the system behaviour, one can restrict the solutions to a finite set of initial conditions. Many methods are available to provide information about the stability of equation (5.5) without resorting to its complete solution.

Lyapunov function: It is a positive definite scalar function $L(x)$ defined to determine the stability of systems. Lyapunov function is unique to each application and is chosen arbitrarily.

The Lyapunov's method involves the following three major theorems:

Theorem 1

Given a system

$$\dot{x} = f(x); f(0) = 0$$

We define a scalar function $L(x)$ for some real number $\varepsilon > 0$, such that it satisfies the following properties, for all x in the region $||x|| \leq \varepsilon$.

  i. $L(x) > 0; x \neq 0$
  ii. $L(0) = 0$
  iii. $L(x)$ has continuous partial derivatives for all values of x.
  iv. $\dfrac{dL}{dt} \leq 0$

For cases (i) and (ii), $L(x)$ is a positive definite scalar function. For case (iv), $L(x)$ is negative semi-definite scalar function. If the above properties are satisfied, then the system is said to be stable at the origin.

Theorem 2

If $\dfrac{dL}{dt} < 0$, $x \neq 0$, then the system is said to be asymptotically stable. Here, $\dfrac{dL}{dt}$ is negative definite scalar function.

Theorem 3

If the condition of theorem 2 is satisfied along with the condition that $L(x)$ must approach infinity as the distance from x to the origin approaches infinity, neglecting the direction, i.e., $L(x) \rightarrow \infty$ as $||x|| \rightarrow \infty$, then the system is said to be asymptotically stable in-the-large at the origin.

This condition is necessary to make $L(x)$ form closed surfaces in the state space.

If this condition is not met, then $L(x)$ does not form closed surfaces, and therefore, the system trajectories tend to infinity.

Theorem 4

Given a system

$$\dot{x} = f(x); \ f(0) = 0$$

We define a scalar function $M(x)$ for some real number $\varepsilon > 0$, such that it satisfies the following properties, for all x in the region $||x|| \leq \varepsilon$.

  i. $M(x) > 0$; $x \neq 0$
  ii. $M(0) = 0$
  iii. $M(x)$ has continuous partial derivatives for all values of x.
  iv. $\dfrac{dM}{dt} \geq 0$. Then the system is unstable at origin. This is the theorem of instability.

Procedure to Comment on the Stability of Non-Linear System Using Lyapunov Stability Criterion

  Step 1: Find an appropriate Lyapunov function for the given problem.
  Step 2: Differentiate the Lyapunov function with respect to the state variables.
  Step 3: Substitute the state variables in the differentiated parts.
  Step 4: Find the describing function (K_N) of given non-linear element.
  Step 5: Investigate the definiteness of $L(x)$.
  Step 6: Comment on the stability of the given non-linear system.

Illustration 6

Comment on the stability of the system represented by the following equations.

$$\dot{x}_1 = -x_2 \tag{i}$$

$$\dot{x}_2 = x_1 - x_2 \tag{ii}$$

Solution

$$\text{Let } L(x) = x_1^2 + x_2^2 \tag{iii}$$

$$\text{Then } \frac{dL(x)}{dt} = \frac{\partial L}{\partial x_1}\dot{x}_1 + \frac{\partial L}{\partial x_2}\dot{x}_2$$

$$\frac{dL(x)}{dt} = 2x_1\dot{x}_1 + 2x_2\dot{x}_2 \tag{iv}$$

Substituting the values of $\dot{x}_1$ and $\dot{x}_2$ from equations (i) and (ii) in equation (iv), we get

$$\frac{dL(x)}{dt} = 2x_1(-x_2) + 2x_2(x_1 - x_2)$$

$$\frac{dL(x)}{dt} = -2x_1x_2 + 2x_2x_1 - 2x_2^2$$

$$\frac{dL(x)}{dt} = -2x_2^2$$

Here,

$L(x) > 0; x \neq 0$

$L(0) = 0$ and also $\dfrac{dL}{dt} < 0$

Hence, the system is asymptotically stable.

Illustration 7

Comment on the stability of the system represented by the following equations

$$\dot{x}_1 = -x_2 \tag{i}$$

$$\dot{x}_2 = x_1 - x_2 x_1^3 \tag{ii}$$

Solution

$$\text{Let } L(x) = x_1^2 + x_2^2 \tag{iii}$$

$$\frac{dL(x)}{dt} = \frac{\partial L}{\partial x_1}\dot{x}_1 + \frac{\partial L}{\partial x_2}\dot{x}_2$$

$$\frac{dL(x)}{dt} = 2x_1\dot{x}_1 + 2x_2\dot{x}_2 \tag{iv}$$

Substitute the values of $\dot{x}_1$ and $\dot{x}_2$ from equations (i) and (ii) in equation (iv), we get

$$\frac{dL(x)}{dt} = 2x_1(-x_2) + 2x_2(x_1 - x_2 x_1^3)$$

$$\frac{dL(x)}{dt} = -2x_1 x_2 + 2x_2 x_1 - 2x_2^2 x_1^3$$

$$\frac{dL(x)}{dt} = -2x_2^2 x_1^3$$

Here,

$$L(x) > 0; \; x \neq 0$$

$$L(0) = 0 \text{ and also } \frac{dL}{dt} < 0. \text{ Also } L(x) \to \infty \text{ as } ||x|| \to \infty$$

Hence, the system is asymptotically stable in thelarge.

Illustration 8

Comment on the stability of the system represented by the following equations

$$\dot{x}_1 = x_2 \tag{i}$$

$$\dot{x}_2 = x_1 + x_2 \tag{ii}$$

Solution

$$\text{Let } L(x) = x_1^2 + x_2^2 \tag{iii}$$

$$\frac{dL(x)}{dt} = \frac{\partial L}{\partial x_1}\dot{x}_1 + \frac{\partial L}{\partial x_2}\dot{x}_2$$

$$\frac{dL(x)}{dt} = 2x_1\dot{x}_1 + 2x_2\dot{x}_2 \tag{iv}$$

Substituting the values of $\dot{x}_1$ and $\dot{x}_2$ from equations (i) and (ii) in equation (iv), we get

$$\frac{dL(x)}{dt} = 2x_1(x_2) + 2x_2(x_1 + x_2)$$

$$\frac{dL(x)}{dt} = 2x_1x_2 + 2x_2x_1 + 2x_2{}^2$$

$$\frac{dL(x)}{dt} = 4x_1x_2 + 2x_2{}^2$$

Here,

$$\frac{dL}{dt} > 0.$$

Hence, the system is unstable at the origin.

Krasovskii Method

Krasovskii's method is used to find the Lyapunov function of non-linear systems. This theorem states about the asymptotic stability of non-linear systems and the proper selection of appropriate Lyapunov function for non-linear systems.

Consider a system

$$\dot{x} = f(x)$$

Assumptions

- $f(0) = 0$
- f is differentiable for all x.

Procedure to Deduce the Stability of Non-Linear System Using Krasovskii's

Method

Step 1: Find the Jacobian matrix $J(x)$ of the given function.

Step 2: Find the Krasovskii's matrix $K(x)$ such that

$$K(x) = J^T(x) + J(x)$$

Step 3: Determine the sign definiteness of $K(x)$, i.e., $K(x)$ must be negative definite/negative semi-definite for the system to be asymptotically stable at the equilibrium point.

Step 4: Determine the Lyapunov function $V(x)$ such that

$$L(x) = f^T(x)f(x)$$

It may be noted that $L(x)$ must be positive definite.

Step 5: If $L(x)$ tends to infinity as x tends to infinity, then the system is said to be asymptotically stable in the large at the equilibrium point. This results in $L(x)$ to be negative definite.

Illustration 9

For the given non-linear system, comment on the stability using the appropriate Lyapunov function.

$$\dot{x}_1 = x_2$$

$$\dot{x}_2 = x_1 - x_2 - x_2{}^3$$

Given

$$\dot{x}_1 = x_2 \tag{i}$$

$$\dot{x}_2 = x_1 - x_2 - x_2{}^3 \tag{ii}$$

The equations (i) and (ii) can be re-written as follows

$$f_1 = \dot{x}_1 = x_2$$

$$f_2 = \dot{x}_2 = x_1 - x_2 - x_2{}^3$$

To obtain the equilibrium points, we set

$$(\dot{x}_1, \dot{x}_2) = (0,0)$$

Clearly, $x_1 = 0$ and $x_2 = 0$.

Step 1: Find the Jacobian matrix $J(x)$ of the given function.
We have

$$\frac{\partial f_1}{\partial x_1} = 0; \frac{\partial f_1}{\partial x_2} = 1$$

$$\frac{\partial f_2}{\partial x_1} = 1; \frac{\partial f_2}{\partial x_2} = -1 - 3x_2{}^2$$

$$J(x) = \begin{bmatrix} \dfrac{\partial f_1}{\partial x_1} & \dfrac{\partial f_1}{\partial x_2} \\ \dfrac{\partial f_2}{\partial x_1} & \dfrac{\partial f_2}{\partial x_2} \end{bmatrix}$$

$$J(x) = \begin{bmatrix} 0 & 1 \\ 1 & -1 - 3x_2{}^2 \end{bmatrix}$$

Step 2: Find the Krasovskii's matrix $K(x)$ such that $K(x) = J^T(x) + J(x)$

$$J^T(x) = \begin{bmatrix} 0 & 1 \\ 1 & -1 - 3x_2{}^2 \end{bmatrix}$$

$$K(x) = J^T(x) + J(x) = \begin{bmatrix} 0 & 1 \\ 1 & -1 - 3x_2{}^2 \end{bmatrix} + \begin{bmatrix} 0 & 1 \\ 1 & -1 - 3x_2{}^2 \end{bmatrix}$$

$$= \begin{bmatrix} 0 & 2 \\ 2 & -2 - 6x_2{}^2 \end{bmatrix}$$

Step 3: Determine the sign definiteness of $K(x)$.

From $K(x)$, it can be seen that $K(x)$ is negative semi-definite. Hence, the system is said to be asymptotically stable at the equilibrium point.

NOTE

The sign definiteness of a given form can be deduced using Sylvester's criterion. The conditions for different sign definiteness are as follows:

$$\text{Suppose } A = \begin{bmatrix} a_{11} & a_{12} & a_{13} \\ a_{21} & a_{22} & a_{23} \\ a_{31} & a_{32} & a_{33} \end{bmatrix} \text{ then, } A \text{ is said to be}$$

i. Positive definite when $a_{11} > 0$; $\begin{vmatrix} a_{11} & a_{12} \\ a_{21} & a_{22} \end{vmatrix} > 0$; $|A| > 0$.

ii. Positive semi-definite when $a_{11} \geq 0$; $\begin{vmatrix} a_{11} & a_{12} \\ a_{21} & a_{22} \end{vmatrix} \geq 0; |A| = 0.$

iii. Negative definite when $a_{11} < 0$; $\begin{vmatrix} a_{11} & a_{12} \\ a_{21} & a_{22} \end{vmatrix} > 0; |A| < 0.$

iv. Negative semi-definite when $a_{11} \leq 0$; $\begin{vmatrix} a_{11} & a_{12} \\ a_{21} & a_{22} \end{vmatrix} \geq 0; |A| = 0.$

Step 4: Determine the Lyapunov function $L(x)$ such that $L(x) = f^T(x)f(x)$.
The Lyapunov function is given by

$$L(x) = f^T(x)f(x)$$

where

$$f^T(x) = \begin{bmatrix} x_2 & x_1 - x_2 - x_2{}^3 \end{bmatrix}$$

$$f(x) = \begin{bmatrix} x_2 \\ x_1 - x_2 - x_2{}^3 \end{bmatrix}$$

$$L(x) = \begin{bmatrix} x_2 & x_1 - x_2 - x_2{}^3 \end{bmatrix} \begin{bmatrix} x_2 \\ x_1 - x_2 - x_2{}^3 \end{bmatrix} = x_2{}^2 + (x_1 - x_2 - x_2{}^3)^2$$

Step 5: Comment on the stability
$L(x)$ is positive definite. Also $L(x) \to \infty$ as $||x|| \to \infty$. Hence, the system is asymptotically stable in the large.

Illustration 10

For the given non-linear system, comment on the stability using the appropriate Lyapunov function.

$$\dot{x}_1 = -x_1 - x_2$$

$$\dot{x}_2 = x_1 - x_2 - x_2{}^3$$

Given

$$\dot{x}_1 = -x_1 - x_2$$

$$\dot{x}_2 = x_1 - x_2 - x_2{}^3$$

The above equation can be re-written as

$$f_1 = \dot{x}_1 = -x_1 - x_2$$

$$f_2 = \dot{x}_2 = x_1 - x_2 - x_2{}^3$$

To obtain the equilibrium points, we set

$$(\dot{x}_1, \dot{x}_2) = (0,0)$$

Clearly, $x_1 = 0$ and $x_2 = 0$.

Hence, the equilibrium point is at the origin.

Step 1: Find the Jacobian matrix $J(x)$ of the given function.
The Jacobian matrix $J(x)$ can be written as follows
We have

$$\frac{\partial f_1}{\partial x_1} = -1; \frac{\partial f_1}{\partial x_2} = -1$$

$$\frac{\partial f_2}{\partial x_1} = 1; \frac{\partial f_2}{\partial x_2} = -1 - 3x_2{}^2$$

$$J(x) = \begin{bmatrix} \dfrac{\partial f_1}{\partial x_1} & \dfrac{\partial f_1}{\partial x_2} \\[2mm] \dfrac{\partial f_2}{\partial x_1} & \dfrac{\partial f_2}{\partial x_2} \end{bmatrix}$$

$$J(x) = \begin{bmatrix} -1 & -1 \\ 1 & -1 - 3x_2{}^2 \end{bmatrix}$$

$$J^T(x) = \begin{bmatrix} -1 & 1 \\ -1 & -1 - 3x_2{}^2 \end{bmatrix}$$

Step 2: Find the Krasovskii's matrix $K(x)$ such that $K(x) = J^T(x) + J(x)$

$$K(x) = J^T(x) + J(x) = \begin{bmatrix} -1 & 1 \\ -1 & -1 - 3x_2{}^2 \end{bmatrix} + \begin{bmatrix} -1 & -1 \\ 1 & -1 - 3x_2{}^2 \end{bmatrix}$$

$$= \begin{bmatrix} -2 & 0 \\ 0 & -2 - 6x_2{}^2 \end{bmatrix}$$

Step 3: Determine the sign definiteness of $K(x)$.

$K(x)$ is negative definite. Hence, the system is said to be asymptotically stable at the equilibrium point.

Step 4: Determine the Lyapunov function $L(x)$ such that $L(x) = f^T(x)f(x)$.

Now, the Lyapunov function is given by

$$L(x) = f^T(x)f(x)$$

$$f^T(x) = \left[-x_1 - x_2 \quad x_1 - x_2 - x_2^3 \right]$$

$$f(x) = \begin{bmatrix} -x_1 - x_2 \\ x_1 - x_2 - x_2^3 \end{bmatrix}$$

$$L(x) = \left[-x_1 - x_2 \quad x_1 - x_2 - x_2^3 \right] \begin{bmatrix} -x_1 - x_2 \\ x_1 - x_2 - x_2^3 \end{bmatrix}$$

$$= \left[(-x_1 - x_2)^2 + (x_1 - x_2 - x_2^3)^2 \right]$$

Step 5: Comment on the stability.

$L(x)$ is positive definite. Also $L(x) \to \infty$ as $||x|| \to \infty$. Hence, the system is asymptotically stable in the large.

Variable Gradient Method

The variable gradient method is used in selecting a Lyapunov function.

Consider the non-linear system

$$\dot{x} = f(x); \, f(0) = 0 \tag{i}$$

Let $L(x)$ be the Lyapunov function

The time derivative of 'L' can be expressed as

$$\dot{L}(x) = \frac{\partial L}{\partial x_1} \dot{x}_1 + \frac{\partial L}{\partial x_2} \dot{x}_2 + \cdots + \frac{\partial L}{\partial x_n} \dot{x}_n \tag{ii}$$

which can be expressed in terms of the gradient of 'V' as

$$\dot{L} = (V_G)^T \dot{x} \tag{iii}$$

$$
\text{where } V_G = \begin{bmatrix} \dfrac{\partial L}{\partial x_1} = V_{G1} \\[2mm] \dfrac{\partial L}{\partial x_2} = V_{G2} \\ \vdots \\ \dfrac{\partial L}{\partial x_n} = V_{Gn} \end{bmatrix} \tag{iv}
$$

The Lyapunov function can be generated by integrating with respect to time on both sides of equation (iii)

$$
L = \int_0^x \frac{dL}{dt} dt = \int_0^x (V_G)^T dx \tag{v}
$$

Illustration 11

For the given non-linear system, comment on the stability

$$
\dot{x}_1 = -x_2^2
$$

$$
\dot{x}_2 = x_1 - x_2 - x_2^3
$$

Given

The given non-linear system is

$$
\dot{x}_1 = -x_2^2
$$

$$
\dot{x}_2 = x_1 - x_2 - x_2^3
$$

The gradient of the system can be written in general as follows:

$$
V_G(x) = \begin{bmatrix} g_1(x) \\ g_2(x) \end{bmatrix} = \begin{bmatrix} a_{11}x_1 + a_{12}x_2 \\ a_{21}x_1 + a_{22}x_2 \end{bmatrix}
$$

Then $\dot{L}(x)$ can be given as follows:

$$
\dot{L}(x) = \begin{bmatrix} g_1(x) & g_2(x) \end{bmatrix} \begin{bmatrix} \dot{x}_1 \\ \dot{x}_2 \end{bmatrix}
$$

$$
\dot{L}(x) = \begin{bmatrix} a_{11}x_1 + a_{12}x_2 & a_{21}x_1 + a_{22}x_2 \end{bmatrix} \begin{bmatrix} -x_2^2 \\ x_1 - x_2 - x_2^3 \end{bmatrix}
$$

$$\therefore \quad \dot{L}(x) = [a_{11}x_1 + a_{12}x_2](-x_2^2) + [a_{21}x_1 + a_{22}x_2](x_1 - x_2 - x_2^3)$$

Now, adjust the parameters of $\dot{V}(x)$ to make it as negative definite.
Substitute $a_{12} = a_{21} = 0$

$$= -a_{11}x_1x_2^2 + a_{22}x_2x_1 - a_{22}x_2^2 - a_{22}x_2^4$$

$$= -a_{11}x_1x_2^2 + a_{22}x_2(x_1 - x_2 - x_2^3)$$

$\dot{L}(x)$ is negative definite. Hence, the system is asymptotically stable at the origin.
  To find the unknown parameters, apply the curl condition to the gradient matrix.

$$V_G(x) = \begin{bmatrix} g_1(x) \\ g_2(x) \end{bmatrix} = \begin{bmatrix} a_{11}x_1 \\ a_{22}x_2 \end{bmatrix}$$

$$\frac{\partial g_1}{\partial x_2} = x_1 \frac{\partial a_{11}}{\partial x_2}$$

$$\frac{\partial g_2}{\partial x_1} = x_2 \frac{\partial a_{22}}{\partial x_1}$$

The condition for curl is given by $\dfrac{\partial g_1}{\partial x_2} = \dfrac{\partial g_2}{\partial x_1}$

$$x_1 \frac{\partial a_{11}}{\partial x_2} = x_2 \frac{\partial a_{22}}{\partial x_1}$$

The curl condition is satisfied only if a_{11} and a_{22} are constants.
  To find $V(x)$, we integrate such that

$$V = \int_0^x \frac{dV}{dt} dt = \int_0^x (V_G)^T dx$$

$$= \int_0^x \begin{bmatrix} a_{11}x_1 \\ a_{22}x_2 \end{bmatrix}^T = \int_0^{x_1} [g_1 dx_1] + \int_0^{x_2} [g_2 dx_2]$$

$$= \int_0^{x_1} [a_{11}x_1] + \int_0^{x_2} [a_{22}x_2]$$

$$= \frac{1}{2}a_{11}x_1^2 + \frac{1}{2}a_{22}x_2^2$$

where a_{11} and a_{22} are strictly greater than zero.

Popov's Stability Criterion

The Popov criterion is a stability criterion by Vasile M. Popov. It illustrates the asymptotic stability of non-linear systems whose nonlinearity is strictly under sector condition. Popov criterion is applicable to sector-type non-linearity and linear time-invariant system only. It can also be applied for systems with delay and higher-order systems.

- It is a frequency domain criterion. It is a modified or extended version of Nyquist criterion.
- No linearization or approximations are involved in Popov criterion.

A feedback system with sector-type nonlinearity and linear system used is as shown in Figure 5.10

Description of the Linear Part G(s)

The following assumptions are made in the linear part when applying the Popov's criterion.

- In the linear part transfer function $G(s)$, the poles of the transfer function must always be greater than the zeros of the transfer function.
- There shouldn't be any condition for pole zero cancellation.

Description of the Non-Linear Part Φ(.)

The following assumptions are made in the non-linear part when applying the Popov's criterion.

- The nonlinearity is sector-type nonlinearity. This means the nonlinearity is bounded in nature. It can be clearly seen from the stability region shown in the Figure 5.11 that, the nonlinearity is bounded between the slopes S_1 and S_2.

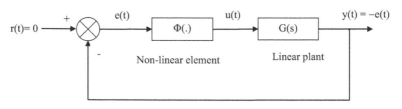

FIGURE 5.10
Feedback system for Popov's stability criterion.

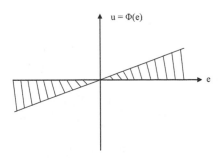

FIGURE 5.11
Stability region.

Popov's Criterion

Popov's criterion states that, 'The closed loop system is asymptotically stable if the Nyquist plot of $G(j\omega)$ does not intersect or encircle the circle given by the intercepts of x and y'.

Procedure

The following steps are the general procedural methods to prove the stability of the system using Popov's criterion.

Step 1: Plot the Nyquist plot of $G(j\omega)$ as shown in Figure 5.12.

Step 2: Draw a tangent to the plot. This line is called the Popov's line. Let it intersect the negative real axis at $(-1/K_1)$.

Step 3: Mark the sector $[0, K_1]$. This is called Popov sector. In this sector, the nonlinearity is asymptotic and output asymptotic.

Step 4: Find the range of K for which the linear system is asymptotically stable. This region $[0, k]$ is called Hurwitz sector.

Illustration 12

For the given system, determine the Popov and Hurwitz region using Popov's stability criterion. $G(s) = \dfrac{4}{(s+1)(0.25s+1)(0.33s+1)}$.

Solution

The given system is $G(s) = \dfrac{4}{(s+1)(0.25s+1)(0.33s+1)}$

Step 1: Plot the Nyquist plot of $G(j\omega)$

$$G(j\omega) = \frac{K}{(j\omega+1)(0.25j\omega+1)(0.33j\omega+1)}$$

Also,

$$G(j\omega) = \frac{4(1 - 0.6625\omega^2)}{(1 - 0.6625\omega^2)^2 + (1.58\omega - 0.0825\omega^3)^2}$$

$$G_R = \frac{4(1 - 0.6625\omega^2)}{(1 - 0.6625\omega^2)^2 + (1.58\omega - 0.0825\omega^3)^2}$$

$$G_I = -\frac{4(1.58\omega - 0.0825\omega^3)}{(1 - 0.6625\omega^2)^2 + (1.58\omega - 0.0825\omega^3)^2}$$

Ω	0	1	5	10	20	100	∞
G_R	4	0.584	−0.62	−0.09	0.002	0	0
G_I	0	−2.5	−0.31	0.05	0.005	0	0

Step 2: Draw a tangent to the Nyquist plot.

This line is called the Popov's line. Let it intersect the negative real axis at $(-1/K_1)$.

Step 3: Mark the sector $[0, K_1]$.

From the Figure 5.13, it can be seen that, the tangent intersects the negative real axis at −0.5. This corresponds to the frequency $\omega = 3.5$. The real part of the system when $\omega = 3.5$ is, $G_R = -0.54$ $G_I = 0.06$.

Hence, the Popov sector is given by $[0,1/0.54]$. This is called Popov sector. In this sector, the nonlinearity is asymptotic and output asymptotic.

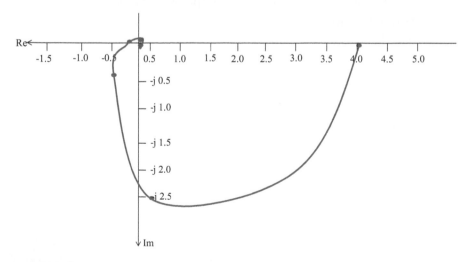

FIGURE 5.12
Nyquist plot of $G(j\omega)$.

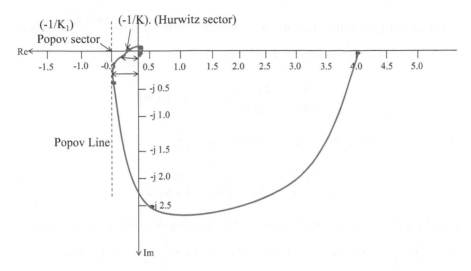

FIGURE 5.13
A tangent drawn to the Nyquist plot.

Step 4: Find the range of K for which the linear system is asymptotically stable.

From Figure 5.13, it can be seen that the Nyquist plot intersects the negative real axis at −0.326. This region [0, k] is called Hurwitz sector. This corresponds to the frequency ω = 4.3. The real part of the system when ω = 4.3 is, G_R = −0.326 G_I = 0.26. Hence, the Hurwitz sector is given by [0,1/0.326].

Circle Criterion

In general, circle criterion is derived from Popov's criterion. It differs from the assumption made when evaluating the Popov criterion, i.e., in Popov criterion, there is no pole-zero cancellation. However, in circle criterion, pole-zero cancellation is considered. Any system which is proved stable using circle criterion can also be called as BIBO stable. It is an extension of Nyquist criterion defined for linear time-invariant systems.

Circle Criterion

Circle criterion states that, the closed loop system is asymptotically stable if the Nyquist plot of $G(j\omega)$ does not intersect the circle given by the intercepts of x and y, except the fact that, the system undergoes pole-zero cancellation.

For investigating the stability, consider the stable system $\dfrac{1+K_2G(s)}{1+K_1G(s)}$ whose real part can be represented as follows:

$$\text{Re}\left[\frac{1+K_2G(j\omega)}{1+K_1G(j\omega)}\right] > 0,\ \forall\ \omega \in [0,\alpha]$$

When $K_1 > 0$, using the Nyquist criterion, we get

$$\frac{1+K_2G(s)}{1+K_1G(s)} = \frac{1}{1+K_1G(s)} + \frac{K_2G(s)}{1+K_1G(s)} \qquad \text{(i)}$$

Equation (i) is stable, if the Nyquist plot of $G(j\omega)$ does not intersect the real axis at $\left(-\dfrac{1}{K_1}+j0\right)$ and also it encircles the point $\left(-\dfrac{1}{K_1}+j0\right)$, n times in the counterclockwise direction, where 'n' is the number of poles of $G(s)$ in the right-half complex plane.

$$\frac{1+K_2G(s)}{1+K_1G(s)} > 0 \ \Rightarrow \ \frac{\dfrac{1}{K_2}+G(j\omega)}{\dfrac{1}{K_1}+G(j\omega)} > 0$$

$$\Rightarrow \text{Re}\left[\frac{\dfrac{1}{K_2}+G(j\omega)}{\dfrac{1}{K_1}+G(j\omega)}\right] > 0,\ \forall\ \omega \in [0,\alpha]$$

is said to be stable, when the Nyquist plot (Figure 5.14) of $G(j\omega)$ does not enter the circle having intercepts $-\dfrac{1}{K_1}, -\dfrac{1}{K_2}$ and encircles it n times in the counterclockwise direction.

When $K_1 = 0$, the stable system becomes $1 + K_2G(s)$

$$\text{Re}[1+K_2G(s)] > 0,\ \forall\omega \in [0,\alpha]$$

$$\text{Re}[G(j\omega)] > -\frac{1}{K_2},\ \forall\omega \in [0,\alpha]$$

In that case, the system is said to be absolutely stable when $G(s)$ is Hurwitz and Nyquist plot of $G(j\omega)$ lies to the right of $-\dfrac{1}{K_2}$ plane.

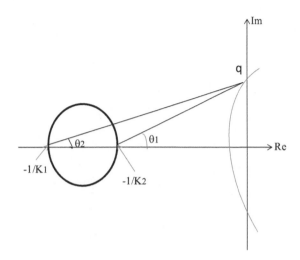

FIGURE 5.14
Nyquist plot.

> ## NOTE
>
> A polynomial function $P(k)$ of any complex variable k is said to be Hurwitz if $P(k)$ is real for all k = real and has roots such that are zero or negative real parts.

When $K_1 < 0 < K_2$, the real part of stable system can be written as follows:

$$\text{Re}\left[\frac{1+K_2G(j\omega)}{1+K_1G(j\omega)}\right] > 0 \;\Rightarrow\; \text{Re}\left[\frac{\dfrac{1}{K_2}+G(j\omega)}{\dfrac{1}{K_1}+G(j\omega)}\right] < 0$$

Inference

a. Here, Nyquist plot $G(j\omega)$ lies inside the circle with intersect $-\dfrac{1}{K_1}$, $-\dfrac{1}{K_2}$.

b. Also, the Nyquist plot does not encircle $-\dfrac{1}{K_1}$.

c. Hence, from Nyquist criterion, it is inferred that $G(s)$ is Hurwitz. The system is absolutely stable if conditions (a) and (b) are satisfied.

6

Bifurcation Behaviour of Non-Linear Systems

This chapter deals with bifurcation behaviour of non-linear systems. It also explains different types of bifurcation namely: Saddle node bifurcation, transcritical bifurcation, pitch fork bifurcation and Hopf bifurcations (both supercritical and subcritical). Later part of the chapter presents the introduction to chaos, Lorentz equations, stability analysis of Lorentz equations and chaos in chemical systems. The above concepts are also illustrated with suitable examples.

Bifurcation Theory

In practice, the dynamical systems with differential equations may contain many parameters. The values of these parameters are often known approximately and are not exactly determined by measurements. Whenever there is a slight variation in the parameter, it can introduce a significant impact on the solution. Thus, it is important to study the behaviour of solutions and examine their dependence on the parameters. This study leads to the area referred to as bifurcation theory. The name bifurcation was first introduced by Henri Poincare in 1885.

Generally, in dynamical systems, a bifurcation occurs when a small change is made to the parameter values of a system causing a sudden qualitative or topological change in its behaviour. At a bifurcation, the local stability properties of equilibria change. The parameter value at which this kind of changes occurs is called as bifurcation value and the parameter which is varied is known as bifurcation parameter.

Bifurcation Analysis

Consider a family of Ordinary Differential Equation (ODE) that depends on one parameter 'α'.

$$x' = f(x, \alpha) \tag{6.1}$$

where $f : \mathbb{R}^{n+1} \to \mathbb{R}^n$ is analytic for $\alpha \in \mathbb{R}$, $x \in \mathbb{R}^n$.

Let $x = x_o(\alpha)$ be a family of equilibrium points of equation (6.1), i.e., $f(x_0(\alpha), \alpha) = 0$.

Now let us consider

$$z = x - x_0(\alpha) \tag{i}$$

Differentiating equation (i), we get

$$z' = A(\alpha)z + 0\left(|z|^2\right) \tag{ii}$$

where $A(\alpha) = \dfrac{\partial f}{\partial x}(x_0(\alpha), \alpha)$

Let $\alpha_1, \alpha_2 \ldots \alpha_n(\alpha)$ be the eigen values of $A(\alpha)$.

If for some 'i', $\operatorname{Re} \alpha_i(\alpha)$ changes sign at $\alpha = \alpha_0$, then we say that 'α_0' is a bifurcation point of equation (6.1).

Bifurcation in One Dimension

Consider, $f : \mathbb{R}^{n+1} \to \mathbb{R}^n$, and let us assume that $n = 1$, then we get $f : \mathbb{R}^2 \to \mathbb{R}^1$ and $x_0(\alpha)$ will be real valued analytic function of α provided

$$\alpha_1(\alpha) = \frac{\partial f}{\partial x}(x_0(\alpha), \alpha) = A(\alpha) = 0$$

Therefore, the equilibrium point is asymptotically stable if $\alpha_1(\alpha) < 0$ and unstable if $\alpha_1(\alpha) > 0$, which implies that α_0 is a bifurcation point if $\alpha_1(\alpha) = 0$.

Hence, the bifurcation points $(x_0(\alpha), \alpha)$ are the solutions of

$$f(x, \alpha) = 0 \text{ and } \frac{\partial f}{\partial x}(x, \alpha) = 0$$

Common Types of Bifurcation

The three common types of bifurcations occurring in scalar differential equations are

  i. Saddle node bifurcation
 ii. Transcritical bifurcation
iii. Pitch fork bifurcation

Saddle Node Bifurcation

The saddle node bifurcation is a collision and disappearance of two equilibria in dynamical systems. This occurs when the critical equilibrium has one zero eigen value. This phenomenon is also called 'fold' or 'limit point' or 'tangent' or 'blue sky' bifurcation.

Consider the dynamical system defined by

$$x' = a - x^2 \tag{6.2}$$

where 'a' is real.

The equilibrium solution is given by $x' = 0$, which gives $x = \pm\sqrt{a}$.

Therefore, if $a < 0$, then we get no real solutions and if $a > 0$, then we get two real solutions.

To examine the stability of the system for each of the two solution (when $a > 0$), let us add a small perturbation

$$x = \bar{x} + \epsilon \tag{6.3}$$

Differentiating equation (6.3), we get

$$\frac{dx}{dt} = x'$$

$$x' = \frac{d}{dt}\bar{x} + \frac{d\epsilon}{dt}$$

$$x' = \frac{d\epsilon}{dt} \quad \left[\because \frac{d\bar{x}}{dt} = 0 \right] \tag{6.4}$$

Substituting equation (6.3) in equation (6.2), we get

$$x' = a - (\bar{x} + \epsilon)^2$$

$$x' = a - \bar{x}^2 - 2\bar{x}\,\epsilon - \epsilon^2 \tag{6.5}$$

From equations (6.4) and (6.5), we get

$$\frac{d\epsilon}{dt} = (a - \bar{x}^2) - 2\bar{x}\,\epsilon - \epsilon^2 \tag{6.6}$$

After simplification, we get

$$\frac{d\epsilon}{dt} = -2\bar{x}\,\epsilon \tag{6.7}$$

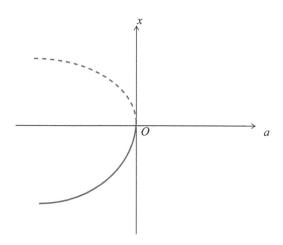

FIGURE 6.1
Bifurcation diagram corresponding to saddle node bifurcation.

The solution of equation (6.7) can be written as follows

$$\in (t) = Ae^{-2\bar{x}t} \tag{6.8}$$

Thus, we have, for $x = +\sqrt{a}$, $|x| \mapsto 0$ as $t \to \infty$ (linear stability)
  and, for $x = -\sqrt{a}$, $|x| \mapsto 0$ as $t \to \infty$ (linear stability)
  The bifurcation diagram corresponding to saddle node bifurcation is
shown in Figure 6.1.

NOTE

The saddle node bifurcation at $a = 0$ corresponds to creation of two new
solution branches. One of these solutions is linearly stable and the other
one is linearly unstable.

Illustration 1

Analyse the bifurcation properties for $x' = r + x^2$.
  Step 1: We should analyse the bifurcation properties for three different
cases, namely, $r < 0, r = 0, r > 0$.
  Step 2: For each of the above three cases, obtain the graph of x versus x' by
taking x on x-axis and x' on y-axis.

Solution

Given $x' = r + x^2$

Case 1: $r < 0$

There are two fixed points given by $x = \pm\sqrt{-r}$. The equilibrium $x = -\sqrt{-r}$ is stable, i.e., solutions beginning near this equilibrium converge to it as time increases. Further, the initial conditions near $+\sqrt{-r}$ diverge from it, as shown in Figure 6.2a.

Case 2: $r = 0$

There is a single fixed point at $x = 0$ and initial conditions less than zero give solutions that converge to zero while positive initial conditions give solutions that increase without bound, as shown in Figure 6.2b.

Case 3: $r > 0$

There are no fixed points at all. For any initial conditions, solutions increase without bound, as shown in Figure 6.2c.

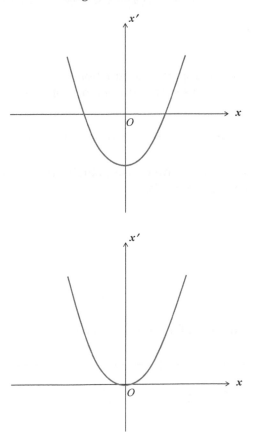

FIGURE 6.2

Bifurcation diagram for $x' = r + x^2$. (a) $r < 0$, (b) $r = 0$, (c) $r > 0$.

(Continued)

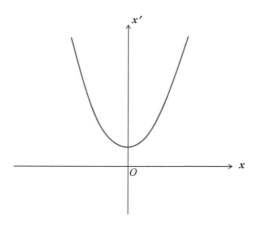

FIGURE 6.2 (CONTINUED)
Bifurcation diagram for $x' = r + x^2$. (a) $r < 0$, (b) $r = 0$, (c) $r > 0$.

Illustration 2

Analyse the bifurcation properties of function $x' = a - x^2$.
   Step 1: We should analyse the bifurcation properties for three different cases, namely,

$$a < 0, (a = -1); \quad a = 0; \quad a > 0, (a = 1)$$

Step 2: For each of the above three cases, obtain the graph of x versus x' by taking x on x-axis and x' on y-axis.

Solution

Given $x' = a - x^2$
   Case 1: $a < 0$

$$\therefore \quad x' = -1 - x^2 \text{ (when } a = -1)$$

From the table giving the values of x' obtained for different values of x.

x	0	1	2	3	−1	−2	−3
x'	−1	−2	−5	−10	−2	−5	−10

Case 2: $a = 0$

$$\therefore \quad x' = -x^2 \text{ (when } a = 0)$$

x	0	1	2	3	−1	−2	−3
x'	0	−1	−4	−9	−1	−4	−9

Case 3: $a > 0$

$$\therefore \quad x' = 1 - x^2 \ (\text{when } a = 1)$$

x	0	1	2	3	−1	−2	−3
x'	1	0	−3	−8	0	−3	−8

The bifurcation diagram for the three cases of $x' = a - x^2$ is as shown in Figure 6.3

Illustration 3

Analyse the bifurcation properties of function $x' = r - \cosh(x)$.

Solution

Given $x' = r - \cosh(x)$

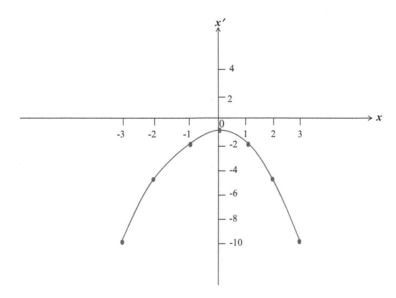

FIGURE 6.3
Bifurcation diagram for $x' = a - x^2$. (a) $a < 0$, (b) $a = 0$, (c) $a > 0$.

(Continued)

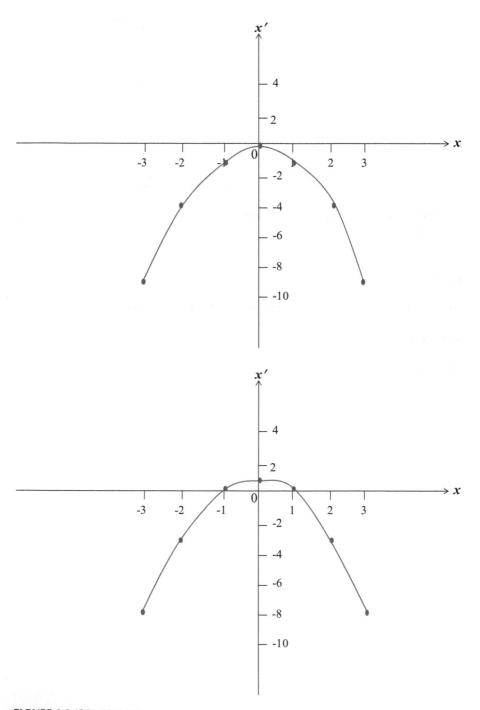

FIGURE 6.3 (CONTINUED)
Bifurcation diagram for $x' = a - x^2$. (a) $a < 0$, (b) $a = 0$, (c) $a > 0$.

Case 1: $r < 0$

$$\therefore \quad x' = -1 - \cosh(x) \text{ (when } r = -1)$$

From the table giving the values of x' obtained for different values of x.

x	0	1	2	3	−1	−2	−3
x'	−2	−2.5	−4.7	−11	−2.5	−4.7	−11

Case 2: $r = 0$

$$\therefore \quad x' = -\cosh(x) \text{ (when } r = 0)$$

x	0	1	2	3	−1	−2	−3
x'	−1	−1.5	−3.7	−10	−1.5	−3.7	−10

Case 3: $r > 0$

$$\therefore \quad x' = 1 - \cosh(x) \text{ (when } r = 1)$$

x	0	1	2	3	−1	−2	−3
x'	0	−0.5	−2.7	−9	−0.5	−2.7	−9

The bifurcation diagram for the three cases of $x' = r - \cosh(x)$ is as shown in Figure 6.4

Transcritical Bifurcation

In this type of bifurcation, two families of fixed points collide and exchange their stability properties. The family which was stable earlier (before bifurcation) will become unstable (after bifurcation) and vice versa.

Consider a system represented by a differential equation

$$x' = \frac{dx}{dt} = ax - bx^2 \tag{6.9}$$

where x, a, b are real and a, b are control parameters.

The two steady states ($x' = 0$) to this system can be given as follows:

$x = \bar{x}_1 = 0$, for any value of a, b

$x = \bar{x}_2 = \dfrac{a}{b}$, for any value of a, b; $b \neq 0$

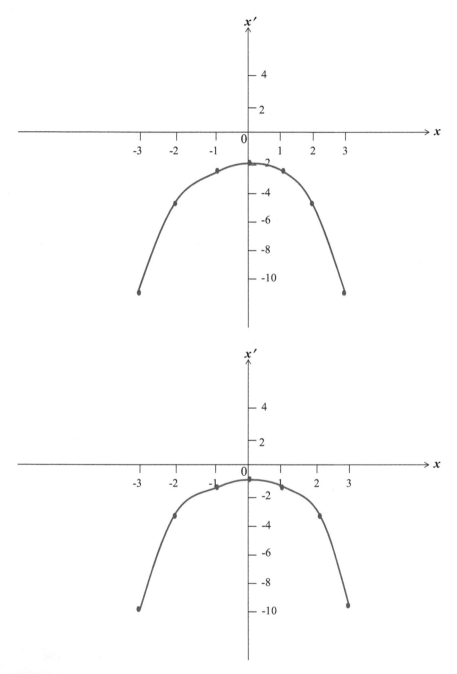

FIGURE 6.4
Bifurcation diagram for $x' = r - \cosh(x)$. (a) $r < 0$, (b) $r = 0$, (c) $r > 0$.

(*Continued*)

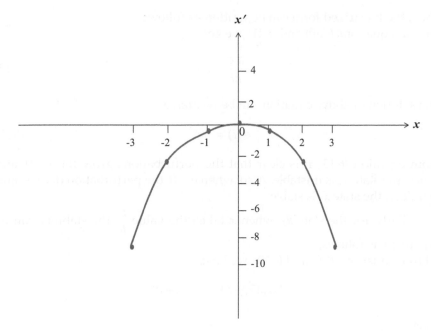

FIGURE 6.4 (CONTINUED)
Bifurcation diagram for $x' = r - \cosh(x)$. (a) $r < 0$, (b) $r = 0$, (c) $r > 0$.

The linear stability of each of these states can be examined as follows:
Let us add a small perturbation ϵ for the state $\bar{x}_1$, i.e.,

$$x = \bar{x}_1 + \epsilon$$

$$\frac{dx}{dt} = \frac{d\bar{x}}{dt} + \frac{d\epsilon}{dt}$$

$$\frac{dx}{dt} = \frac{d\epsilon}{dt} \quad \left(\because \frac{d\bar{x}}{dt} = 0 \right) \tag{6.10}$$

We have

$$\frac{dx}{dt} = x' = a(\bar{x}_1 + \epsilon) - b(\bar{x}_1 + \epsilon)^2$$

$$= a\bar{x}_1 + a\,\epsilon - b\bar{x}_1{}^2 - 2b\bar{x}_1\,\epsilon - b\,\epsilon^2$$

$$\frac{dx}{dt} = a\,\epsilon - b\,\epsilon^2 \quad (\bar{x}_1 = 0, \text{ equilibrium state})$$

$$\frac{dx}{dt} = a\,\epsilon \quad (\because \epsilon^2 \text{ is very small}) \tag{6.11}$$

Thus, the linearized form can be written as follows:
From equations (6.10) and (6.11), we get

$$\frac{d\,\epsilon}{dt} = a\,\epsilon \tag{6.12}$$

The solution of above equation can be written as

$$\epsilon(t) = Ae^{at} \tag{6.13}$$

From equation (6.13), it is clear that the perturbations grow for $a > 0$, and hence, the state $\bar{x}_1$ is unstable. Also, when $a < 0$, the perturbation decays, and therefore, the state $\bar{x}_1$ is stable.

Similarly, for the state $\bar{x}_2$, when x takes the value $\dfrac{a}{b}$, the stability can be explained as follows.
From equations (6.9) and (6.10), we have

$$x' = a(\bar{x}_1 + \epsilon) - b(\bar{x}_1 + \epsilon)^2$$

when $\bar{x} = \dfrac{a}{b}$

$$x' = a\left(\frac{a}{b} + \epsilon\right) - b\left(\frac{a}{b} + \epsilon\right)^2$$

$$x' = \frac{a^2}{b} + a\,\epsilon - \frac{ba^2}{b^2} - b\,\epsilon^2 - \frac{2a\,\epsilon\,b}{b}$$

$$x' = \frac{a^2}{b} + a\,\epsilon - \frac{a^2}{b} - b\,\epsilon^2 - 2a\,\epsilon$$

$$= -a\,\epsilon - b\,\epsilon^2$$

$$x' = -a\,\epsilon \quad (\because \epsilon^2 \text{ is very small})$$

Thus, the linearized form can be written as follows:

$$\frac{d\,\epsilon}{dt} = -a\,\epsilon \tag{6.14}$$

The solution of the equation (6.14) can be obtained as follows:

$$\epsilon(t) = Ae^{-at} \tag{6.15}$$

From equation (6.15), it is clear that the state $x = \dfrac{a}{b}$ is linearly stable for $a > 0$ and linearly unstable for $a < 0$. The bifurcation point $a = 0$ corresponds to

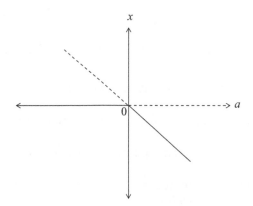

FIGURE 6.5
Bifurcation diagram corresponding to transcritical bifurcation.

exchange of stabilities between the two solution branches. The bifurcation diagram for transcritical bifurcation is shown in Figure 6.5.

Illustration 4

Analyse the bifurcation properties of the system represented by $x' = rx - x^2$

Solution

Given $x' = rx - x^2$
  Case 1: $r < 0$

$$\therefore \quad x' = -x - x^2 \text{ (when } r = -1)$$

x	0	1	2	3	−1	−2	−3
x'	0	−2	−6	−12	0	−2	−6

  Case 2: $r = 0$

$$\therefore \quad x' = -x^2$$

x	0	1	2	3	−1	−2	−3
x'	0	−1	−4	−9	−1	−4	−9

  Case 3: $r > 0$

$$\therefore \quad x' = x - x^2 \text{ (when } r = 1)$$

x	0	1	2	3	−1	−2	−3
x'	0	1	−2	−6	−2	−6	−12

The bifurcation diagram for the three cases of $x' = rx - x^2$ is as shown in Figure 6.6.

When $r < 0$, the non-zero fixed point is towards left and the system is unstable.

When $r > 0$, the non-zero fixed point is towards right and the system is stable.

When $r = 0$, the fixed point in $x = 0$, which is semi-stable. (Stable from the right and unstable from the left.)

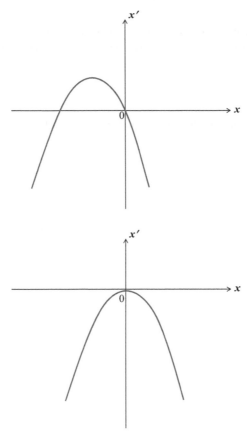

FIGURE 6.6
(a–c) Bifurcation diagrams of $x' = rx - x^2$.

(*Continued*)

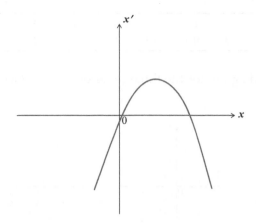

FIGURE 6.6 (CONTINUED)
(a–c) Bifurcation diagrams of $x' = rx - x^2$.

Illustration 5

Analyse the bifurcation properties of the system represented by $x' = rx - \ln(1 + x)$.

Solution

Given $x' = rx - \ln(1 + x)$
  Case 1: $r < 0$

$$\therefore \quad x' = -x - \ln(1 + x) \text{ (when } r = -1)$$

x	0	1	2	3	4	5	6
x'	0	−1.6	−3	−4.3	−5.6	−6.7	−7.9

Case 2: $r = 0$

$$\therefore \quad x' = -\ln(1 + x)$$

x	0	1	2	3	4	5	6
x'	0	−0.6	−1	−1.3	−1.6	−1.7	−1.9

Case 3: $r > 0$

$$\therefore \quad x' = x - x^2 \text{ (when } r = 1)$$

x	0	1	2	3	4	5	6
x'	0	0.3	0.9	1.6	2.3	3.2	4

The bifurcation diagram for the three cases of $x' = rx - \ln(1 + x)$ is as shown in Figure 6.7.

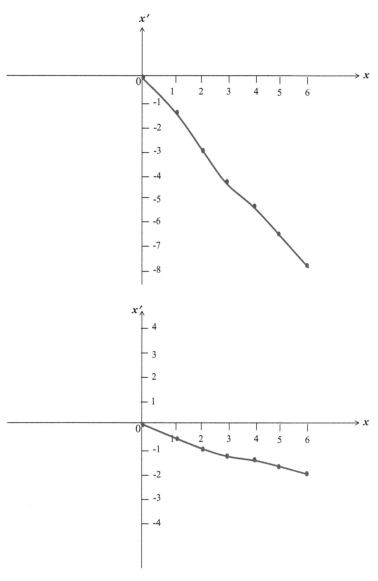

FIGURE 6.7
Bifurcation diagrams of $x' = rx - \ln(1 + x)$. (a) $r < 0$, (b) $r = 0$, (c) $r > 0$.

(Continued)

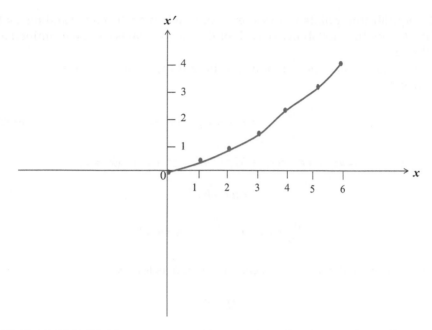

FIGURE 6.7 (CONTINUED)
Bifurcation diagrams of $x' = rx - \ln(1 + x)$. (a) $r < 0$, (b) $r = 0$, (c) $r > 0$.

Pitchfork Bifurcation

In pitchfork bifurcation, one family of fixed point transfers its stability properties to two families after or before the bifurcation point. If the transfer takes place after the bifurcation point, it is called as supercritical bifurcation, whereas if the transfer takes place before the bifurcation point, then it is called as subcritical bifurcation.

Now, consider the dynamical system

$$x' = ax - bx^3 \tag{6.16}$$

where a, b are real and are external control parameters.

The three steady states to this dynamical system can be given as follows:

$$x = \bar{x}_1 = 0, \text{ for any value of } a, \, b$$

$$x = \bar{x}_2 = -\sqrt{\frac{a}{b}}, \text{ for } a/b > 0$$

$$x = \bar{x}_3 = +\sqrt{\frac{a}{b}}, \text{ for } a/b > 0$$

The equilibrium points $\bar{x}_2$ and $\bar{x}_3$ exist only when $a > 0$, if $b > 0$ and for $a < 0$ if $b < 0$. The linear stability of each of the steady states can be examined as follows:

Let us add a small perturbation for the state $\bar{x}_1 = 0$ as $x = \bar{x}_1 + \in$

We get

$$\frac{dx}{dt} = x' = a(\bar{x}_1 + \in) - b(\bar{x}_1 + \in)^3 \tag{6.17}$$

$$= a\bar{x}_1 + a \in -b(\bar{x}_1^3 + 3\bar{x}_1^2 \in +3\bar{x}_1 \in^2 + \in^3) \quad (\bar{x}_1 = 0)$$

$$= a \in -b \in^3$$

$$\frac{dx}{dt} = a \in \quad (\because \in^2 \text{ is very small})$$

The solution of above equation can be written as follows:

$$\in (t) = Ae^{at}$$

Hence, the state $\bar{x}_1 = 0$ is linearly unstable when $a > 0$ and the state $\bar{x}_1 = 0$ is linearly stable when $a < 0$.

Similarly, for the states $x = \bar{x}_2$ and $x = \bar{x}_3$, set $\bar{x}_1 = \pm\sqrt{\frac{a}{b}} + \in$

After simplification, we get

$$\frac{d \in}{dt} = a \in -3b\bar{x}_1^2 \in$$

The solution of above equation can be written as follows:

$$\in (t) = Ae^{ct}$$

where $c = -2a$.

Thus, the states $\bar{x}_2$ and $\bar{x}_3$ are linearly stable when $a > 0$ and linearly unstable when $a < 0$. The bifurcation diagram corresponding to pitchfork bifurcation is as shown in Figure 6.8.

Illustration 6

Analyse the bifurcation properties of the system represented by $x' = rx - x^3$

Solution

Given $x' = rx - x^3$

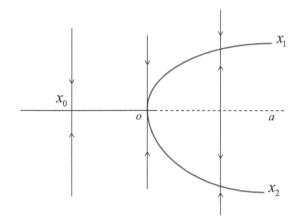

FIGURE 6.8
Bifurcation diagram corresponding to pitchfork bifurcation.

Case 1: $r < 0$

$$\therefore \quad x' = -x - x^3 \text{ (when } r = -1)$$

x	0	1	2	3	-1	-2	-3
x'	0	-2	-10	-30	2	10	30

Case 2: $r = 0$

$$\therefore \quad x' = -x^3 \text{ (when } r = 0)$$

x	0	1	2	3	-1	-2	-3
x'	0	-1	-8	-27	1	8	27

Case 3: $r > 0$

$$\therefore \quad x' = x - x^3 \text{ (when } r = 1)$$

x	0	1	2	3	-1	-2	-3
x'	0	0	-6	-24	0	6	24

The bifurcation diagram for the three cases of $x' = rx - x^3$ is as shown in Figure 6.9.

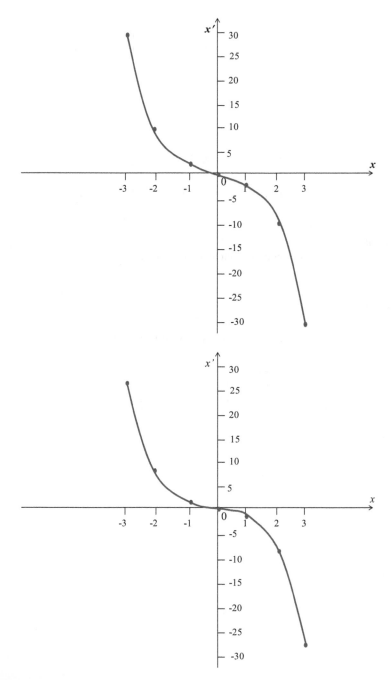

FIGURE 6.9
Bifurcation diagram of $x' = rx - x^3$. (a) $r < 0$, (b) $r = 0$, (c) $r > 0$.

(Continued)

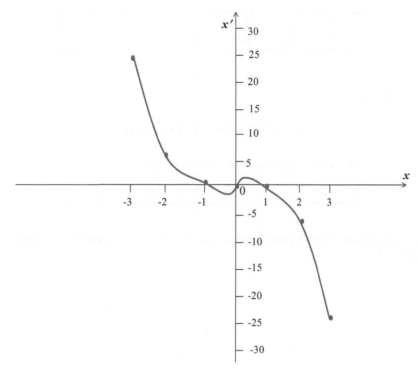

FIGURE 6.9 (CONTINUED)
Bifurcation diagram of $x' = rx - x^3$. (a) $r < 0$, (b) $r = 0$, (c) $r > 0$.

Illustration 7

Analyse the bifurcation properties of the system represented by $x' = rx - \sinh x$.

Solution

Given $x' = rx - \sinh x$
   Case 1: $r < 0$

$$\therefore \quad x' = -x - \sinh x \text{ (when } r = -1)$$

x	0	1	2	3	−1	−2	−3
x'	0	−2.1	−5.6	−13	2.1	5.6	13

Case 2: $r = 0$

$$\therefore \quad x' = -\sinh x \text{ (when } r = 0)$$

x	0	1	2	3	−1	−2	−3
x'	0	−1.1	−3.6	−10	1.1	3.6	10

Case 3: $r > 0$

$$\therefore \quad x' = x - \sinh x \ (\text{when } r = 1)$$

x	0	1	2	3	−1	−2	−3
x'	0	−0.1	−1.6	−7	0.1	1.6	7

The bifurcation diagram for the three cases of $x' = rx - \sinh x$ is as shown in Figure 6.10.

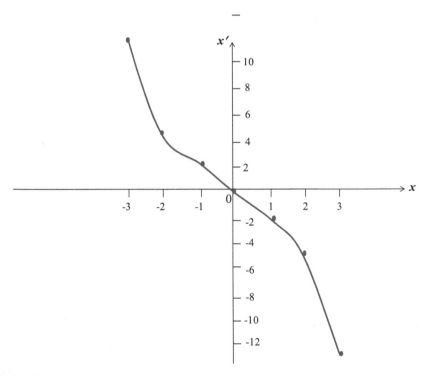

FIGURE 6.10
Bifurcation diagram of $x' = rx - \sinh x$. (a) $r < 0$, (b) $r = 0$, (c) $r > 0$.

(Continued)

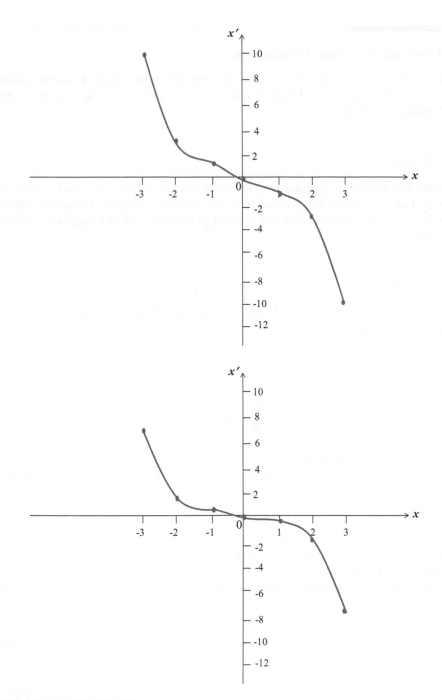

FIGURE 6.10 (CONTINUED)
Bifurcation diagram of $x' = rx - \sinh x$. (a) $r < 0$, (b) $r = 0$, (c) $r > 0$.

Bifurcation in Two Dimension

The Hopf bifurcation falls under two-dimension bifurcation. It is also called as Poincare–Andronov–Hopf. It appears in systems having two or more differential equations.

Definition

Hopf bifurcation is defined as a local bifurcation in which a fixed point of a dynamical system loses stability when a pair of complex conjugate eigen values of linearization around the fixed point crosses the imaginary axis of complex plane.

Theorem

Consider a two-dimensional system represented by the following equations (6.18) and (6.19)

$$\frac{dx}{dt} = f(x,y,\lambda) \tag{6.18}$$

$$\frac{dy}{dt} = g(x,y,\lambda) \tag{6.19}$$

where 'λ' is the parameter.

Let $(\bar{x}(\lambda), \bar{y}(\lambda))$ be the coordinates of the equilibrium point and let $\alpha(\lambda) \pm j\beta(\lambda)$ be the eigen values of Jacobian matrix evaluated at the equilibrium point.

Let us also assume that the change in the stability of the equilibrium point occurs at $\lambda = \lambda^*$ where $\alpha(\lambda^*) = 0$. Let us first transform the above system so that equilibrium is at the origin and the parameter λ at $\lambda^* = 0$ provides pure imaginary eigen values.

Thus, the system represented in equations (6.18) and (6.19) can be rewritten as follows:

$$\frac{dx}{dt} = a_{11}(\lambda)x + a_{12}(\lambda)y + f_1(x,y,\lambda) \tag{6.20}$$

$$\frac{dy}{dt} = a_{21}(\lambda)x + a_{22}(\lambda)y + g_1(x,y,\lambda) \tag{6.21}$$

where f_1 and g_1 have continuous third-order partial derivatives in x and y.

The linearization of system represented in equations (6.18) and (6.19) about the origin is given by

$$\frac{dX}{dt} = J(\lambda)X \tag{6.22}$$

where $X = \begin{bmatrix} x \\ y \end{bmatrix}$ and $J(\lambda) = \begin{bmatrix} a_{11}(\lambda) & a_{12}(\lambda) \\ a_{21}(\lambda) & a_{22}(\lambda) \end{bmatrix}$ which will be the Jacobian

matrix evaluated at origin.

Let us also assume that the origin is an equilibrium point of equations (6.20) and (6.21) so that the Jacobian matrix $J(\lambda)$ given above is valid for all sufficiently small $|\lambda|$.

Also, assume that the eigen values of matrix $J(\lambda)$ are $\alpha(\lambda) \pm j\beta(\lambda)$ where $\alpha(0) = 0$, $\beta(0) \neq 0$ such that the eigen values cross the imaginary axis, with a non-zero speed.

That is,

$$\left.\frac{d\alpha}{dt}\right|_{\lambda=0} \neq 0 \tag{6.23}$$

Then, in any open set 'U' containing the origin in $\mathbb{R}^2$ and for any $\lambda_0 \geq 0$, there exists a value $\bar{\lambda}, |\lambda| \leq \lambda_0$ such that the system of differential equations given in (6.20) and (6.21) has a periodic solution for $\lambda = \bar{\lambda}$ in U.

_____

Supercritical Hopf Bifurcation

The Hopf bifurcation becomes 'supercritical' if the equilibrium point (0,0) is asymptotically stable at the bifurcation point, i.e., when $\lambda = 0$. In supercritical bifurcation, the limit cycle grows out of the equilibrium point. In other words, the limit cycle has zero amplitude at the parameters of Hopf bifurcation, and this amplitude grows as the parameters move further into the limit cycle as shown in Figure 6.11.

Illustration 8

Analyse the bifurcation property of the given system

$$\frac{dx}{dt} = \lambda x - y - x\left(x^2 + y^2\right)$$

$$\frac{dy}{dt} = x + \lambda y - y\left(x^2 + y^2\right)$$

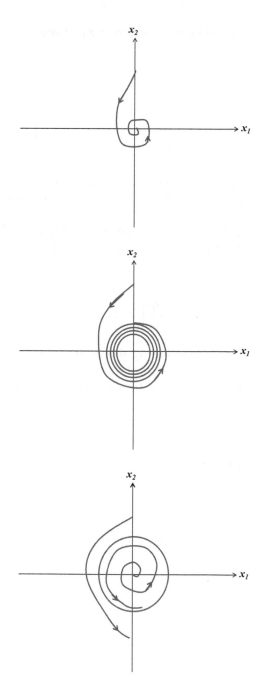

FIGURE 6.11
Bifurcation diagram corresponding to supercritical Hopf bifurcation. (a) $r < 0$, (b) $r = 0$, (c) $r > 0$.

Solution

Given

$$\frac{dx}{dt} = \lambda x - y - x\left(x^2 + y^2\right)$$

$$\frac{dy}{dt} = x + \lambda y - y\left(x^2 + y^2\right)$$

Step 1: Deduce the equilibrium point

The equilibrium of the above system can be written as $(\bar{x}(\lambda), \bar{y}(\lambda)) = (0,0)$.

Step 2: Evaluate the Jacobian matrix

$$J(\lambda) = \begin{bmatrix} \lambda - 3x^2 - y^2 & -1 - 2xy \\ 1 - 2xy & \lambda - x^2 - 3y^2 \end{bmatrix}$$

$$J(\lambda)_{(0,0)} = \begin{bmatrix} \lambda & -1 \\ 1 & \lambda \end{bmatrix}$$

Step 3: Find the eigen values

The eigen values of this system are $\lambda \pm i$. For values of λ, it can be seen that the eigen values cross the imaginary axis.

Step 4: Find the polar coordinates of the system by substituting $x = r\cos\theta$ and $y = r\sin\theta$

We know that

$$\dot{x} = \lambda x - y - x\left(x^2 + y^2\right) \tag{i}$$

$$\dot{y} = x + \lambda y - y\left(x^2 + y^2\right) \tag{ii}$$

For stability at equilibrium $\lambda = \lambda 0 = 0$, the equations (i) and (ii) become,

$$\dot{x} = -y - x\left(x^2 + y^2\right)$$

$$\dot{y} = x - y\left(x^2 + y^2\right)$$

In polar coordinates, we have

$$r^2 = x^2 + y^2 \tag{iii}$$

On differentiating the equation (iii), we get

$$2r\dot{r} = 2x\dot{x} + 2y\dot{y}$$

$$\Rightarrow r\dot{r} = x\dot{x} + y\dot{y} \tag{iv}$$

Substitute $\dot{x}$ and $\dot{y}$ in the equation (iv), we get

$$r\dot{r} = x\left(-y - x\left(x^2 + y^2\right)\right) + y\left(x - y\left(x^2 + y^2\right)\right)$$

$$r\dot{r} = x\left(-y - xr^2\right) + y\left(x - yr^2\right)$$

$$r\dot{r} = -xy - x^2 r^2 + yx - y^2 r^2$$

$$r\dot{r} = -r^2\left(x^2 + y^2\right)$$

$$r\dot{r} = -r^2 . r^2$$

$$r\dot{r} = -r^4$$

Step 5: Analyse the bifurcation for the three cases ($r < 0$, $r = 0$, $r > 0$)

$$\therefore \dot{r} = -r^3$$

r	-3	-2	-1	0	1	2	3
$\dot{r}$	27	8	1	0	-1	-8	-27

The bifurcation diagram for the three cases of the given system $\dot{r} = -r^3$ is as shown in Figure 6.12.

Here, the radius r becomes smaller over time. Also, $\dot{\theta} = 1$. It means that the rotation rate is changing at the same rate.

From the above two inferences, the phase diagram can be drawn as shown in Figure 6.13.

Thus, the system exhibits a supercritical Hopf bifurcation.

Subcritical Hopf Bifurcation

The Hopf bifurcation becomes 'subcritical' if the equilibrium point (0,0) is negatively asymptotically stable (as $t \to \infty$). The bifurcation diagram corresponding to subcritical Hopf bifurcation is shown in Figure 6.14.

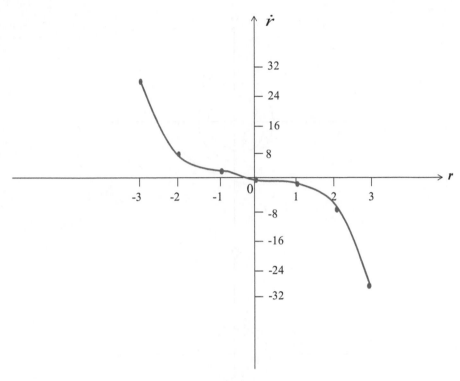

FIGURE 6.12
Bifurcation diagram of $\dot{r} = -r^3$.

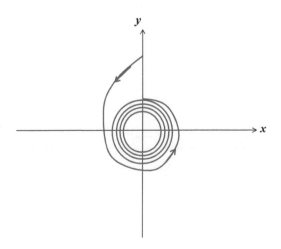

FIGURE 6.13
Phase portrait of $\dot{r} = -r^3$.

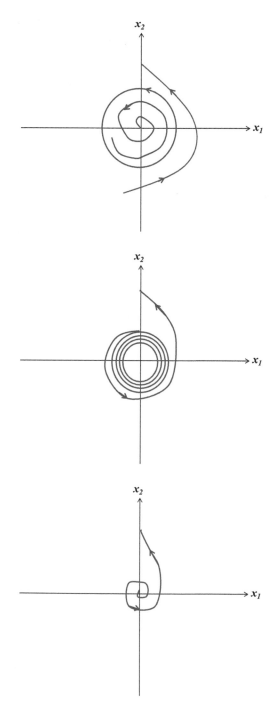

FIGURE 6.14
Bifurcation diagram corresponding to subcritical Hopf bifurcation. (a) $r < 0$, (b) $r = 0$, (c) $r > 0$.

Illustration 9

Consider the two-dimensional system

$$\frac{dx}{dt} = x^2 - \beta \tag{i}$$

$$\frac{dy}{dt} = -(x^2 + 5)y \tag{ii}$$

where 'β' is parameter. Comment on the behaviour of the system.

Solution

From equation (i), we have at equilibrium point $x' = 0$.

$$\therefore \frac{dx}{dt} = 0 \Rightarrow x^2 - \beta = 0$$

$$x^2 = \beta$$

$$x = \pm\sqrt{\beta}$$

$$x = \sqrt{\beta}, x = -\sqrt{\beta}$$

- If $\beta < 0$, the system has no equilibrium point
- If $\beta = 0$, the system has one equilibrium point at $(0,0)$
- If $\beta > 0$, the system has two equilibrium points $(-\sqrt{\beta},0)$ and $(\sqrt{\beta},0)$.

The Jacobian matrix is given by

$$J = \begin{bmatrix} 2x & 0 \\ -2xy & -(x^2 + 5) \end{bmatrix}$$

At the equilibrium point, $J(0,0) = \begin{bmatrix} 0 & 0 \\ 0 & -5 \end{bmatrix}$

$$J(-\sqrt{\beta},0) = \begin{bmatrix} -2\sqrt{-\beta} & 0 \\ 0 & -\beta - 5 \end{bmatrix}$$

$$J(\sqrt{\beta},0) = \begin{bmatrix} 2\sqrt{\beta} & 0 \\ 0 & -\beta - 5 \end{bmatrix}$$

For $\beta = 0$, there will be a line equilibrium (since one of the eigen values is zero) and for $\beta > 0$, the point $(-\sqrt{\beta},0)$ is a sink and $(\sqrt{\beta},0)$ is a saddle point so that $\beta = 0$ is the bifurcation point for the given differential equation system.

Illustration 10

Analyse the bifurcation property of the given system

$$\dot{x} = \lambda x - y + x(x^2 + y^2)(2 - x^2 - y^2)$$

$$\dot{y} = x + \lambda y + y(x^2 + y^2)(2 - x^2 - y^2)$$

Solution

Given

$$\dot{x} = \lambda x - y + x(x^2 + y^2)(2 - x^2 - y^2)$$

$$\dot{y} = x + \lambda y + y(x^2 + y^2)(2 - x^2 - y^2)$$

Step 1: Deduce the equilibrium point

The equilibrium of the above system can be written as $(\bar{x}(\lambda), \bar{y}(\lambda)) = (0,0)$.

Step 2: Evaluate the Jacobian matrix

$$J(\lambda)_{(0,0)} = \begin{bmatrix} \lambda & -1 \\ 1 & \lambda \end{bmatrix}$$

Step 3: Find the eigen values

The eigen values of this system are $\lambda \pm i$. For values of λ, it can be seen that the eigen values cross the imaginary axis.

Step 4: Find the polar coordinates of the system by substituting $x = r\cos\theta$ and $y = r\sin\theta$

We know that

$$\dot{x} = \lambda x - y + x\left(x^2 + y^2\right)\left(2 - x^2 - y^2\right) \tag{i}$$

The equation (i) can be rewritten as given in equation (ii)

$$\dot{x} = \lambda x - y + x\left(r^2\right)\left(2 - x^2 - y^2\right) \tag{ii}$$

$$\Rightarrow \dot{x} = \lambda x - y + 2xr^2 - xr^4$$

Also,

$$\dot{y} = x + \lambda y + y\left(x^2 + y^2\right)\left(2 - x^2 - y^2\right) \tag{iii}$$

The equation (iii) can be rewritten as follows

$$\dot{y} = x + \lambda y + y\left(r^2\right)\left(2 - x^2 - y^2\right)$$

$$\Rightarrow \dot{y} = x + \lambda y + 2yr^2 - yr^4$$

In polar coordinates, we know that

$$r^2 = x^2 + y^2 \qquad\qquad\text{(iv)}$$

On differentiating the equation (iv), we get

$$2r\dot{r} = 2x\dot{x} + 2y\dot{y}$$

$$\Rightarrow r\dot{r} = x\dot{x} + y\dot{y} \qquad\qquad\text{(v)}$$

Substitute $\dot{x}$ and $\dot{y}$ in the equation (v), we get

$$r\dot{r} = x\left(\lambda x - y + 2xr^2 - xr^4\right) + y\left(x + \lambda y + 2yr^2 - yr^4\right)$$

$$r\dot{r} = \lambda x^2 - xy + 2x^2r^2 - x^2r^4 + yx + \lambda y^2 + 2y^2r^2 - y^2r^4$$

$$r\dot{r} = \lambda x^2 + 2x^2r^2 - x^2r^4 + \lambda y^2 + 2y^2r^2 - y^2r^4$$

$$r\dot{r} = \lambda\left(x^2 + y^2\right) + 2r^2\left(x^2 + y^2\right) - r^4\left(x^2 + y^2\right)$$

$$r\dot{r} = \lambda r^2 + 2r^2\left(r^2\right) - r^4\left(r^2\right)$$

$$r\dot{r} = r^2\left(\lambda + 2r^2 - r^4\right)$$

$$\Rightarrow r\dot{r} = \lambda r + 2r^3 - r^5$$

Step 5: Analyse the bifurcation for the three cases ($\lambda < 0, \lambda = 0, \lambda > 0$)
   Case 1: $\lambda < 0$

$$\therefore \quad \dot{r} = -r + 2r^3 - r^5 \text{ (when } \lambda = -1)$$

r	-3	-2	-1	0	1	2	3
$\dot{r}$	192	18	0	0	0	-18	-192

Case 2: $\lambda = 0$

$$\therefore \quad \dot{r} = 2r^3 - r^5 \text{ (when } \lambda = 0)$$

r	-3	-2	-1	0	1	2	3
$\dot{r}$	189	16	-1	0	1	-16	-189

Case 3: $\lambda > 0$

$$\therefore \quad \dot{r} = r + 2r^3 - r^5 \text{ (when } \lambda = 1)$$

r	-3	-2	-1	0	1	2	3
$\dot{r}$	186	14	-2	0	2	-14	186

The bifurcation diagram for the three cases of $\dot{r} = \lambda r + 2r^3 - r^5$ is as shown in Figure 6.15.

Here, $\dot{\theta} = 1$. This means that the rotation rate is changing at the same rate. From the above two inferences, the phase diagram can be drawn as shown in Figure 6.16.

Thus, the system exhibits a subcritical Hopf bifurcation.

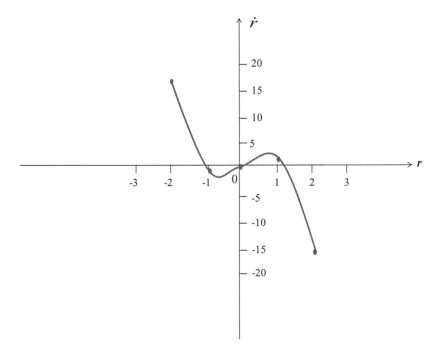

FIGURE 6.15
Bifurcation diagram for $\dot{r} = \lambda r + 2r^2 - r^3$. (a) $\lambda < 0$, (b) $\lambda = 0$, (c) $\lambda > 0$.

(Continued)

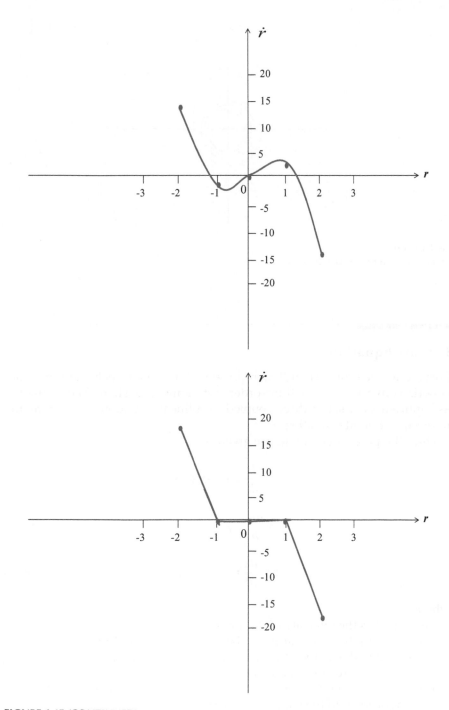

FIGURE 6.15 (CONTINUED)
Bifurcation diagram for $\dot{r} = \lambda r + 2r^2 - r^3$. (a) $\lambda < 0$, (b) $\lambda = 0$, (c) $\lambda > 0$.

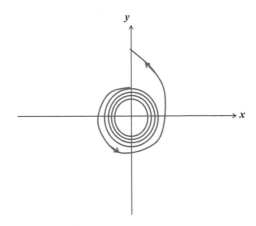

FIGURE 6.16
Phase diagram for $\dot{r} = \lambda r + 2r^2 - r^3$.

Lorentz Equation

Lorentz, a meteorologist, in 1963, represented the chaotic behaviour of atmospheric convections as a third-order autonomous system. The dynamics as captured by a set of three coupled non-linear equations are commonly referred as Lorentz equation.

Thus, the Lorentz equations are given by

$$\dot{x} = \frac{dx}{dt} = \sigma(-x+y) \tag{6.24}$$

$$\dot{y} = \frac{dy}{dt} = \gamma x - y - xz \tag{6.25}$$

$$\dot{z} = \frac{dz}{dt} = -\beta z + xy \tag{6.26}$$

where
  x is related to the intensity of fluid motion.
  y is related to the temperature variation in horizontal direction.
  z is related to the temperature variation in vertical direction.
  σ, γ and β are three real and positive parameters.
  σ and β depend on the material and geometrical properties of fluid layer.
  γ is proportional to the temperature difference, ΔT.

Typical values are: $\sigma = 10$, $\gamma = 28$ and $\beta = 8/3$.

Solutions for Lorentz Equations

From Lorentz's first equation, we have

$$\dot{x} = \sigma(-x + y)$$

$$\sigma(-x + y) = 0$$

$$\sigma = 0$$

(or)

$$(-x + y) = 0 \Rightarrow x = y \qquad \text{(i)}$$

From Lorentz's second equation, we have

$$\dot{y} = \gamma x - y - xz$$

$$\gamma x - y - xz = 0$$

$$\gamma x - x - xz = 0 \quad (\because x = y)$$

$$x(\gamma - 1 - z) = 0$$

$$x = 0 \qquad \text{(ii)}$$

(or)

$$(\gamma - 1 - z) = 0$$

$$z = (\gamma - 1) \qquad \text{(iii)}$$

From Lorentz's third equation, we have

$$\dot{z} = -\beta z + xy$$

$$-\beta z + xy = 0$$

$$-\beta z + x \cdot x = 0 \quad (\because x = y)$$

$$-\beta z + x^2 = 0 \qquad \text{(iv)}$$

We have from equation (i)

$$x = y$$

From equation (ii), we have

$$x = 0 \tag{v}$$

$\therefore$ From equations (i) and (ii), we get

$$y = 0 \tag{vi}$$

Also, in equation (iv), when we substitute $x = 0$, we get

$$-\beta z = 0$$

$$z = 0 \tag{vii}$$

Substituting equation (iii) in equation (iv), we get

$$-\beta z + x^2 = 0$$

$$-\beta(\gamma - 1) + x^2 = 0$$

$$x = \pm\sqrt{\beta(\gamma - 1)} \tag{viii}$$

Thus, from equations (ii, vi and vii), for $\gamma < 1$, the critical point P_1 can be given as P_1 (0,0,0).

Similarly, from equations (iii and viii), for $\gamma > 1$, the critical point P_2 and P_3 can be given as follows:

$$P_2 \left(\sqrt{\beta(\gamma - 1)}, \ \sqrt{\beta(\gamma - 1)}, \ (\gamma - 1) \right)$$

and

$$P_3 \left(-\sqrt{\beta(\gamma - 1)}, \ -\sqrt{\beta(\gamma - 1)}, \ (\gamma - 1) \right)$$

Stability Analysis of Lorentz Equations

The stability analysis of Lorentz equations can be performed for two cases, namely, critical point P_1 and for critical point P_2.

Case 1: For Critical Point, P_1 (0,0,0)

We have the Lorentz set of equations given by

$$\dot{x} = \frac{dx}{dt} = \sigma(-x + y)$$

$$\dot{y} = \frac{dy}{dt} = \gamma x - y - xz$$

$$\dot{z} = \frac{dz}{dt} = -\beta z + xy$$

The Jacobian matrix of the above set of Lorentz equations at critical point $P_1(0,0,0)$ can be obtained as follows:

$$J_{P_1} = \begin{bmatrix} \left\{\dfrac{\partial}{\partial x}(-\sigma x + \sigma y)\right\} & \left\{\dfrac{\partial}{\partial y}(-\sigma x + \sigma y)\right\} & \left\{\dfrac{\partial}{\partial z}(-\sigma x + \sigma y)\right\} \\[2ex] \left\{\dfrac{\partial}{\partial x}(\gamma x - y - xz)\right\} & \left\{\dfrac{\partial}{\partial y}(\gamma x - y - xz)\right\} & \left\{\dfrac{\partial}{\partial z}(\gamma x - y - xz)\right\} \\[2ex] \left\{\dfrac{\partial}{\partial x}(-\beta z + xy)\right\} & \left\{\dfrac{\partial}{\partial y}(-\beta z + xy)\right\} & \left\{\dfrac{\partial}{\partial z}(-\beta z + xy)\right\} \end{bmatrix}$$

After simplification, we get

$$J_{P_1} = \begin{bmatrix} -\sigma & \sigma & 0 \\ \gamma & -1 & 0 \\ 0 & 0 & -\beta \end{bmatrix}$$

The characteristic polynomial is given by

$$\left|\lambda I - J_{P_1}\right| = 0$$

$$\begin{vmatrix} (\lambda + \sigma) & -\sigma & 0 \\ -\gamma & (\lambda + 1) & 0 \\ 0 & 0 & (\lambda + \beta) \end{vmatrix} = 0$$

$$(\lambda + \sigma)\left[(\lambda + 1)(\lambda + \beta)\right] - (-\sigma)\left[(-\gamma)(\lambda + \beta)\right] = 0$$

After simplification, we get

$$\lambda^3 + \lambda^2(\sigma + 1 + \beta) + \lambda\left[\sigma(1 - \gamma) + \beta(\sigma + 1)\right] + \beta\sigma(1 - \gamma) = 0 \qquad (6.27)$$

Thus, the generalized form of the above characteristic polynomial is given by

$$f(\lambda) = \lambda^n + a_1\lambda^{n-1} + a_2\lambda^{n-2} + a_3\lambda^{n-3} + \cdots + a_{n-1}\lambda + a_n \qquad (6.28)$$

The Hurwitz matrix of the above characteristic polynomial is given by

$$H = \begin{bmatrix} a_1 & 1 & 0 & 0 & 0 & 0 & . & . & . & 0 \\ a_3 & a_2 & a_1 & 1 & 0 & 0 & . & . & . & 0 \\ a_5 & a_4 & a_3 & a_2 & a_1 & 1 & 0 & . & . & 0 \\ \vdots & \vdots & \vdots & \vdots & \vdots & \vdots & \vdots & \vdots & \vdots & \vdots \\ 0 & 0 & 0 & 0 & 0 & 0 & 0 & 0 & 0 & a_n \end{bmatrix} \qquad (6.29)$$

For the system to be stable, the following conditions have to be satisfied.

$$\left. \begin{aligned} \Delta_1 &= |a_1| > 0 \\[2mm] \Delta_2 &= \begin{vmatrix} a_1 & 1 \\ a_3 & a_2 \end{vmatrix} > 0 \\[2mm] \Delta_3 &= \begin{vmatrix} a_1 & 1 & 0 \\ a_3 & a_2 & a_1 \\ a_5 & a_4 & a_3 \end{vmatrix} > 0 \\[2mm] &\vdots \\ \Delta_n &> 0 \end{aligned} \right\} \qquad (6.30)$$

Illustration 11

Investigate the stability of the following Lorentz differential system, at the critical point, P_1 (0,0,0).

$$\dot{x} = -10x + 10y$$

$$\dot{y} = 0.5x - y - xz$$

$$\dot{z} = -\frac{8}{3}z + xy$$

Solution

The characteristic polynomial at critical point P_1 (0,0,0) is given by

$$\lambda^3 + \lambda^2(\sigma + 1 + \beta) + \lambda[\sigma(1-\gamma) + \beta(\sigma+1)] + \beta\sigma(1-\gamma) = 0$$

From the given Lorentz differential systems, we have

$$\sigma = 10, \ \gamma = 0.5, \ \beta = 8/3$$

Substituting the above values in the characteristic polynomial, we get

$$\lambda^3 + \lambda^2 \left(10 + 1 + \frac{8}{3}\right) + \lambda \left[10(1 - 0.5) + \frac{8}{3}(10 + 1)\right] + \frac{8}{3} \times 10(1 - 0.5) = 0$$

$$\lambda^3 + \lambda^2 \left(\frac{41}{3}\right) + \lambda \left[\frac{103}{3}\right] + \frac{40}{3} = 0 \qquad \text{(i)}$$

The generalized form of characteristic polynomial is given by

$$\lambda^3 + a_1 \lambda^2 + a_2 \lambda + a_3 = 0 \qquad \text{(ii)}$$

Comparing equation (i) with the generalized form given in equation (ii), we get

$$a_1 = \frac{41}{3}; \ a_2 = \frac{103}{3}; \ a_3 = \frac{40}{3}$$

Thus, the Hurwitz matrix of equation (i) can be formed as follows

$$H = \begin{bmatrix} a_1 & 1 & 0 \\ a_3 & a_2 & a_1 \\ 0 & 0 & a_3 \end{bmatrix} = \begin{bmatrix} \dfrac{41}{3} & 1 & 0 \\ \dfrac{40}{3} & \dfrac{103}{3} & \dfrac{41}{3} \\ 0 & 0 & \dfrac{40}{3} \end{bmatrix}$$

For the Lorentz system to be stable, the following conditions are to be satisfied.

i. $\Delta_1 = |a_1| = \left|\dfrac{41}{3}\right| > 0$

ii. $\Delta_2 = \begin{vmatrix} a_1 & 1 \\ a_3 & a_2 \end{vmatrix} = \begin{vmatrix} \dfrac{41}{3} & 1 \\ \dfrac{40}{3} & \dfrac{103}{3} \end{vmatrix}$

$$= \left| \left[\left(\frac{41}{3}\right)\left(\frac{103}{3}\right)\right] - \left(\frac{40}{3}\right) \right|$$

$$= 458.88 > 0$$

$$\text{iii. } \Delta_3 = \begin{vmatrix} a_1 & 1 & 0 \\ a_3 & a_2 & a_1 \\ a_5 & a_4 & a_3 \end{vmatrix} = \begin{vmatrix} \dfrac{41}{3} & 1 & 0 \\ \dfrac{40}{3} & \dfrac{103}{3} & \dfrac{41}{3} \\ 0 & 0 & \dfrac{40}{3} \end{vmatrix}$$

$$= \left| \left(\frac{40}{3} \right) \left[\left(\frac{41}{3} \times \frac{103}{3} \right) - \left(\frac{40}{3} \right) \right] \right|$$

$$= 6078.5 > 0$$

$\therefore$ The given Lorentz differential system is stable.

Case 2: For Critical Point, $P_2 \left(\sqrt{\beta(\gamma-1)}, \sqrt{\beta(\gamma-1)}, (\gamma-1) \right)$

We have the Lorentz set of equations given by

$$\dot{x} = \frac{dx}{dt} = \sigma(-x+y)$$

$$\dot{y} = \frac{dy}{dt} = \gamma x - y - xz$$

$$\dot{z} = \frac{dz}{dt} = -\beta z + xy$$

The Jacobian matrix of the above set of Lorentz equations at critical point, $P_2 \left(\sqrt{\beta(\gamma-1)}, \sqrt{\beta(\gamma-1)}, (\gamma-1) \right)$ can be obtained as follows.

$$J_{P_2} = \begin{bmatrix} \left\{ \dfrac{\partial}{\partial x}(-\sigma x + \sigma y) \right\} & \left\{ \dfrac{\partial}{\partial y}(-\sigma x + \sigma y) \right\} & \left\{ \dfrac{\partial}{\partial z}(-\sigma x + \sigma y) \right\} \\ \left\{ \dfrac{\partial}{\partial x}(\gamma x - y - xz) \right\} & \left\{ \dfrac{\partial}{\partial y}(\gamma x - y - xz) \right\} & \left\{ \dfrac{\partial}{\partial z}(\gamma x - y - xz) \right\} \\ \left\{ \dfrac{\partial}{\partial x}(-\beta z + xy) \right\} & \left\{ \dfrac{\partial}{\partial y}(-\beta z + xy) \right\} & \left\{ \dfrac{\partial}{\partial z}(-\beta z + xy) \right\} \end{bmatrix}$$

After simplification, we get

$$J_{P_2} = \begin{bmatrix} -\sigma & \sigma & 0 \\ 1 & -1 & -\sqrt{\beta(\gamma-1)} \\ \sqrt{\beta(\gamma-1)} & \sqrt{\beta(\gamma-1)} & -\beta \end{bmatrix}$$

The characteristic polynomial is given by

$$\left| \lambda I - J_{P_2} \right| = 0$$

$$\begin{vmatrix} (\lambda + \sigma) & \sigma & 0 \\ -1 & (\lambda + 1) & \sqrt{\beta(\gamma - 1)} \\ -\sqrt{\beta(\gamma - 1)} & -\sqrt{\beta(\gamma - 1)} & (\lambda + \beta) \end{vmatrix} = 0$$

$$(\lambda + \sigma)\left\{(\lambda + 1)(\lambda + \beta) - \left(-\sqrt{\beta(\gamma - 1)}\right)\left(-\sqrt{\beta(\gamma - 1)}\right)\right\}$$

$$- \sigma\left\{(-1)(\lambda + \beta) - \left(-\sqrt{\beta(\gamma - 1)}\right)\left(-\sqrt{\beta(\gamma - 1)}\right)\right\} = 0$$

After simplification, we get

$$\lambda^3 + \lambda^2(\sigma + 1 + \beta) + \lambda\left(\beta(\sigma + \gamma)\right) + 2\beta\sigma(\gamma - 1) = 0$$

The generalized form of the above characteristic polynomial is given by

$$f(\lambda) = \lambda^n + a_1\lambda^{n-1} + a_2\lambda^{n-2} + a_3\lambda^{n-3} + \cdots + a_{n-1}\lambda + a_n$$

The procedure for formulation of Hurwitz matrix and the conditions to be satisfied for the Lorentz system to be stable remain same as that described earlier for the critical point P_1 (0,0,0).

Illustration 12

Investigate the stability of the following Lorentz differential system, at the critical point, $P_2\left(\sqrt{\beta(\gamma - 1)}, \sqrt{\beta(\gamma - 1)}, (\gamma - 1)\right)$.

$$\dot{x} = -10x + 10y$$

$$\dot{y} = 28x - y - xz$$

$$\dot{z} = xy - \frac{8}{3}z$$

Solution

The characteristic polynomial at critical point $P_2\left(\sqrt{\beta(\gamma - 1)}, \sqrt{\beta(\gamma - 1)}, (\gamma - 1)\right)$ is given by

$$\lambda^3 + \lambda^2(\sigma + 1 + \beta) + \lambda[\sigma(1 - \gamma) + \beta(\sigma + 1)] + \beta\sigma(1 - \gamma) = 0$$

From the given Lorentz differential systems, we have

$$\sigma = 10, \gamma = 28, \beta = 8/3$$

Substituting the above values in the characteristic polynomial, we get

$$\lambda^3 + \lambda^2\left(10+1+\frac{8}{3}\right) + \lambda\left[\frac{8}{3}(10+28)\right] + 2\times\frac{8}{3}\times10(28-1) = 0$$

After simplification, we get

$$\lambda^3 + \lambda^2\left(\frac{41}{3}\right) + \lambda\left[\frac{304}{3}\right] + 1440 = 0 \tag{i}$$

The generalized form of characteristic polynomial is given by

$$\lambda^3 + a_1\lambda^2 + a_2\lambda + a_3 = 0 \tag{ii}$$

Comparing equation (i) with the generalized form given in equation (ii), we get

$$a_1 = \frac{41}{3}; a_2 = \frac{304}{3}; a_3 = 1440$$

$\therefore$ Hurwitz matrix is given by

$$H = \begin{bmatrix} a_1 & 1 & 0 \\ a_3 & a_2 & a_1 \\ 0 & 0 & a_3 \end{bmatrix} = \begin{bmatrix} \dfrac{41}{3} & 1 & 0 \\ 1440 & \dfrac{304}{3} & \dfrac{41}{3} \\ 0 & 0 & 1440 \end{bmatrix}$$

For the Lorentz system to be stable, the following conditions are to be satisfied.

i. $\Delta_1 = |a_1| = \left|\dfrac{41}{3}\right| > 0$

ii. $\Delta_2 = \begin{vmatrix} a_1 & 1 \\ a_3 & a_2 \end{vmatrix} = \begin{vmatrix} \dfrac{41}{3} & 1 \\ 1440 & \dfrac{304}{3} \end{vmatrix} = -5511 < 0$

$$
\text{iii. } \Delta_3 = \begin{vmatrix} a_1 & 1 & 0 \\ a_3 & a_2 & a_1 \\ a_5 & a_4 & a_3 \end{vmatrix} = \begin{vmatrix} \dfrac{41}{3} & 1 & 0 \\ 1440 & \dfrac{304}{3} & \dfrac{41}{3} \\ 0 & 0 & 1440 \end{vmatrix} = -79,360 < 0
$$

Thus, the given Lorentz system at critical point P_2 is unstable.

Chaos Theory

Chaos theory is useful in understanding the unpredictable fluctuations in the dynamic behaviour of chemical processes. Henri Poincare, a French mathematician, in 1890, was perhaps the first person to study the unpredictable behaviour of non-linear systems. Much later, in 1963, the meteorologist, Edward Lorentz, carried out research on chaos in the atmosphere to forecast the weather. In the study of chaos theory, the dynamics of non-linear systems are described by a set of non-linear equations which may be differential equations or difference equations.

Once, it was envisaged that the deterministic input applied to a deterministic system will result deterministic output and stochastic input applied to a system can result stochastic output. Also it was believed that, small change in initial condition results only small change in output. Now, it is clear that the deterministic input applied to a deterministic system can produce a stochastic or chaotic output. Also, small change in initial condition may result entirely different output, as time elapses.

Definitions in the Study of Chaos Theory

Trajectory: The solution to a differential equation is called a trajectory.

Flow: The collection of all such solutions of differential equation is called flow.

Map: The collection of solutions of difference equation is called the map.

Dissipative system: It is a system where energy loss takes place. (May be due to friction or damping, etc.)

Hamiltonian system: It is a system where the total mechanical energy is preserved.

Characteristics of Chaos

- Chaos theory investigates the behaviour of dynamical systems which are highly sensitive to initial conditions, an effect referred to as 'butterfly effect'.

- In the chaotic systems, small differences in initial conditions result in widely diverging outcomes. Hence, it is not possible to make long-term prediction.
- Chaos happens in feedback systems, in which past events affect the current events and the current events affect the future events.

Application of Chaos Theory

The chaos theory applications include the following:

- Controlling oscillations in chemical reaction
- Increasing the power of lasers
- Stabilizing the erratic beat of unhealthy animal hearts
- Encoding electronic messages for secure communication
- Synchronizing the output of electronic circuits.

Chaos in Chemical Process

The chaotic behaviour is observed in most of the chemical processes as they are inherent by nonlinear and exhibit structurally unstable dynamics. More specifically, the occurrence of chaos is seen in exothermic chemical reactor, fluidized bed catalytic reactor and continuous stirred tank reactor (CSTR), as they deal with highly non-linear interactions, fluid flow and heat and mass transfer.

Chaos in PI-Controlled CSTR

Let us consider a Proportional–Integral (PI)-controlled CSTR where an isothermal reaction takes place, as shown in Figure 6.17. The control objective is to maintain the output concentration (c) at the desired value.

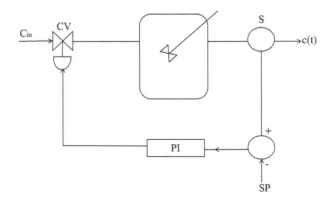

FIGURE 6.17
PI-controlled CSTR.

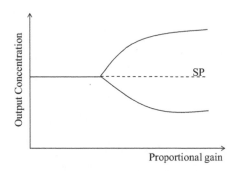

FIGURE 6.18
Variation of output concentration with proportional gain.

The sensor 'S' sends the discrete sampled values of the outlet concentration to the PI controller. The Control Valve (CV) receives the difference signal (difference between set point and sampled concentration) from the controller.

Now, by keeping the integral action in null position, the proportional gain is increased. A plot of the variation of concentration with changes in proportional gain is drawn as shown in Figure 6.18 which depicts chaotic behaviour of PI-controlled CSTR.

7

Optimal Control

This chapter provides an insight into the optimal control branch of engineering. The need for optimal control in modern control theory is explained by comparing optimal control with classical control theory. The basic definitions including: Function, maximum and minima of a function, functional, increment and variation of functional and the performance indices related to optimal problems are initially explained. This chapter covers the topics related to time-varying cases. Both finite time and infinite time problems are discussed. The design of Linear Quadratic Regulator (LQR) using Hamiltonian–Jacobi equation, Pontryagin's principle, Linear Quadratic Tracking (LQT) and Linear Quadratic Gaussian (LQG) are also presented. The matrix and Algebraic Riccati Equations (AREs) are derived and their numerical solutions are put forth. Application examples to illustrate finite time problem, infinite time problem and LQR design are meticulously solved.

Introduction

Optimal control is an important branch of control system, which mainly deals with extensive calculus of variations applied on systems to achieve desired performance. Optimal control can be applied to linear and non-linear systems by imposing proper constraints on them. In optimal control, a control law is formulated to achieve a certain optimal criterion (maximum or minimum) by satisfying some physical constraints.

If a plant is under steady-state conditions, it can be represented by algebraic equations and the optimal control can be designed using calculus, Lagrange multipliers, linear and non-linear programming. On the other hand, if the system variables are changing (dynamic state), it can be represented by differential equations, and the optimal control can be designed using dynamic programming and variational calculus.

Classical Control versus Optimal Control

The need for optimal control can be justified by comparing the salient features of optimal control with classical control scheme.

Classical Control

- It mostly involves the control of linear, time-invariant, Single-Input Single-Output (SISO) systems.
- Classical control approach is a frequency domain approach. Laplace transform is used here.
- The control and its solution are robust.
- Computationally easy.

Optimal Control

- Optimal control mostly involves the control of nonlinear, time-varying, Multi-Input Multi-Output (MIMO) system.
- Control is applied on a constrained (both on states and the control inputs) environment.
- It is a time domain approach. State space representation is used here.
- It always involves an optimization procedure.
- The control and its solution are not robust. Since an optimal control is applied to non-linear systems, the solution to the system is either tedious or does not exist.
- Computationally tedious.

Objective

The key objective of an optimal control is to develop a control signal, which allows a plant (process) to achieve a proposed performance criterion by means of satisfying some physical constraints.

The procedural steps of optimal control problem are listed below:

1. Formulation of mathematical model of the process which has to be controlled.
2. Identifying a suitable performance index.
3. Specifying the constraints and boundary conditions on the states/ control.

Case 1: Linear System

Consider the linear plant model

$$\dot{x} = Ax(t) + Bu(t) \tag{7.1}$$

where A and B are $n \times n$ state and $n \times p$ input real matrices, $x(t)$ is $n \times 1$ state vector and u is $p \times 1$ input vector.

The optimal control problem is to find the optimal control law $u^*(t)$ to give the trajectory $x^*(t)$ that optimizes the performance index (J).

$$J = x^T(t_f)Hx(t_f) + \int_{t_i}^{t_f} (x^T(t)Qx(t) + u^T(t)Ru(t))dt \tag{7.2}$$

where H and Q are real symmetric positive semi-definite $n \times n$ matrices and R is a real symmetric positive definite $p \times p$ matrix. Here, t_i represents initial time and t_f represents the final time.

Case 2: Non-Linear System

Consider the non-linear plant model

$$\dot{x}(t) = f(x(t), u(t), t) \tag{7.3}$$

The optimal control problem is to find the optimal control law $u^*(t)$ to give the trajectory $x^*(t)$ that optimizes the performance index

$$J = S\left(x(t_f), t_f\right) + \int_{t_i}^{t_f} V(x(t), u(t), t)dt \tag{7.4}$$

where S is real symmetric positive semi-definite $n \times n$ matrix and V is a real symmetric positive definite $p \times p$ matrix.

Let us first introduce few basic concepts that help us in understanding optimal control in a better way.

Function

Single Variable

A variable x is a function of a variable quantity t, i.e., $x(t) = f(t)$ supposed to be continuous for all values of this variable. In other words, for every value of t over a certain range of t, there corresponds a value x. Here, t may be any independent variable and it need not always be time.

Two Independent Variables

Let $f(x,y)$ be any function of two independent variables x and y supposed to be continuous for all values of these variables in the neighbourhood of their values a and b, respectively.

Maxima and Minima for Functions of Two Variables

Maximum Value

A function $f(a,b)$ is said to be maximum value $f(x,y)$, if there exists some neighbourhood of the point (a,b) such that for every point $(a+h, b+k)$ of the neighbourhood is less than $f(a,b)$.

$$\text{i.e., } f(a,b) > f(a+h, b+k)$$

Minimum Value

A function $f(a,b)$ is said to be minimum value of $f(x,y)$, if there exists some neighbourhood of the point (a,b) such that for every point $(a+h, b+k)$ of the neighbourhood is greater than $f(a,b)$.

$$\text{i.e., } f(a,b) < f(a+h, b+k)$$

Extreme Values

A function $f(a,b)$ is said to be an extremum value of $f(x,y)$, if it is either a maximum or a minimum. In other words, the maximum and minimum values of a function are together called extreme values or turing values. The points at which they are attained are called points of maxima and minima. The points at which a function has extreme values are called turning points.

Necessary Conditions

The necessary conditions that $f(x,y)$ should be maximum or minimum at $x = a, y = b$ are that

$$\left.\frac{\partial f}{\partial x}\right|_{\substack{x=a\\y=b}} = 0 \text{ or } f_x(a,b) = 0$$

and

$$\left.\frac{\partial f}{\partial y}\right|_{\substack{x=a \\ y=b}} = 0 \text{ or } f_y(a,b) = 0$$

NOTE

$$\frac{\partial f}{\partial x} = f_x, \frac{\partial f}{\partial y} = f_y, \frac{\partial^2 f}{\partial x^2} = f_{xx}, \frac{\partial^2 f}{\partial y^2} = f_{yy}, \frac{\partial f}{\partial x \partial y} = f_{xy}$$

Sufficient Conditions

  i. Let $f_x(a,b) = 0$, $f_y(a,b) = 0$, $f_{xx}(a,b) = r$, $f_{yy}(a,b) = t$ and $f_{xy}(a,b) = s$, then $f(x,y)$ will have a maximum or a minimum value at $x = a$, $y = b$ if $rt > s^2$. Further, $f(x,y)$ is maximum or minimum accordingly as r is negative or positive.

 ii. If $f(x,y)$ will have neither a maximum or a minimum value at $x = a$, $y = b$ if $rt < s^2$, i.e., $x = a$, $y = b$ is a saddle point.

iii. If $rt = s^2$, this case is doubtful case and further advanced investigation is needed to determine whether $f(x,y)$ is a maximum or minimum at $x = a$, $y =$ or not. This case is very rarely used.

Properties of Relative Maxima and Minima

• Maximum and minimum values must occur alternatively.

• There may be several maximum or minimum values of same function.

• Atleast one maximum or one minimum must be between two equal values of a function.

• A function $y = f(x)$ is maximum at $x = a$, if $\frac{dy}{dx}$ changes sign from positive to negative as x passes through a.

• A function $y = f(x)$ is minimum at $x = a$, if $\frac{dy}{dx}$ changes sign from negative to positive as x passes through a.

• If the sign of $\frac{dy}{dx}$ does not change while x passes through a, then y is neither maximum nor minimum at $x = a$.

Definition of Stationary Values

Single Variable

A function $f(x)$ is said to be stationary at $x = a$, if $f'(a) = 0$. Thus, for a function $f(x)$ to be a maximum or minimum at $x = a$, it must be stationary at $x = a$.

Two Variables

A function $f(x,y)$ is said to be stationary at (a,b) or $f(a,b)$ and is said to be a stationary value of $f(x,y)$, if $f_x(a,b) = 0$ and $f_y(a,b) = 0$.

NOTE

Every extremum value is a stationary value; however, a stationary value need not be an extremum value.

Conditions for Maximum or Minimum Values

The necessary condition for $f(x)$ to be a maximum or minimum at $x = a$ is that, $f'(a) = 0$.

Sufficient Conditions of Maximum and Minimum Values

The sufficient condition for $f(x)$ to be a maximum at $x = a$ is that, $f''(a)$ should be negative. On the other hand, the sufficient condition for $f(x)$ to be a minimum at $x = a$ is that, $f''(a)$ should be positive.

Procedural Steps for Solving Maxima and Minima of Function f(x)

  i. Find $f'(x)$ and equate it to zero.
  ii. Solve the resulting equation of x. Let its roots be $a_1, a_2 \ldots a_n$. Then $f(x)$ is stationary at $x = a_1, a_2 \ldots$ Thus, $x = a_1, a_2 \ldots a_n$ are the only points at which $f(x)$ can be maximum or minimum.
  iii. Find $f''(x)$ and substitute the values of $x = a_1, a_2 \ldots a_n$ when $f''(x)$ is negative, then $f(x)$ will be maximum, and when $f''(x)$ is positive, then $f(x)$ will be minimum.

Illustration 1

A function is given as $f(x,y) = 4x^2 + 6y^2 - 8x - 4y + 8$. Find the optimal value of $f(x,y)$.

Solution

Given that

$$f(x,y) = 4x^2 + 6y^2 - 8x - 4y + 8$$

$$\frac{\partial f}{\partial x} = 8x - 8$$

$$\frac{\partial f}{\partial y} = 12y - 4$$

Substitute $\frac{\partial f}{\partial x} = 0$ and $\frac{\partial f}{\partial y} = 0$

$$8x - 8 = 0$$

$$8x = 8$$

$$x = 1$$

$$12y - 4 = 0$$

$$y = \frac{4}{12}$$

$$y = \frac{1}{3}$$

Therefore, $\left(1, \frac{1}{3}\right)$ is the only stationary point.

$$r = \left[\frac{\partial^2 f}{\partial x^2}\right]_{x=1} = 8$$

$$s = \left[\frac{\partial^2 f}{\partial x \partial y}\right]_{\substack{x=1 \\ y=1/3}} = 0$$

$$t = \left[\frac{\partial^2 f}{\partial y^2}\right]_{y=1/3} = 12$$

$$rt = 8 \times 12 = 96$$

$$s^2 = 0$$

Since $rt > s^2$, we have either a maxima or minima at $\left(1, \dfrac{1}{3}\right)$. Also since $r = 8 > 0$,

the point $\left(1, \dfrac{1}{3}\right)$ is a point of minima.

The minimum value is

$$f\left(1, \frac{1}{3}\right) = 4(1)^2 + 6\left(\frac{1}{3}\right)^2 - 8(1) - 4\left(\frac{1}{3}\right) + 8$$

$$= \left(\frac{10}{3}\right)$$

Therefore, the optimal value of $f(x,y)$ is minimum and is equal to $\dfrac{10}{3}$.

Functional

A functional J is a variable quantity which is dependent on a function $f(x)$ and it can be represented as

$$J = J(f(x))$$

For each function $f(x)$, there corresponds a value J.

Functional depends on several functions, i.e., a function of a function.

$$J(f(x)) = \int_{x_1}^{x_2} f(x)\,dx$$

Illustration 2

Consider $x(t) = 2t^2 + 1$. Then

$$J(x(t)) = \int_0^1 x(t)\,dt$$

$$= \int_0^1 (2t^2 + 1)\,dt = \frac{2}{3} + 1$$

$$J(x(t)) = \frac{5}{3} \text{ is the area under the curve } x(t).$$

If $v(t)$ is the velocity of a vehicle, then

$$J(v(t)) = \int_{t_0}^{t_f} v(t)\,dt \text{ is the path traversed by the vehicle.}$$

Thus, here $x(t)$ and $v(t)$ are functions of t and J is a functional of $x(t)$ or $v(t)$.

Increment of a Function

The increment of a function f is represented as Δf, and it is defined by the following equation (7.5):

$$\Delta f \triangleq f(t + \Delta t) - f(t) \tag{7.5}$$

From equation (7.5), it is clear that Δf depends on both the independent variable t and increment of the independent variable Δt.

$\therefore$ The increment of a function can be represented as

$$\Delta f(t, \Delta t) \tag{7.6}$$

Illustration 3

If $f(t) = t^2 - t + 1$, find the increment of the function $f(t)$.

Solution

$$\begin{aligned} \Delta f &\triangleq f(t + \Delta t) - f(t) \\ &= \left[(t + \Delta t)^2 - (t + \Delta t) + 1\right] - (t^2 - t + 1) \\ &= t^2 + \Delta t^2 + 2t\Delta t - t - \Delta t + 1 - t^2 + t - 1 \\ \Delta f &= \Delta t^2 + 2t\Delta t - \Delta t \end{aligned}$$

Increment of a Functional

The increment of a functional J is represented as ΔJ, and it is defined by equation (7.7).

$$\Delta J \triangleq J(x(t) + \delta x(t)) - J(x(t)) \tag{7.7}$$

where $\delta x(t)$ is the variation of the function $x(t)$.

From equation (7.7), it is clear that the increment of a functional is mainly dependent upon the function $x(t)$ and its variation $\delta x(t)$.

∴The increment of a functional can also be represented as

$$\text{Increment of } J = \Delta J(x(t), \delta x(t)) \tag{7.8}$$

Illustration 4

Find the increment of the functional

$$J = \int_{t_0}^{t_f} [5x^2(t) + 4] dt$$

Solution

The increment of J is given by

$$\Delta J \triangleq J(x(t) + \delta x(t)) - J(x(t))$$

$$\Delta J = \int_{t_0}^{t_f} \left[5(x(t) + \delta x(t))^2 + 4 \right] dt - \int_{t_0}^{t_f} \left[5x^2(t) + 4 \right] dt$$

$$= \int_{t_0}^{t_f} \left[5\left(x^2(t) + 2x(t)\delta x(t) + (\delta x(t))^2\right) + 4 \right] dt - \int_{t_0}^{t_f} \left[5x^2(t) + 4 \right] dt$$

$$= \int_{t_0}^{t_f} \left[5x^2(t) + 10x(t)\delta x(t) + 5(\delta x(t))^2 + 4 \right] dt - \int_{t_0}^{t_f} \left[5x^2(t) + 4 \right] dt$$

$$\Delta J = \int_{t_0}^{t_f} \left[10x(t)\delta x(t) + 5(\delta x(t))^2 \right] dt$$

Differential of a Functional

The increment of the function f at a point t^* can be represented as

$$\Delta f \triangleq f(t^* + \Delta t) - f(t^*) \tag{7.9}$$

Using Taylor series, expanding $f(t* + \Delta t)$ about $t*$, we get

$$\Delta f = f(t^*) + \left(\frac{df}{dt}\right)_* \Delta t + \frac{1}{2!}\left(\frac{d^2 f}{dt^2}\right)_* (\Delta t)^2 + \ldots - f(t^*) \tag{7.10}$$

By neglecting the higher-order terms in Δt, we get

$$\Delta f = \left(\frac{df}{dt}\right)_* \Delta t = \dot{f}(t^*)\Delta t = df \tag{7.11}$$

where
df is called the differential of f at the point $t*$.
$\dot{f}(t^*)$ is the derivative or slope of f at $t*$.

The differential df is the first-order approximation to increment Δt.

Illustration 5

Let $f(t) = t^2 - t + 1$. Find the increment and derivative of the function $f(t)$.

Solution

Given that $f(t) = t^2 - t + 1$
    We know that the increment of $f(t)$ is given by

$$\Delta f = f(t^*) + \left(\frac{df}{dt}\right)_* \Delta t + \frac{1}{2!}\left(\frac{d^2 f}{dt^2}\right)_* (\Delta t)^2 + \cdots - f(t^*) \tag{i}$$

After neglecting the higher-order terms, we can rewrite the above equation as follows:

$$\Delta f = \dot{f}(t)\Delta t \tag{ii}$$

where $\dot{f}(t)$ is the derivative.

$$\frac{\partial f}{\partial t} = 2t - 1$$

$$\frac{\partial^2 f}{\partial t^2} = 2$$

$$\therefore \quad \Delta f = (2t - 1)\Delta t + \frac{1}{2!}2(\Delta t)^2$$

$$\Delta f = (2t - 1)\Delta t \quad \text{(neglecting higher order terms)}$$

$$\dot{f}(t) = (2t - 1)$$

Variation of a Functional

Small changes in the functional's value due to the small changes in the function that is its argument are known as variations of a functional.

The first variation is defined as the linear part of the change in the functional.

The second variation is defined as the quadratic part of the change in the functional.

Consider the increment of a functional

$$\Delta J \triangleq J(x(t) - \delta x(t)) - J(x(t))$$

Using Taylor series, expanding $J(x(t) - \delta x(t))$, we get

$$\Delta J = J(x(t)) + \frac{\partial J}{\partial x}\delta x(t) + \frac{1}{2!}\frac{\partial^2 J}{\partial x^2}\left(\delta x(t)^2 + \cdots - J(x(t))\right)$$

$$= \frac{\partial J}{\partial x}\delta x(t) + \frac{1}{2!}\frac{\partial^2 J}{\partial x^2}\left(\delta x(t)^2 + \cdots\right)$$

$$= \delta J + \delta^2 J + \cdots$$

where

$$\partial J = \frac{\partial J}{\partial x}\delta x(t) \text{ is the first variation of the functional, } J \qquad (7.12a)$$

and

$$\partial^2 J = \frac{1}{2!}\frac{\partial^2 J}{\partial x^2}\left(\delta x(t)\right)^2 \text{ is the second variation of the functional, } J \quad (7.12b)$$

The second variation $\partial^2 J(x(t))$ is said to be strongly positive if $\partial^2 J(x(t)) \geq k\|x(t)\|^2$ for all $x(t)$ and for some constant $k > 0$.

Using the above definitions, for first variation, second variation, and strongly positive, the sufficient condition for a minimum of a functional can be stated as follows.

Sufficient Condition for a Minimum

The functional $J(x(t))$ has a minimum of $x(t) = \hat{x}(t)$ if its first variation $\delta J = 0$ at $x(t) = \hat{x}(t)$ and its second variation $\delta^2 J$ is strongly positive at $x(t) = \hat{x}(t)$.

Illustration 6

Given the functional

$$J(x(t)) = \int_{t_0}^{t_f} [7x^2(t) + 2x(t) + 1]dt$$

Evaluate the first and second variations of the functional.

Solution

We know that the increment of a function is given by

$$\Delta J \triangleq J(x(t) + \delta x(t)) - J(x(t))$$

After expanding it, using Taylor series, we get

$$\Delta J = J(x(t) + \frac{\partial J}{\partial x}\delta x(t) + \frac{1}{2!}\frac{\partial^2 J}{\partial x^2}(\delta x(t))^2 + \cdots - J(x(t))$$

$$\Delta J = \int_{t_0}^{t_f} [7x^2(t) + 2x(t) + 1]dt + \int_{t_0}^{t_f} [14x(t) + 2]dt\,\delta x(t)$$

$$+ \frac{1}{2!}\int_{t_0}^{t_f} [14\delta^2 x(t)]dt - \int_{t_0}^{t_f} [7x^2(t) + 2x(t) + 1]dt$$

$$= \int_{t_0}^{t_f} [14x(t) + 2]dt\,\delta x(t) + \frac{1}{2!}\int_{t_0}^{t_f} [14\delta^2 x(t)]dt$$

$$= \int_{t_0}^{t_f} [14x(t) + 2]dt\,\delta x(t) + \int_{t_0}^{t_f} [7\delta^2 x(t)]dt$$

By only considering the first-order terms, we get the first variation as

$$\delta J(x(t), \delta x(t)) = \int_{t_0}^{t_f} [14x(t) + 2]\delta x(t)dt$$

By only considering the second-order terms, we get the second variation as

$$\delta^2 J(x(t), \delta x(t)) = \int_{t_0}^{t_f} [7\delta^2 x(t)] dt$$

Performance Measures in Optimal Control

Performance measure can be viewed as a mathematical representation of the objective of a problem in which optimization has to be performed in a particular situation. It is also called as performance index or cost function. Some of the performance indices used in optimal control problem are listed here. The choice of the performance index is to optimize a particular problem lies in the application of the problem and the constraints associated with the problem.

Minimum Time Criterion

The main objective of the minimum time criterion is 'to transfer the system from a given initial state $x(t_i)$ at given time t_i to a final state $x(t_f)$ or to a specified target set, T in minimal time'. Hence, the criterion is a time minimization problem. The cost function associated with the minimum time criteria can be written as in equation (7.13).

$$J = t_f - t_i$$

$$J = \int_{t_i}^{t_f} dt \tag{7.13}$$

where J is the cost function, t_i is the initial time and t_f is the first instance that the state $x(t_i)$ reached the target set T.

Few applications that can employ minimum time criteria are listed below.

- Inverted pendulum control
- Travelling salesman problem, where an object/person tries to reach a particular target in a minimum time by travelling along the least distance path
- Cascade control loop problems
- Drug reaction to a particular disease

Minimum Energy Criterion

The main objective of the minimum energy criterion is 'to transfer the system from a given initial state $x(t_i)$ to a particular point situated in a target set T by utilizing minimal energy'. Hence, the criterion is an energy minimization problem. The cost function associated with the minimum energy criterion can be written as given in equation (7.14)

$$J = \int_{t_i}^{t_f} u^2(t)\,dt \qquad (7.14)$$

where J is the cost function, t_i is the initial time and t_f is the first instance that the state $x(t_i)$ reached the target set T, $u^2(t)$ denotes the measure of instantaneous rate of expenditure of energy.

For a multi-class problem, the cost function of minimum energy criterion can be given as in equation (7.15).

$$J = \int_{t_i}^{t_f} (u^T(t)u(t))\,dt \qquad (7.15)$$

A more generalized way to represent the cost function along with weights to visualize which control variable from the set $u_i(t)$ contributes to the minimum energy consumption condition, a weighing matrix called R is introduced as given in equation (7.16)

$$J = \int_{t_i}^{t_f} (u^T(t)Ru(t))\,dt \qquad (7.16)$$

where R is a positive definite matrix.

Few applications that can employ minimum energy criterion are listed below.

- Missile-launching system
- Wind mill application
- Turbine start-up operation

Minimum Fuel Criterion

The main objective of the minimum fuel criterion is 'to transfer the system from a given initial state $x(t_i)$ to a specified target set T by utilizing minimal fuel'. Hence, the criterion is a fuel consumption minimization problem.

The cost function associated with the minimum fuel criterion can be written as in equation (7.17a)

$$J = \int_{t_i}^{t_f} |u(t)| \, dt \qquad (7.17a)$$

Here, the rate of fuel consumption of an engine is proportional to the thrust developed.

For a multi-jet system, the generalized form of cost function of minimum fuel criterion can be written as given in equation (7.17b)

$$J = \int_{t_i}^{t_f} (K_1 |u_1(t)| + K_2 |u_2(t)| + \cdots) \, dt \qquad (7.17b)$$

Few applications that can employ minimum fuel criterion are listed below

- Rocket fuel consumption problem
- Battery consumption rate in e-vehicles
- Nuclear power plants.

State Regulator Criterion

The main objective of the state regulator criterion is 'to transfer the system from a given initial state $x(t_i)$ to another state say, $x(t_f)$ in a target set T with minimum integral-square error'. The state $x(t_f)$ can also be an equilibrium state. In another words, the aim of a state regulator criterion is to maintain all the components of a state vector $x(t)$ as small as possible. Hence, the problem is again a minimization problem. The cost function associated with the state regulator criterion can be written as given in equation (7.18).

$$J = \int_{t_i}^{t_f} \left[\sum_{i=1}^{n} (x_t(t))^2 \right] dt \qquad (7.18)$$

$$J = \int_{t_i}^{t_f} (x^T(t)x(t)) \, dt \qquad (7.19)$$

A more generalized way to represent the cost function along with weights is to add a weighing matrix called Q as given in equation (7.20).

$$J = \int_{t_i}^{t_f} (x^T(t)Qx(t)) \, dt \qquad (7.20)$$

where Q is a positive semi-definite and symmetric matrix.

When any changes in the final state is to be considered, then the cost function assumes a terminal constraint as given in equation (7.21).

$$J = x^T(t_f)Hx(t_f) + \int_{t_i}^{t_f} (x^T(t)Qx(t))\,dt \qquad (7.21)$$

where H is a positive semi-definite, real, symmetric, constant matrix. Also considering the input constraints, the cost function can be modified as in equation (7.22)

$$J = \frac{1}{2}x^T(t_f)Hx(t_f) + \frac{1}{2}\int_{t_i}^{t_f} (x^T(t)Qx(t) + u^T(t)Ru(t))\,dt \qquad (7.22)$$

where R is a real, symmetric, positive definite matrix.

When time t tends to infinity, then the cost function ignores the terminal constraint, since the system reaches an equilibrium state. Then, the performance index is as given in equation (7.23)

$$J = \frac{1}{2}\int_{t_i}^{\infty} (x^T(t)Qx(t) + u^T(t)Ru(t))\,dt \qquad (7.23)$$

In other words, if the optimal control aims at maintaining the state $x(t)$ near origin, then the optimal problem is called state regulator problem.

Few applications include the following:

- Boiler drum-level control system
- Motor load disturbance rejection, where a sudden disturbance occurs, the state of the motor must be regulated to the nominal state
- Actuator position control.

Output Regulator Criterion

The main objective of the output regulator criterion is 'to maintain all the components of a state vector $x(t)$ as small as possible'. Hence, the problem is again a minimization problem. The cost function associated with the output regulator criterion can be written as follows:

$$J = \frac{1}{2}y^T(t_f)Hy(t_f) + \frac{1}{2}\int_{t_i}^{t_f} (y^T(t)Qy(t) + u^T(t)Ru(t))\,dt \qquad (7.24)$$

In other words, if the optimal control aims at maintaining the output $y(t)$ near origin, then the optimal problem is called output regulator problem.
    Few applications include

- Flow control through a restriction
- Noise rejection in output.

Servo or Tracking Criterion

The main objective of the servo or tracking criterion is 'to maintain the system $x(t)$ as close as possible to a desired state $r(t)$ in a given time interval $[t_i, t_f]$. The cost function associated with the tracking problem can be written as given in equation (7.25).

$$J = \frac{1}{2} e^T(t_f) He(t_f) + \frac{1}{2} \int_{t_i}^{t_f} (e^T(t) Qe(t) + u^T(t) Ru(t)) dt \qquad (7.25)$$

where $e(t)$ is an error value, i.e., $e(t) = x(t) - r(t)$
    Few applications include

- Satellite tracking control problem
- Line follower robot problem.

Time-Varying Optimal Control

Optimal control extensively deals with non-linear time-varying systems. Here, the control law is formulated within a control interval defined by the initial time and final time. Based on the constraints imposed on the final time, time-varying optimal control problems can be classified as finite horizon problem and infinite horizon problem.

- **Finite horizon:** When an optimal problem is defined by specifying the final time (t_f) as a bounded or finite entity, i.e., $t_f \rightarrow$ (bounded value), then the optimal problem is called a finite horizon optimal problem.
- **Infinite horizon:** When an optimal problem is defined by specifying the final time (t_f) as a free or infinite entity, i.e., $t_f \rightarrow \infty$, then the optimal problem is called an infinite horizon optimal problem.

Continuous Time-Varying Optimal Regulator

The continuous time-varying optimal regulator solves a quadratic cost function J for a given continuous Linear Time Varying (LTV) system represented by the state and output equations as given in equations (7.26) and (7.27), respectively

$$\dot{x} = A(t)x(t) + B(t)u(t) \tag{7.26}$$

$$y(t) = C(t)x(t) \tag{7.27}$$

where A is $n \times n$ state matrix, B is $n \times p$ control matrix, C is $q \times n$ output matrix, $x(t)$ is a $n \times 1$ state vector, $u(t)$ is a $p \times 1$ input vector and $y(t)$ is a $q \times 1$ output vector. For this continuous LTV system, the control interval is given as $[t_i, t_f]$.

If the objective of the optimal control problem is to find an appropriate control law $u^*(t)$ that maintains the state $x(t)$ near the equilibrium or at the origin, then the problem is called as state regulator optimal control problem. The quadratic cost function is given in equation (7.22) and is reproduced here.

$$J = \frac{1}{2}x^T(t_f)Hx(t_f) + \frac{1}{2}\int_{t_i}^{t_f}(x^T(t)Qx(t) + u^T(t)Ru(t))dt$$

where J is the cost function, $x(t)$ is the state vector, H is a real, symmetric, positive semi-definite $n \times n$ matrix, Q is real, symmetric, positive semi-definite $n \times n$ matrix and R is a real, symmetric, positive definite $p \times p$ matrix.

If the objective of the optimal control problem is to find an appropriate law $u^*(t)$ that maintains the output $y(t)$ near the equilibrium or the origin, then the problem is called as output regulator optimal control problem. The quadratic cost function is given in equation (7.24) which is reproduced here.

$$J = \frac{1}{2}y^T(t_f)Hy(t_f) + \frac{1}{2}\int_{t_i}^{t_f}(y^T(t)Qy(t) + u^T(t)Ru(t))dt$$

where J is the cost function, $y(t)$ is the output vector, H is a real symmetric, positive semi-definite $n \times n$ matrix, Q is real, symmetric positive semi-definite $n \times n$ matrix and R is a real, symmetric, positive definite $p \times p$ matrix.

LQR Optimal Regulator Design Using Hamiltonian–Jacobi Equation

Consider the LTV system as given in equations (7.26) and (7.27), respectively, which is reproduced here.

$$\dot{x} = A(t)x(t) + B(t)u(t)$$

$$y(t) = C(t)x(t)$$

Objective

To obtain the value of feedback gain $K(t)$ and thereby formulate the control law.

$$u*(t) = -K(t)x(t) \tag{7.28}$$

Case 1: Finite Horizon

The cost function for this optimal control problem is given by equation (7.22), which is reproduced here.

$$J = \frac{1}{2}x^T(t_f)Hx(t_f) + \frac{1}{2}\int_{t_i}^{t_f}(x^T(t)Qx(t) + u^T(t)Ru(t))dt$$

The auxiliary cost function is a combination of the above cost function along with the co-state vector and is given by equation (7.29).

$$J_\lambda = \lambda^T(t)[A(t)x(t) + B(t)u(t)] \tag{7.29}$$

Assumptions

1. The control input $u(t)$ is an unconstrained input.
2. The final time t_f is specified, i.e., t_f is fixed.
3. The initial condition $x(t_0) = x_0$ is given. However, the final state $x(t_f)$ is not given, i.e., $x(t_f)$ is free.

Assumptions (2) and (3) are called the boundary conditions.

From the Hamiltonian–Jacobi equation, we have

$$H\left(x(t), u(t), \frac{\partial J^{*}}{\partial x}(x(t), t), t\right) = \frac{1}{2} x^{T}(t)Q(t)x(t) + \frac{1}{2} u^{T}(t)R(t)u(t)$$

$$+\left(\frac{\partial J^{*}}{\partial x}(x(t), t)\right)^{T} [A(t)x(t) + B(t)u(t)]$$

For the purpose of simplicity, let us replace $\frac{\partial J^{*}}{\partial x}(x(t), t)$ with $\Delta J^{*}(t)$, we get

$$H(x(t), u(t), \Delta J^{*}(t), t) = \frac{1}{2} x^{T}(t)Q(t)x(t) + \frac{1}{2} u^{T}(t)R(t)u(t)$$

$$+\left(\Delta J^{*}(t)\right)^{T} [A(t)x(t) + B(t)u(t)] \qquad (7.30)$$

To formulate the control equation,

$$\frac{\partial H}{\partial u}(x(t), u(t), \Delta J^{*}(t), t) = 0 \qquad (7.31a)$$

$$R(t)u(t) + B^{T}(t)\Delta J^{*}(t) = 0 \qquad (7.31b)$$

Solving equation (7.31b) for $u(t)$, we get

$$\therefore \quad u^{*}(t) = -R^{-1}(t)B^{T}(t)\Delta J^{*}(t) \qquad (7.32)$$

By substituting equation (7.32) in equation (7.30), we get

$$H(x(t), \Delta J^{*}(t), t) = \frac{1}{2} x^{T}(t)Q(t)x(t) + \frac{1}{2}[\Delta J^{*}(t)]^{T} B(t)R^{-1}(t)R(t)R^{-1}(t)B^{T}(t)\Delta J^{*}(t)$$

$$+[\Delta J^{*}(t)]^{T}\left[A(t)x(t) - \left[B(t)\left(R^{-1}(t)B^{T}(t)\Delta J^{*}(t)\right)\right]\right] \qquad (7.33)$$

$$H(x(t), \Delta J^{*}(t), t) = \frac{1}{2} x^{T}(t)Q(t)x(t) - \frac{1}{2}(\Delta J^{*}(t))^{T} B(t)R^{-1}(t)B^{T}(t)\Delta J^{*}(t)$$

$$+[\Delta J^{*}(t)]^{T} A(t)x(t) \qquad (7.34)$$

Thus, the Hamiltonian–Jacobi equation is given as follows:

$$0 = \Delta J^{*}(t) + \frac{1}{2} x^{T}(t)Q(t)x(t) - \frac{1}{2}(\Delta J^{*}(t))^{T} [B(t)R^{-1}(t)B^{T}(t)\Delta J^{*}(t)]$$

$$+[\Delta J^{*}(t)]^{T} A(t)x(t) \qquad (7.35)$$

From the boundary conditions, we know that

$$J^*(x(t_f), t_f) = \frac{1}{2} x^T(t_f) H x(t_f).$$

Since, the performance index is a time-varying quadratic function of state, it can also be written as follows:

$$J^*(x(t_f), t_f) = \frac{1}{2} x^T(t_f) P(t_f) x(t_f) \quad (\because P(t_f) = H) \tag{7.36a}$$

$$\therefore \quad \frac{\partial H}{\partial X} = P(t)X(t) \tag{7.36b}$$

$$\Delta J^*(t) = \frac{\partial H}{\partial X} = P(t)X(t) \tag{7.37}$$

By substituting equation (7.37) in equation (7.32), we get

$$\therefore \quad u^*(t) = -R^{-1}(t)B^T(t)P(t)x(t)$$

$$u^*(t) = -K(t)x(t) \tag{7.38}$$

where $K(t)$ is the gain.

Thus, from the control law, the gain can be expressed as given in equation (7.39).

$$K(t) = R^{-1}(t)B^T(t)P(t) \tag{7.39}$$

Using the derived Kalman gain and control law, we can solve the objective function J. However, this requires $P(t)$ value.

From equation (7.37), we know that

$$\Delta J^*(t) = P(t)x(t)$$

Let, $\Delta J^*(t) = \lambda(t)$

$$\dot{\lambda}(t) = \dot{P}(t)x(t) + P(t)\dot{x}(t) \tag{7.40}$$

By substituting (7.40) in the co-state equation and simplifying, we get

$$\dot{P}(t)x(t) + P(t)\dot{x}(t) = -Q(t)x(t) - A^T(t)\lambda(t)$$

Substituting $\dot{x}(t) = A(t)x(t) + B(t)u(t)$ and $\lambda(t) = \Delta J^*(t) = P(t)x(t)$, we get

$$\dot{P}(t)x(t) + P(t)[A(t)x(t) + B(t)u(t)] = -Q(t)x(t) - A^T(t)(P(t)x(t))$$

$$\dot{P}(t)x(t) + P(t)A(t)x(t) + P(t)B(t)u(t) = -Q(t)x(t) - A^T(t)P(t)x(t))$$

Now, substituting $u(t) = -R^{-1}(t)B^T(t)P(t)x(t)$ and simplifying, we get

$$\dot{P}(t) = -Q(t) - A^T(t)P(t) - P(t)A(t) + P(t)B(t)R^{-1}B(t)P(t) \qquad (7.41)$$

The above expression given in equation (7.41) is called Matrix Riccati Equation (MRE). In this equation, $Q(t)$, $A(t)$, $B(t)$, $R(t)$ all are known. The value of $P(t)$ is unknown. To obtain the values of $P(t)$ within the control interval, equation (7.41) is integrated backwards.

Now, solve for the optimal state equation

$$\dot{x}^* = [A(t) - B(t)R^{-1}B^T(t)P(t)]x(t) \qquad (7.42)$$

The block diagram for LQR design using Hamilton–Jacobi approach is shown in Figure 7.1.

Now, substitute the values in J and optimize the performance index.

$$J = \frac{1}{2}x^T(t)Px(t) \qquad (7.43)$$

Case 2: Infinite Horizon

For the infinite time linear state regulator, the objective function is chosen as given in equation (7.23) and is reproduced here.

$$J = \frac{1}{2}\int_{t_i}^{\infty}(x^T(t)Qx(t) + u^T(t)Ru(t))dt$$

where the final time t_f tends to infinity ($t_f \to \infty$).

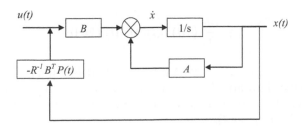

FIGURE 7.1
Block diagram for LQR design (Hamilton–Jacobi approach).

Assumptions

1. The control input $u(t)$ is a unconstrained input.
2. The final time t_f is ∞.
3. The initial condition $x(t_0) = x_0$ is given. However, the final state $x(t_f)$ is not given, i.e., $x(t_f)$ is free.

Assumptions (2) and (3) are called as boundary conditions.

The design procedure is similar to that of finite horizon LQR.

Thus, an LQR is designed using optimal control law by applying Hamilton–Jacobi equation. The step-by-step procedure is explained below.

Procedural Steps for Hamilton–Jacobi Method for Solving LQR Optimal Control Problem

Step 1: Represent the plant by linear continuous time state equation

$$\dot{x} = A(t)x(t) + B(t)u(t)$$

Step 2: Represent the performance index 'J' as

$$J = \frac{1}{2}x^T(t_f)Hx(t_f) + \frac{1}{2}\int_{t_i}^{t_f}(x^T(t)Qx(t) + u^T(t)Ru(t))dt$$

Step 3: Form the Hamiltonian equation

$$H(x(t), u(t), \Delta J*(t), t) = \frac{1}{2}x^T(t)Q(t)x(t) + \frac{1}{2}u^T(t)R(t)u(t)$$

$$+ (\Delta J*(t))^T[A(t)x(t) + B(t)u(t)]$$

where $\Delta J*(t) = \dfrac{\partial J*}{\partial x}(x(t), t)$

Step 4: Minimize $H(x(t), u(t), \Delta J*(t), t)$

$$\text{i.e., } \frac{\partial H}{\partial u}(x(t), u(t), \Delta J*(t), t) = 0$$

and hence, find $u*(t) = -R^{-1}(t)B^T(t)\Delta J*(t) = -R^{-1}(t)B^T(t)P(t)X(t)$

Step 5: Find the Hamilton–Jacobi equation by substituting $u*$ for u (in Step 3).

Step 6: Solve the Hamilton–Jacobi equation with appropriate boundary conditions to obtain $J*(x, t)$.

Step 7: Substitute the results of Step 6 in $u*(t)$, and hence, find the optimal control.

Illustration 7

For a given system $\dot{x} = 1.5x(t) + u(t)$, construct an optimal control law using Hamiltonian–Jacobi method that minimizes the performance index

$$J = \frac{1}{2}\int_{0}^{t_f} \left(2x^2 + \frac{1}{9}u^2 \right) dt.\ \text{Take}\ t_f = 1\ \text{second}.$$

Solution

Given that

$$\dot{x} = 1.5x(t) + u(t) \qquad\qquad\qquad (i)$$

$$J = \frac{1}{2}\int_{0}^{t_f} \left(2x^2 + \frac{1}{9}u^2 \right) dt \qquad\qquad\qquad (ii)$$

t_f is fixed as 1 second. (Finite horizon)
  We have

$$\dot{x} = Ax(t) + Bu(t) \qquad\qquad\qquad (iii)$$

and

$$J = \frac{1}{2}x^T(t_f)Hx(t_f) + \frac{1}{2}\int_{t_i}^{t_f} (x^T(t)Qx(t) + u^T(t)Ru(t))dt \qquad\qquad\qquad (iv)$$

This is a scalar case. From equations (i) to (iv), the following parameters can be deduced: $A = 1.5$, $B = 1$, $H = 0$, $Q = 2$, $R = 1/9$
  Step 1: Find the value of $\dot{P}(t)$ from the MRE.
  The MRE for a scalar case can be given as follows:

$$\dot{p}(t) + Q - PBR^{-1}B^T P + PA + A^T P = 0$$

$$\dot{p}(t) + 2 - 9P^2 + 3P = 0$$

$$\frac{dp(t)}{dt} + 2 - 9P^2 + 3P = 0$$

$$\frac{dp(t)}{dt} = 9P^2 - 3P - 2$$

This can be factored as follows:

$$\frac{dp(t)}{dt} = (3P+1)(3P-2)$$

$$\frac{dp(t)}{dt} = 3(P+1/3)3(P-2/3)$$

$$\frac{dp(t)}{dt} = 9(P+1/3)(P-2/3)$$

Step 2: Obtain the expression for $P(t)$

$$\text{Solve } \int_{t_i}^{t_f} \frac{dp}{9\left(p+\dfrac{1}{3}\right)\left(p-\dfrac{2}{3}\right)} = \int_{t_i}^{t_f} dt \text{ for } p(t) \tag{i}$$

By applying partial fraction, we get

$$\frac{1}{\left(P+\dfrac{1}{3}\right)\left(P-\dfrac{2}{3}\right)} = \frac{M}{\left(P+\dfrac{1}{3}\right)} + \frac{N}{\left(P-\dfrac{2}{3}\right)}$$

$$1 = M\left(P-\frac{2}{3}\right) + N\left(P+\frac{1}{3}\right)$$

$$1 = MP - \frac{2}{3}M + NP + N\frac{1}{3}$$

On comparing the P values, we get

$$0 = M+N$$

$$M = -N \tag{ii}$$

On comparing the constant terms, we get

$$1 = -\frac{2}{3}M + N\frac{1}{3} \tag{iii}$$

Substituting $M = -N$ in equation (iii), we get

$$1 = \frac{2}{3}N + N\frac{1}{3}$$

$$N = 1.$$

or from equation (ii), we can simply write $M = -N$, $M = -1$.

By substituting the values in equation (i), we get

$$\frac{1}{9}\int_{t_i}^{t_f}\left[-\frac{1}{\left(p+\frac{1}{3}\right)}+\frac{1}{\left(p-\frac{2}{3}\right)}\right]dp=\int_{t_i}^{t_f}dt$$

The above equation can be rewritten as

$$\frac{1}{9}\int_{t_i}^{t_f}\left[\frac{1}{\left(p-\frac{2}{3}\right)}-\frac{1}{\left(p+\frac{1}{3}\right)}\right]dp=\int_{t_i}^{t_f}dt$$

On integrating, the above equation becomes

$$\frac{1}{9}\left\{\ln\left[\frac{\left(p(t)-\frac{2}{3}\right)\left(p(t_f)+\frac{1}{3}\right)}{\left(p(t_f)-\frac{2}{3}\right)\left(p(t)+\frac{1}{3}\right)}\right]\right\}=t_i-t_f$$

Now, substitute $p(t_f)=0$ and take exponential

$$\frac{1}{9}\left\{\ln\left[\frac{-\frac{1}{3}\left(p(t)-\frac{2}{3}\right)}{\frac{2}{3}\left(p(t)+\frac{1}{3}\right)}\right]\right\}=t_i-t_f$$

$$\frac{\frac{1}{3}\left(p(t)-\frac{2}{3}\right)}{-\frac{2}{3}\left(p(t)+\frac{1}{3}\right)}=\exp 9(t_i-t_f)$$

$$p(t)=\frac{\frac{-1}{3}\left[1-\exp 9(t_i-t_f)\right]}{1+\frac{1}{2}\exp 9(t_i-t_f)}$$

Step 3: Find the optimal control law, $u^*(t)=-R^{-1}B^T P\,x(t))$

$$u^*(t)=-(1/9)^{-1}p(t)x(t)$$

$$u^*(t)=-9p(t)x(t)$$

Thus, the structure of the optimal controller designed is as shown in Figure 7.2.

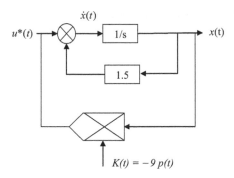

FIGURE 7.2
Block diagram for the controller design using Hamilton–Jacobi method.

LQR Optimal Regulator Design Using Pontryagin's Principle

Consider the LTV system as given in equations (7.26) and (7.27), respectively, which is reproduced here.

$$\dot{x} = A(t)x(t) + B(t)u(t)$$

$$y(t) = C(t)x(t)$$

Objective

To obtain the value of feedback gain $K(t)$ and thereby formulate the control law $u^*(t) = -K(t)\, x(t)$.

Case 1: Finite Horizon

The cost function for this optimal control problem is as given in equation (7.22).

$$J = \frac{1}{2}x^T(t_f)Hx(t_f) + \frac{1}{2}\int_{t_i}^{t_f}(x^T(t)Qx(t) + u^T(t)Ru(t))dt$$

The auxiliary cost function is a combination of the above cost function along with the co-state vector as given in equation (7.29).

$$J_\lambda = \lambda^T(t)[A(t)x(t) + B(t)u(t)]$$

Here also, the same assumptions made under case 1 for LQR optimal regulator design using Hamilton–Jacobi equation hold good.

From the Pontryagin's equation, we can write

$$H(x,u,\lambda,t) = \frac{1}{2}x^T(t)Q(t)x(t) + \frac{1}{2}u^T(t)R(t)u(t) + \lambda^T(t)[A(t)x(t) + B(t)u(t)] \quad (7.44)$$

Along the optimal trajectory, we have

$$\frac{\partial H}{\partial u}(x,u,\lambda,t) = 0 \quad (7.45)$$

$$R(t)u(t) + B^T(t)\lambda(t) = 0$$

and

$$u^*(t) = -R^{-1}(t)B^T(t)\lambda(t) \quad (7.46)$$

Since, $\frac{\partial^2 H}{\partial u^2}(x,u,\lambda,t) = R(t)$ is positive definite, $u^*(t)$ given in equation (7.46) minimizes the Pontryagin's function, $H(x,u,\lambda,t)$. Substituting equation (7.46) in equation (7.44) and simplifying, we get

$$H^*(x,\lambda,t) = \frac{1}{2}x^T(t)Q(t)x(t) + \lambda^T(t)A(t)x(t) - \frac{1}{2}\lambda^T(t)B(t)R^{-1}(t)B^T(t)\lambda(t)$$

The state and co-state equations can be obtained as follows:

$$\text{State equation } \dot{x}(t) = \frac{\partial H^*}{\partial \lambda} = A(t)x(t) + B(t)(-R^{-1}(t)B^T(t)\lambda(t)) \quad (7.47)$$

$$\text{Co-state equation } \dot{\lambda}(t) = -\frac{\partial H^*}{\partial x} = -Q(t)x(t) - A^T(t)\lambda(t) \quad (7.48)$$

$$\begin{bmatrix} \dot{x}(t) \\ \dot{\lambda}(t) \end{bmatrix} = \begin{bmatrix} A(t) & -B(t)R^{-1}(t)B^T(t) \\ -Q(t) & -A^T(t) \end{bmatrix} \begin{bmatrix} x(t) \\ \lambda(t) \end{bmatrix} \quad (7.49)$$

From the boundary conditions, we have

$$\frac{\partial H}{\partial X} - \lambda^T(t) = 0$$

$$\lambda^T(t) = \frac{\partial H}{\partial x} \quad (7.50)$$

$$\therefore \quad \lambda(t) = \frac{\partial H}{\partial x}\bigg|_{t=t_f} \quad \text{or } \lambda(t_f) = Hx(t_f) \quad (7.51)$$

Let the state transition matrix be given as in equation (7.52).

$$\Phi(t,t_i) = \begin{bmatrix} \Phi_{11}(t,t_i) & \Phi_{12}(t,t_i) \\ \Phi_{21}(t,t_i) & \Phi_{22}(t,t_i) \end{bmatrix} \tag{7.52}$$

The solution of the above state model can be given as in equation (7.53)

$$\begin{bmatrix} x(t) \\ \lambda(t) \end{bmatrix} = \Phi(t,t_i) \begin{bmatrix} x(t_i) \\ \lambda(t_i) \end{bmatrix} \tag{7.53}$$

Similarly, for the finial time t_f, the following equation holds satisfactorily.

$$x(t) = \Phi_{11}(t,t_f)x(t_f) + \Phi_{12}(t,t_f)\lambda(t_f) \tag{7.54}$$

$$\lambda(t) = \Phi_{21}(t,t_f)x(t_f) + \Phi_{22}(t,t_f)\lambda(t_f) \tag{7.55}$$

Now, substituting equation (7.51) in equation (7.55), we get

$$x(t) = [\Phi_{11}(t,t_f)x(t_f) + \Phi_{12}(t,t_f)H]x(t_f) \tag{7.56}$$

$$\lambda(t) = [\Phi_{21}(t,t_f)x(t_f) + \Phi_{22}(t,t_f)H][\Phi_{21}(t,t_f)x(t_f) + \Phi_{22}(t,t_f)H]^{-1}x(t_f) \tag{7.57}$$

We know that

$$H = \frac{1}{2}x^T(t_f)Hx(t_f)$$

From equation (7.57), it is inferred that, the co-state $\lambda(t)$ and the state $x(t)$ can be related as given in equation (7.58).

$$\lambda(t) = P(t)x(t) \tag{7.58}$$

where

$$P(t) = [\Phi_{21}(t,t_f) + \Phi_{22}(t,t_f)H][\Phi_{11}(t,t_f) + \Phi_{12}(t,t_f)H]^{-1} \tag{7.59}$$

The value of $P(t)$ given in equation (7.59) is the solution of Riccati equation given in equation (7.41).

From the control law $u(t) = -R^{-1}B^T(t)\lambda(t)$

By substituting equation (7.58) in control law $u(t)$, we get

$$u(t) = -R^{-1}B^T(t)P(t)x(t)$$

$$u(t) = -K(t)x(t)$$

From the control law, the gain can be expressed as given in equation (7.60)

$$K(t) = R^{-1}B^T(t)P(t) \tag{7.60}$$

To solve the objective function J, it is required to use the derived Kalman gain and control law. This requires $P(t)$ value. To get the values of $P(t)$, perform direct integration on the MRE.

Solve for the optimal state equation given in equation (7.61)

$$\dot{x}^* = [A(t) - B(t)R^{-1}B^T(t)P(t)]x(t) \tag{7.61}$$

The block diagram for LQR design using Pontryagin's principle approach is shown in Figure 7.3.

Now, substitute the values in J and optimize the performance index.

$$J = \frac{1}{2}x^T(t)Px(t) \tag{7.62}$$

Case 2: Infinite Horizon

For the infinite time linear state regulator, the objective function is as given in equation (7.63)

$$J = \frac{1}{2}\int_{t_i}^{\infty}(x^T(t)Qx(t) + u^T(t)Ru(t))dt \tag{7.63}$$

where the final time t_f tends to infinity $(t_f \to \infty)$.

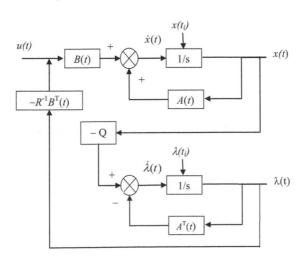

FIGURE 7.3
Block diagram for LQR design (Pontryagin's principle approach).

Assumptions

Here too, the same assumptions made under case 2 for LQR optimal regulator design using Hamilton–Jacobi equation holds good.

The step-by-step design procedure for solving an LQR optimal control problem using Pontryagin's principle is explained below.

Procedural Steps for Pontryagin's Principle Approach for Solving LQR Optimal Control Problem

Step 1: Represent the plant by linear continuous time equation

$$\dot{x} = A(t)x(t) + B(t)u(t)$$

Step 2: Represent the performance index 'J' as

$$J = \frac{1}{2}x^T(t_f)Hx(t_f) + \frac{1}{2}\int_{t_i}^{t_f}(x^T(t)Qx(t) + u^T(t)Ru(t))dt$$

Step 3: Form the Pontryagin's equation

$$H(x,u,\lambda,t) = \frac{1}{2}x^T(t)Q(t)x(t) + \frac{1}{2}u^T(t)R(t)u(t) + \lambda^T(t)[A(t)x(t) + B(t)u(t)]$$

Step 4: Minimize $H(x,u,\lambda,t)$

$$\text{i.e., } \frac{\partial H}{\partial u}(x,u,\lambda,t) = 0$$

and hence, find $u*(t) = -R^{-1}(t)B^T(t)\lambda(t)$

Step 5: Find the Pontryagin's function

$$H*(x,\lambda,t) = \min H(x,u,\lambda,t)$$

Step 6: Solve the state and co-state equations using the given boundary conditions

$$\dot{x} = \frac{\partial H*}{\partial \lambda} \quad \text{(State equation)}$$

$$\dot{\lambda} = -\frac{\partial H*}{\partial x} \quad \text{(Co-state equation)}$$

Step 7: Substitute the results of Step 6 in $u*(t)$, and hence, find the optimal control.

Illustration 8

For a given system $\dot{x} = 1.5x(t) + u(t)$, construct an optimal control law using Pontryagin's principle method that minimizes the performance index

$$J = \frac{1}{2} \int_0^{t_f} \left(2x^2 + \frac{1}{9}u^2 \right) dt.$$ Here $t_f = 1$ second.

Solution

This is a scalar case. From the given data, the following parameters can be deduced. $A = 1.5, B = 1, H = 0, Q = 2, R = 1/9$.

Step 1: Obtain the combined co-state and state equations using equation (7.42).

$$\left[\begin{array}{c} \dot{x}(t) \\ \dot{\lambda}(t) \end{array} \right] = \left[\begin{array}{cc} A(t) & -B(t)R^{-1}B^T(t) \\ -Q(t) & -A^T(t) \end{array} \right] \left[\begin{array}{c} x(t) \\ \lambda(t) \end{array} \right]$$

$$\therefore \left[\begin{array}{c} \dot{x}(t) \\ \dot{\lambda}(t) \end{array} \right] = \left[\begin{array}{cc} 1.5 & -9 \\ -2 & -1.5 \end{array} \right] \left[\begin{array}{c} x(t) \\ \lambda(t) \end{array} \right]$$

Step 2: Find the state transition matrix using equation (7.45)

$$\Phi(t,t_i) = \left[\begin{array}{cc} \Phi_{11}(t,t_i) & \Phi_{12}(t,t_i) \\ \Phi_{21}(t,t_i) & \Phi_{22}(t,t_i) \end{array} \right]$$

Find the matrix sI

$$I = \left[\begin{array}{cc} 1 & 0 \\ 0 & 1 \end{array} \right]$$

$$sI = \left[\begin{array}{cc} s & 0 \\ 0 & s \end{array} \right]$$

Find the matrix $[sI - A_C]$ where $A_C = \left[\begin{array}{cc} 1.5 & -9 \\ -2 & -1.5 \end{array} \right]$

$$[sI - A_C] = \left[\begin{array}{cc} s & 0 \\ 0 & s \end{array} \right] - \left[\begin{array}{cc} 1.5 & -9 \\ -2 & -1.5 \end{array} \right]$$

$$= \left[\begin{array}{cc} s - 1.5 & 9 \\ 2 & s + 1.5 \end{array} \right]$$

Find the inverse of the matrix $[sI - A_C]$

$$[sI - A_C]^{-1} = \frac{Adj(sI - A_C)}{|sI - A_C|}$$

$$|sI - A_C| = (s - 1.5)(s - 1.5) - 18$$

$$= s^2 + 1.5s - 1.5 - 18 = s^2 - 18$$

$$= (s - 4.2)(s + 4.2)$$

$$Adj(sI - A_C) = \begin{bmatrix} s + 1.5 & -9 \\ -2 & s - 1.5 \end{bmatrix}$$

$$[sI - A_C]^{-1} = \frac{Adj(sI - A_C)}{|sI - A_C|} = \frac{\begin{bmatrix} s + 1.5 & -9 \\ -2 & s - 1.5 \end{bmatrix}}{(s - 4.2)(s + 4.2)}$$

$$= \begin{bmatrix} \dfrac{s + 1.5}{(s - 4.2)(s + 4.2)} & \dfrac{-9}{(s - 4.2)(s + 4.2)} \\[4mm] \dfrac{-2}{(s - 4.2)(s + 4.2)} & \dfrac{s - 1.5}{(s - 4.2)(s + 4.2)} \end{bmatrix}$$

Find the Laplace inverse of $[sI - A_C]^{-1}$

$$e^{At} = L^{-1}[sI - A_C]^{-1}$$

$$= L^{-1} \begin{bmatrix} \dfrac{s + 1.5}{(s - 4.2)(s + 4.2)} & \dfrac{-9}{(s - 4.2)(s + 4.2)} \\[4mm] \dfrac{-2}{(s - 4.2)(s + 4.2)} & \dfrac{s - 1.5}{(s - 4.2)(s + 4.2)} \end{bmatrix}$$

Finding the partial fraction

$$\frac{s + 1.5}{(s + 4.2)(s - 4.2)} = \frac{M}{(s + 4.2)} + \frac{N}{(s - 4.2)}$$

$$s + 1.5 = M(s - 4.2) + N(s + 4.2)$$

When $s = 4.2$	When $s = -4.2$
$4.2 + 1.5 = N(4.2 + 4.2)$	$-4.2 + 1.5 = M(8.4) + N(0)$
$5.7 = 8.4N$	$-2.7 = -8.4M$
$N = 6/8$ (approx)	$M = 3/8$ (approx)

$$\frac{-9}{(s+4.2)(s-4.2)} = \frac{M}{(s+4.2)} + \frac{N}{(s-4.2)}$$

$$-9 = M(s-4.2) + N(s+4.2)$$

When $s = 4.2$

$-9 = N(4.2 + 4.2)$

$-9 = 8.4N$

$N = -9/8$ (approx)

When $s = -4.2$

$-9 = M(8.4) + N(0)$

$-9 = -8.4M$

$M = 9/8$ (approx)

$$\frac{-2}{(s+4.2)(s-4.2)} = \frac{M}{(s+4.2)} + \frac{N}{(s-4.2)}$$

$$-2 = M(s-4.2) + N(s+4.2)$$

When $s = 4.2$

$-2 = N(4.2 + 4.2)$

$-2 = 8.4N$

$N = -2/8$ (approx)

When $s = -4.2$

$-2 = M(8.4) + N(0)$

$-2 = -8.4M$

$M = 2/8$ (approx)

$$\frac{s-1.5}{(s+4.2)(s-4.2)} = \frac{M}{(s+4.2)} + \frac{N}{(s-4.2)}$$

$$s - 1.5 = M(s-4.2) + N(s+4.2)$$

When $s = 4.2$

$4.2 - 1.5 = N(4.2 + 4.2)$

$2.7 = 8.4N$

$N = 3/8$ (approx)

When $s = -4.2$

$4.2 + 1.5 = M(8.4) + N(0)$

$-5.7 = -8.4M$

$M = 6/8$ (approx)

$$= L^{-1}\begin{bmatrix} \left(\dfrac{3/8}{(s+4.2)} + \dfrac{6/8}{(s-4.2)}\right) & \left(\dfrac{9/8}{(s+4.2)} - \dfrac{9/8}{(s-4.2)}\right) \\ \left(\dfrac{2/8}{(s+4.2)} - \dfrac{2/8}{(s-4.2)}\right) & \left(\dfrac{6/8}{(s+4.2)} + \dfrac{3/8}{(s-4.2)}\right) \end{bmatrix}$$

$$\Phi(t,t_i) = \begin{bmatrix} \dfrac{3}{8}\left(e^{4.2(t_i-t_f)} + 2e^{-4.2(t_i-t_f)}\right) & \dfrac{9}{8}\left(e^{4.2(t_i-t_f)} - e^{-4.2(t_i-t_f)}\right) \\ \dfrac{2}{8}\left(e^{4.2(t_i-t_f)} - e^{-4.2(t_i-t_f)}\right) & \dfrac{3}{8}\left(2e^{4.2(t_i-t_f)} + e^{-4.2(t_i-t_f)}\right) \end{bmatrix}$$

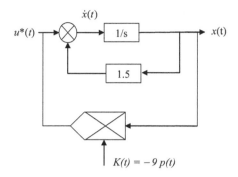

FIGURE 7.4
Block diagram for the controller design using Pontryagin's principle approach.

Step 3: Find $P(t)$ using the equation (7.52)

$$P(t) = [\Phi_{21}(t, t_f)x(t_f) + \Phi_{22}(t, t_f)H][\Phi_{21}(t, t_f)x(t_f) + \Phi_{22}(t, t_f)H]^{-1}$$

($\because H = 0$) the above equation reduces to

$$P(t) = [\Phi_{21}(t, t_f)x(t_f)][\Phi_{21}(t, t_f)x(t_f)]^{-1}$$

$$P(t) = \left[\frac{2}{8}\left(e^{4.2(t_i - t_f)} - e^{-4.2(t_i - t_f)}\right)\right]\left[\frac{3}{8}\left(e^{4.2(t_i - t_f)} + 2e^{-4.2(t_i - t_f)}\right)\right]^{-1}$$

$$p(t) = \frac{\dfrac{-1}{3}\left[1 - \exp 9(t_i - t_f)\right]}{1 + \dfrac{1}{2}\exp 9(t_i - t_f)}$$

Step 4: Find the optimal control law $u^*(t) = -R^{-1}B^T P x(t)$

$$u^*(t) = -9p(t)x(t)$$

The structure of the optimal controller thus designed is as shown in Figure 7.4.
   For the ease of the reader, a more generalized way to understand the design of steady-state LQR is as follows.

LQR Steady-State Optimal Regulator

Consider a linear time-variant system represented in state variable form as given in equations (7.64) and (7.65)

$$\dot{x} = Ax(t) + Bu(t) \tag{7.64}$$

$$y(t) = Cx(t) \tag{7.65}$$

and the control law is

$$u = -Kx + v \tag{7.66}$$

where A and B are state and input matrices, respectively, and v is the noise.
∴By substituting equation (7.66) in (7.64), we get

$$\dot{x} = Ax + B(-Kx + v)$$

$$\dot{x} = Ax - BKx + Bv$$

$$\dot{x} = (A - BK)x + Bv$$

$$\dot{x} = A_C x + Bv \tag{7.67}$$

where $A_C = (A - BK)$

Assumptions

  i. The system is completely observable and controllable.
 ii. The initial condition is zero.

To design an optimal state variable feedback, the performance index or cost function, in general, is defined as follows:

$$J = \frac{1}{2} \int_0^\infty x^T Q x \tag{7.68}$$

Also, by considering the effect of control input 'u', the above equation can be rewritten as follows:

$$J = \frac{1}{2} \int_0^\infty \left(x^T Q x + u^T R u \right) dt \tag{7.69}$$

By substituting the value $u = kx$ in equation (7.69), we get

$$J = \frac{1}{2} \int_0^\infty \left(x^T Q x + x^T K^T R K x \right) dt \tag{7.70}$$

$$J = \frac{1}{2} \int_0^\infty x^T \left(Q + K^T R K \right) x \, dt \qquad (7.71)$$

Here, the plant is linear and the performance index is quadratic. The problem of finding a suitable feedback gain 'K' value to minimize 'J' such that 'K', i.e., the feedback gain regulates the states to zero. Thus the control problem is called as a LQR problem.

Objective

The objective of optimal design is to select the state variable feedback gain 'K' such that it minimizes the performance measure J. In this problem, the cost function can be assumed to be a minimum energy problem.
   Here,

   Q is positive semi-definite ($\therefore x^T Q x$ is positive or zero).
   R is positive definite ($\therefore u^T Q u$ is positive).

To find optimal feedback K, consider a constant matrix P, such that

$$-\frac{d}{dt}(x^T P x) = x^T (Q + K^T R K) x \qquad (7.72)$$

$$J = -\frac{1}{2} \int_0^\infty \frac{d}{dt}(x^T P x) = \frac{1}{2} x^T (0) P x(0) \qquad (7.73)$$

From equation (7.73), it can be seen that, J is dependent on the constant matrix P and initial conditions only. It is independent of 'K'.
   From equation (7.72), it can be derived that

$$\dot{x}^T P x + x^T P \dot{x} + x^T Q x + x^T K^T R K x = 0 \qquad (7.74)$$

$$x^T A_C^T P x + x^T P A_C x + x^T Q x + x^T K^T R K x = 0 \quad (\because \dot{x} = A_C x)$$

$$x^T (A_C^T P + P A_C + Q + K^T R K) x = 0 \quad (\because \dot{x}^T = x^T A_C) \qquad (7.75)$$

From equation (7.75), we get

$$x^T x = 0$$

$$A_C^T P + P A_C + Q + K^T R K = 0$$

$$(A - BK)^T P + P(A - BK) + Q + K^T R K = 0 \quad (\because A_C = A - BK)$$

$$A^T P + PA + Q + K^T RKP - K^T B^T P - PBK = 0 \qquad (7.76)$$

By taking optimal gain as $K = R^{-1} B^T P$ \qquad (7.77)

$$A^T P + PA + Q + (R^{-1} B^T P)^T R(R^{-1} B^T P)P - (R^{-1} B^T P)^T B^T P - PB(R^{-1} B^T P) = 0$$

After simplification, we get

$$Q - PBR^{-1} B^T P + PA + A^T P = 0 \qquad (7.78)$$

Equation (7.78) is called the Matrix Riccati Equation (MRE).

An MRE at steady state is called Algebraic Riccati Equation (ARE).

The steps involved in designing the LQR are as follows:

i. Select appropriate cost function
ii. Choose Q and R matrices
iii. Find the solution of the Riccati equation such that the constant P matrix is derived
iv. Apply P in the optimal gain formula $K = R^{-1} B^T P$.

The LQR thus designed is as shown in Figure 7.5.

Illustration 9

For a given system $\dot{x} = \begin{bmatrix} 0 & 10 \\ 0 & 0 \end{bmatrix} x + \begin{bmatrix} 0 \\ 10 \end{bmatrix} u$, construct an optimal con-

trol law that minimizes the performance index $J = \int_{0}^{\infty} (x^2 + u^2) du$. Assume

$Q = \begin{bmatrix} 2 & 0 \\ 0 & 0 \end{bmatrix}$ and $R = [2]$

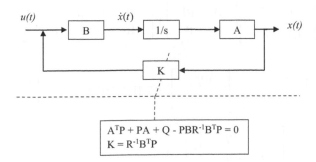

FIGURE 7.5
Block diagram for LQR design.

Solution

Given that

$$A = \begin{bmatrix} 0 & 10 \\ 0 & 0 \end{bmatrix}, B = \begin{bmatrix} 0 \\ 10 \end{bmatrix}$$

Step 1: Check for controllability (since, the given performance index is a state regulator problem).

$$Q_C = \begin{bmatrix} B & AB \end{bmatrix}$$

$$Q_C = \begin{bmatrix} 0 & 100 \\ 10 & 0 \end{bmatrix} = 0 - 1000 = -1000 \neq 0$$

Hence, the given system is controllable.
  Step 2: Find $P(t)$ using reduced MRE.
  We have the reduced MRE as $Q - PBR^{-1}B^TP + PA + A^TP = 0$

$$\begin{bmatrix} 2 & 0 \\ 0 & 0 \end{bmatrix} - \begin{bmatrix} p_{11} & p_{12} \\ p_{21} & p_{22} \end{bmatrix} \begin{bmatrix} 0 \\ 10 \end{bmatrix} [1/2] \begin{bmatrix} 0 & 10 \end{bmatrix} \begin{bmatrix} p_{11} & p_{12} \\ p_{21} & p_{22} \end{bmatrix}$$

$$+ \begin{bmatrix} p_{11} & p_{12} \\ p_{21} & p_{22} \end{bmatrix} \begin{bmatrix} 0 & 10 \\ 0 & 0 \end{bmatrix} + \begin{bmatrix} 0 & 0 \\ 10 & 0 \end{bmatrix} \begin{bmatrix} p_{11} & p_{12} \\ p_{21} & p_{22} \end{bmatrix} = 0$$

Due to symmetric property, $p_{12} = p_{21}$

$$\begin{bmatrix} 2 & 0 \\ 0 & 0 \end{bmatrix} - \begin{bmatrix} p_1 & p_2 \\ p_2 & p_3 \end{bmatrix} \begin{bmatrix} 0 \\ 10 \end{bmatrix} [1/2] \begin{bmatrix} 0 & 10 \end{bmatrix} \begin{bmatrix} p_1 & p_2 \\ p_2 & p_3 \end{bmatrix}$$

$$+ \begin{bmatrix} p_1 & p_2 \\ p_2 & p_3 \end{bmatrix} \begin{bmatrix} 0 & 10 \\ 0 & 0 \end{bmatrix} + \begin{bmatrix} 0 & 0 \\ 10 & 0 \end{bmatrix} \begin{bmatrix} p_1 & p_2 \\ p_2 & p_3 \end{bmatrix} = 0$$

$$\begin{bmatrix} 2 & 0 \\ 0 & 0 \end{bmatrix} - \begin{bmatrix} 10p_2 \\ 10p_3 \end{bmatrix} [1/2] \begin{bmatrix} 0 & 10 \end{bmatrix} \begin{bmatrix} p_1 & p_2 \\ p_2 & p_3 \end{bmatrix}$$

$$+ \begin{bmatrix} 0 & 10p_1 \\ 0 & 10p_2 \end{bmatrix} + \begin{bmatrix} 0 & 0 \\ 10p_1 & 10p_2 \end{bmatrix} = 0$$

$$\begin{bmatrix} 2 & 0 \\ 0 & 0 \end{bmatrix} - \begin{bmatrix} 50p_2^2 & 50p_2p_3 \\ 50p_1p_3 & 50p_3^2 \end{bmatrix} + \begin{bmatrix} 0 & 10p_1 \\ 0 & 10p_2 \end{bmatrix} + \begin{bmatrix} 0 & 0 \\ 10p_1 & 10p_2 \end{bmatrix} = 0$$

After simplification, we get

$$\begin{bmatrix} 2 - 50p_2^2 & -50p_2p_3 + 10p_1 \\ -50p_1p_3 + 10p_1 & -50p_3^2 + 10p_2 + 10p_2 \end{bmatrix} = 0$$

The 'P' matrix can be obtained by solving the following equations.

$$2 - 50p_2^2 = 0$$

$$-50p_2p_3 + 10p_1 = 0$$

$$-50p_1p_3 + 10p_1 = 0$$

$$-50p_3^2 + 20p_2 = 0$$

By solving the above equations, we get

$$p_1 = \frac{\sqrt{2}}{5}, \; p_2 = \frac{1}{5}, \; p_3 = \frac{\sqrt{2}}{5}$$

Thus, the P matrix can be written as follows:

$$P = \begin{bmatrix} \dfrac{\sqrt{2}}{5} & \dfrac{1}{5} \\ \dfrac{1}{5} & \dfrac{\sqrt{2}}{5} \end{bmatrix}$$

Step 3: Find the control law, $u^*(t) = -R^{-1}B^T P x(t) = K x(t)$

$$u^*(t) = -[1/2]\begin{bmatrix} 0 & 10 \end{bmatrix}\begin{bmatrix} \dfrac{\sqrt{2}}{5} & \dfrac{1}{5} \\ \dfrac{1}{5} & \dfrac{\sqrt{2}}{5} \end{bmatrix} x(t)$$

$$u^*(t) = -\begin{bmatrix} 1 & \sqrt{2} \end{bmatrix} x(t)$$

where $K_1 = -1$ and $K_2 = -\sqrt{2}$.
   Hence, the control law can be written as $u^*(t) = -1\, x(t) - \sqrt{2}\, x(t)$.

Illustration 10

For a given system $\dot{x} = \begin{bmatrix} 0 & 1 \\ 0 & -1 \end{bmatrix} x + \begin{bmatrix} 0 \\ 10 \end{bmatrix} u$ and $y = \begin{bmatrix} 1 & 0 \end{bmatrix} x$ construct an

optimal control law that minimizes the performance index $J = \int_0^\infty (y^2 + u^2) du$.

Assume $Q = \begin{bmatrix} 1 & 0 \\ 0 & 0 \end{bmatrix}$ and $R = \begin{bmatrix} 1 \end{bmatrix}$

Solution

Given that

$$A = \begin{bmatrix} 0 & 1 \\ 0 & -1 \end{bmatrix}, B = \begin{bmatrix} 0 \\ 10 \end{bmatrix}, C = \begin{bmatrix} 1 & 0 \end{bmatrix}$$

Step 1: Check for observability (since, the given performance index is an output regulator problem).

$$Q_O = \begin{bmatrix} C^T & C^T A^T \end{bmatrix}$$

$$Q_O = \begin{bmatrix} 1 & 0 \\ 0 & 1 \end{bmatrix} = 1 - 0 = 1 \neq 0$$

Hence, the given system is observable.

Step 2: Find $P(t)$ using the reduced MRE.

We have the reduced MRE as $Q - PBR^{-1}B^TP + PA + A^TP = 0$

$$\begin{bmatrix} 1 & 0 \\ 0 & 0 \end{bmatrix} - \begin{bmatrix} p_{11} & p_{12} \\ p_{21} & p_{22} \end{bmatrix} \begin{bmatrix} 0 \\ 10 \end{bmatrix} [1][0 \quad 10] \begin{bmatrix} p_{11} & p_{12} \\ p_{21} & p_{22} \end{bmatrix}$$

$$+ \begin{bmatrix} p_{11} & p_{12} \\ p_{21} & p_{22} \end{bmatrix} \begin{bmatrix} 0 & 1 \\ 0 & -1 \end{bmatrix} + \begin{bmatrix} 0 & 0 \\ 1 & -1 \end{bmatrix} \begin{bmatrix} p_{11} & p_{12} \\ p_{21} & p_{22} \end{bmatrix} = 0$$

Due to symmetric property, $p_{12} = p_{21}$

$$\begin{bmatrix} 1 & 0 \\ 0 & 0 \end{bmatrix} - \begin{bmatrix} p_1 & p_2 \\ p_2 & p_3 \end{bmatrix} \begin{bmatrix} 0 \\ 10 \end{bmatrix} [1][0 \quad 10] \begin{bmatrix} p_1 & p_2 \\ p_2 & p_3 \end{bmatrix}$$

$$+ \begin{bmatrix} p_1 & p_2 \\ p_2 & p_3 \end{bmatrix} \begin{bmatrix} 0 & 1 \\ 0 & -1 \end{bmatrix} + \begin{bmatrix} 0 & 0 \\ 1 & -1 \end{bmatrix} \begin{bmatrix} p_1 & p_2 \\ p_2 & p_3 \end{bmatrix} = 0$$

$$\begin{bmatrix} 1 & 0 \\ 0 & 0 \end{bmatrix} - \begin{bmatrix} 10p_2 \\ 10p_3 \end{bmatrix} \begin{bmatrix} 10p_2 & 10p_3 \end{bmatrix} + \begin{bmatrix} 0 & p_1 - p_2 \\ 0 & p_2 - p_3 \end{bmatrix}$$

$$+ \begin{bmatrix} 0 & 0 \\ p_1 - p_2 & p_2 - p_3 \end{bmatrix} = 0$$

$$\begin{bmatrix} 1 & 0 \\ 0 & 0 \end{bmatrix} - \begin{bmatrix} 100p_2{}^2 & 100p_2 p_3 \\ 100p_2 p_3 & 100p_3{}^2 \end{bmatrix} + \begin{bmatrix} 0 & p_1 - p_2 \\ p_1 - p_2 & 2p_2 - 2p_3 \end{bmatrix} = 0$$

$$\begin{bmatrix} 1 - 100p_2{}^2 & -100p_2 p_3 + p_1 - p_2 \\ -100p_2 p_3 + p_1 - p_2 & -100p_3{}^2 + 2p_2 - 2p_3 \end{bmatrix} = 0$$

By solving the above equations, we get

$$p_1 = 0.45,\ p_2 = 0.1,\ p_3 = 0.035$$

Thus, the P matrix can be written as follows:

$$P = \begin{bmatrix} 0.45 & 0.1 \\ 0.1 & 0.035 \end{bmatrix}$$

Step 3: Find the control law, $u^*(t) = -R^{-1} B^T P\, x(t) = K\, x(t)$

$$u^*(t) = -[1][0\ \ 10] \begin{bmatrix} 0.45 & 0.1 \\ 0.1 & 0.035 \end{bmatrix} x(t)$$

$$u^*(t) = -[1\ \ 0.35] x(t)$$

where $K_1 = -1$ and $K_2 = -0.35$.

Hence, the control law can be written as $u^*(t) = -1\, x(t) - 0.35\, x(t)$.

Illustration 11

For a given system $\dot{x} = \begin{bmatrix} 0 & 1 \\ -3 & -4 \end{bmatrix} x + \begin{bmatrix} 0 \\ 1 \end{bmatrix} u$, construct an optimal con-

trol law that minimizes the performance index $J = \dfrac{1}{2} \displaystyle\int_0^\infty (x_1{}^2 + x_2{}^2)\,dt$. Assume

$Q = \begin{bmatrix} 1 & 0 \\ 0 & 0 \end{bmatrix}$ and $R = [1]$. Also find the minimum value of J when $x_1(0) = 1$,

$x_2(0) = 1$.

Solution

Given that

$$A = \begin{bmatrix} 0 & 1 \\ -3 & -4 \end{bmatrix}, B = \begin{bmatrix} 0 \\ 1 \end{bmatrix}$$

Step 1: Check for controllability

$$Q_C = \begin{bmatrix} B & AB \end{bmatrix}$$

$$Q_C = \begin{bmatrix} 0 & 1 \\ 1 & -4 \end{bmatrix} = 0 - 1 = -1 \neq 0$$

Hence, the given system is controllable.

Step 2: Find $P(t)$ using the reduced MRE.

We have the reduced MRE as $Q - PBR^{-1}B^T P + PA + A^T P = 0$

$$\begin{bmatrix} 1 & 0 \\ 0 & 0 \end{bmatrix} - \begin{bmatrix} p_{11} & p_{12} \\ p_{21} & p_{22} \end{bmatrix} \begin{bmatrix} 0 \\ 1 \end{bmatrix} [1][0 \ 1] \begin{bmatrix} p_{11} & p_{12} \\ p_{21} & p_{22} \end{bmatrix}$$

$$+ \begin{bmatrix} p_{11} & p_{12} \\ p_{21} & p_{22} \end{bmatrix} \begin{bmatrix} 0 & 1 \\ -3 & -4 \end{bmatrix} + \begin{bmatrix} 0 & -3 \\ 1 & -4 \end{bmatrix} \begin{bmatrix} p_{11} & p_{12} \\ p_{21} & p_{22} \end{bmatrix} = 0$$

Due to symmetric property, $p_{12} = p_{21}$

$$\begin{bmatrix} 1 & 0 \\ 0 & 0 \end{bmatrix} - \begin{bmatrix} p_1 & p_2 \\ p_2 & p_3 \end{bmatrix} \begin{bmatrix} 0 \\ 1 \end{bmatrix} [1][0 \ 1] \begin{bmatrix} p_1 & p_2 \\ p_2 & p_3 \end{bmatrix}$$

$$+ \begin{bmatrix} p_1 & p_2 \\ p_2 & p_3 \end{bmatrix} \begin{bmatrix} 0 & 1 \\ -3 & -4 \end{bmatrix} + \begin{bmatrix} 0 & -3 \\ 1 & -4 \end{bmatrix} \begin{bmatrix} p_1 & p_2 \\ p_2 & p_3 \end{bmatrix} = 0$$

$$\begin{bmatrix} 1 & 0 \\ 0 & 0 \end{bmatrix} - \begin{bmatrix} p_2 \\ p_3 \end{bmatrix} [p_2 \ p_3] + \begin{bmatrix} -3p_2 & p_1 - 4p_2 \\ -3p_3 & p_2 - 4p_3 \end{bmatrix}$$

$$+ \begin{bmatrix} -3p_2 & -3p_3 \\ p_1 - 4p_2 & p_2 - 4p_3 \end{bmatrix} = 0$$

$$\begin{bmatrix} 1 & 0 \\ 0 & 0 \end{bmatrix} - \begin{bmatrix} p_2^2 & p_2 p_3 \\ p_2 p_3 & p_3^2 \end{bmatrix} + \begin{bmatrix} -6p_2 & p_1 - 4p_2 - 3p_3 \\ p_1 - 4p_2 - 3p_3 & 2p_2 - 8p_3 \end{bmatrix} = 0$$

$$\begin{bmatrix} 1 - p_2{}^2 - 6p_2 & -p_2 p_3 + p_1 - 4p_2 - 3p_3 \\ -p_2 p_3 + p_1 - 4p_2 - 3p_3 & -p_3{}^2 + 2p_2 - 8p_3 \end{bmatrix} = 0$$

By solving the above equations, we get

$$p_1 = 0.75, \; p_2 = 0.16, \; p_3 = 0.03$$

Thus, the P matrix can be written as follows:

$$P = \begin{bmatrix} 0.75 & 0.16 \\ 0.16 & 0.03 \end{bmatrix}$$

Step 3: Find the control law, $u^*(t) = -R^{-1} B^T P \, x(t) = K \, x(t)$

$$u^*(t) = -[1][0 \quad 1]\begin{bmatrix} 0.75 & 0.16 \\ 0.16 & 0.03 \end{bmatrix} x(t)$$

$$u^*(t) = -[0.16 \quad 0.03] x(t)$$

where $K_1 = -0.16$ and $K_2 = -0.03$.
   Hence, the control law can be written as $u^*(t) = -0.16 \, x(t) - 0.03 \, x(t)$.
Step 4: Find the minimum value of J for the given performance index.
We have

$$J_{min} = \frac{1}{2}\left[x(0)^T P x(0) \right]$$

$$= \frac{1}{2}[1 \quad 1]\begin{bmatrix} 0.75 & 0.16 \\ 0.16 & 0.03 \end{bmatrix}\begin{bmatrix} 1 \\ 1 \end{bmatrix} \quad \because\left(x(0) = \begin{pmatrix} x_1(0) \\ x_2(0) \end{pmatrix} \right)$$

$$= \begin{bmatrix} \dfrac{1}{2} & \dfrac{1}{2} \end{bmatrix}\begin{bmatrix} 0.75 & 0.16 \\ 0.16 & 0.03 \end{bmatrix}\begin{bmatrix} 1 \\ 1 \end{bmatrix}$$

$$= \begin{bmatrix} \dfrac{1}{2} & \dfrac{1}{2} \end{bmatrix}\begin{bmatrix} 0.91 \\ 0.19 \end{bmatrix}$$

$$= [0.455 + 0.095] = 0.55$$

$$J_{min} = 0.55$$

Symmetric Property of Matrix Riccati Equation

An MRE of an optimal problem is a mathematical representation of the system. Hence, from the symmetric property of the Riccati equation, it is possible to extract qualitative measures about the system. It also makes the computations of solutions much easier.

To show the symmetric property of the MRE, let us recall equations (7.35) and (7.36a).

$$0 = \Delta J * (t) + \frac{1}{2} x^T (t) Q(t) x(t) - \frac{1}{2} \left(\Delta J * (t) \right)^T [B(t) R^{-1}(t) B^T (t) \Delta J * (t)]$$

$$+ \left[\Delta J * (t) \right]^T A(t) x(t) \tag{i}$$

Since, in the linear regulator problem, the minimum value of the performance index is a time-varying quadratic function of state, it can be assumed that

$$J * (x(t), t) = \frac{1}{2} x^T (t) P(t) x(t) \tag{ii}$$

Substituting equation (ii) in equation (i), we get

$$0 = \frac{1}{2} x^T (t) P(t) x(t) + \frac{1}{2} x^T (t) Q(t) x(t) - \frac{1}{2} x^T (t) P(t) B(t) R^{-1}(t) B^T (t) P(t) x(t)$$

$$+ x^T (t) P(t) A(t) x(t)$$

$$0 = x^T \left[2P(t) A(t) - P(t) B(t) R^{-1}(t) B^T (t) P(t) + Q(t) + P(t) \right] x(t) \tag{7.79}$$

Equation (7.79) yields a solution of $x(t)$ only when equation (7.80) is satisfied.

$$2P(t) A(t) - P(t) B(t) R^{-1}(t) B^T (t) P(t) + Q(t) + P(t) = 0 \tag{7.80}$$

In equation (7.80), all the terms except $2P(t)A(t)$ are symmetric.

To make the solution symmetric, we know that, for a function representing $x^T Z x$, the symmetric part of 'Z' is only important, i.e., $Z_S = \dfrac{Z + Z^T}{2}$.

Thus, the symmetric part of $2P(t)A(t)$ can be written as given in equation (7.81).

$$2P(t) A(t) = 2 \frac{P(t) A(t) + A^T (t) P(t)}{2} = P(t) A(t) + A^T (t) P(t) \tag{7.81}$$

By substituting equation (7.81) in equation (7.80), we get

$$P(t)A(t) + A^T(t)P(t) - P(t)B(t)R^{-1}(t)B^T(t)P(t) + P(t) + Q(t) = 0 \qquad (7.82)$$

Equation (7.82) is satisfied when $P(t_f) = H$. Also, it is said to be symmetric, and hence, matrix H is also symmetric. Therefore, the solution $P(t)$ for all values of $t_i \le t \le t_f$ is also symmetric. This proves the symmetric property of the MRE.

Numerical Solution of Matrix Riccati Equation

The numerical solution to MRE is important because they help in quantitatively deducing/converging a solution for any optimal control problem. The numerical solution of MRE can be arrived using three different methods, namely, (1) direct integration, (2) negative exponential method and (3) Lyapunov method. Some other methods used in the literatures are Davison-Maki method, Schur method and Chandrasekhar method.

Direct Integration Method

The direct integration method can be applied to both time-varying and time-invariant cases. In this method, the MRE as given in equations (7.83) and (7.84) is expressed as a set of n^2 simultaneous non-linear first-order differential equations.

$$-\dot{P}(t) = Q(t) - P(t)B(t)R^{-1}(t)B^T(t)P(t) + P(t)A(t) + A^T(t)P(t) \qquad (7.83)$$

$$P(t_1) = H \qquad (7.84)$$

These n^2 simultaneous non-linear first-order differential equations are integrated backwards from t_1 to get the solution of the Riccati equation. To preserve the symmetrical property of the Riccati equation $P(t)$, these equations are reduced to $\dfrac{n(n+1)}{2}$ instead of n^2.

For time-varying case, integrate the Riccati equation with $P(t_1) = 0$ such that t_1 has a larger value. This leads to a constant matrix P^0.

For time-invariant case, techniques such as negative exponential method and Lyapunov methods are applied.

Advantages of Direct Integration Method

i. It is simple and can be implemented using a computer program.

ii. The derivation of P^0 matrix is accurate.

Disadvantages of Direct Integration Method

i. This method involves more number of steps to achieve accuracy. This is because time is discretized into smaller intervals to achieve greater accuracy.

ii. The symmetry of $P(t)$ may be destroyed due to numerical errors. This is compensated by reducing the n^2 equations or simply replacing $P(t)$ with $\dfrac{P^T(t) + P(t)}{2}$

Negative Exponential Method

The negative exponential method is employed for time-invariant systems and is very effective in evaluating the constant matrix P^0. From the derivation of LQR steady-state optimal control problem, the state and co-state matrices $\Phi(t,t_0)$ can be obtained as follows:

From equations (7.58) and (7.59), the relationship between state $x(t)$ and co-state $\lambda(t)$ can be written as follows:

$$\lambda(t) = P(t)X(t)$$

where

$$P(t) = [\Phi_{21}(t_f) + \Phi_{22}(t_f)H][\Phi_{11}(t_f) + \Phi_{12}(t_f)H]^{-1}$$

The matrices $\Phi_{ij}(t,t_0)$ are obtained using equation (7.49), which is reproduced here.

$$\begin{bmatrix} \dot{x}(t) \\ \dot{\lambda}(t) \end{bmatrix} = \begin{bmatrix} A & -BR^{-1}B^T \\ -Q & -A^T \end{bmatrix} \begin{bmatrix} x(t) \\ \lambda(t) \end{bmatrix}$$

$$\begin{bmatrix} \dot{x}(t) \\ \dot{\lambda}(t) \end{bmatrix} = [M] \begin{bmatrix} x(t) \\ \lambda(t) \end{bmatrix}$$

The solution for $\Phi(t_{i+1}, t_i)$ is given by $e^{-M\Delta t}$. Using power series algorithm, this exponential can be evaluated and can be converged to a constant matrix P^0.

Advantages of Negative Exponential Method

i. Computational complexity is less.

ii. The steady-state solution can be easily obtained.

Disadvantages of Negative Exponential Method

    i. Computation of the transformation matrix M, is unreliable.

Lyapunov Method

This method is an iterative method. Here, the Lyapunov-type equation is continuously iterated to get the desired solution. Consider a reduced MRE derived by substituting the constant matrix P^0 as in equation (7.85).

$$0 = A^T P^0 + P^0 A - P^0 B R^{-1} B^T P^0 + Q \qquad (7.85)$$

$$0 = A^T P^0 + P^0 A - P^0 S P^0 + Q \qquad (7.86)$$

where $S = BR^{-1}B^T$

    Let $f(P^0)$ be a matrix function as given in equation (7.87)

$$f(P^0) = Q - P^0 S P^0 + A^T P^0 + P^0 A \qquad (7.87)$$

An iterative procedure is employed to find P^0 such that, $f(P^0) = 0$. To derive the matrix P^0, assume that during ith iteration, solution P_i is much similar to P^0. Substituting $P_0 = P_i + \tilde{P}$ in equation (7.87), we get

$$f(P^0) = A^T(P_i + \tilde{P}) + (P_i + \tilde{P})A - (P_i + \tilde{P})S(P_i + \tilde{P}) + Q \qquad (7.88)$$

$$f(P^0) = A^T(P_i + \tilde{P}) + (P_i + \tilde{P})A - P_i S P_i - P_i S \tilde{P} - \tilde{P} S P_i + Q \qquad (7.89)$$

In equation (7.89), if $\tilde{P}_i$ produces $f(P^0) = 0$, then it can be rewritten as follows:

$$0 = A_i^T P_{i+1} + P_{i+1} A_i - P_i S P_i + Q \qquad (7.90)$$

$$\text{where } A_i = A - S P_i \text{ and } P_{i+1} = P_i + \tilde{P}_i \qquad (7.91)$$

Equation (7.90) resembles a Lyapunov direct method equation. Iterating this equation repeatedly yields the constant matrix P^0. Hence, this method is also called as iterative method. Choosing an appropriate initial value for P^0 promises the convergence.

Advantages of Lyapunov Method

    i. No updation of exponential terms is required. Hence, this method is computationally simpler than negative exponential method.

    ii. A proper choice of P^0 yields easy convergence.

Disadvantages of Lyapunov Method

i. For higher-order systems, the iterative method becomes tedious.

Discrete LQR Optimal Regulator Design

The LQR design for a discrete, linear, time-varying system is extended here. Consider the discrete LTV system as given in equations (7.92) and (7.93).

$$x(k+1) = A(k)x(k) + B(k)u(k) \tag{7.92}$$

$$y(k) = C(k)x(k) \tag{7.93}$$

where x is a $n \times 1$ state vector, u is the $p \times 1$ input vector, y is a $q \times 1$ output vector. A, B and C are, respectively, $n \times n$ state matrix, $n \times p$ input matrix and $q \times n$ output matrix. All the three matrices are time-varying real matrices.

Objective

To obtain the value of feedback gain $K(k)$ and thereby formulate the control law $u^*(k) = -K(k)x(k)$.

Case 1: Finite Horizon

The cost function for this optimal control problem is as given in equation (7.94).

$$J = \frac{1}{2}x^T(k_f)H(k_f)x(k_f) + \frac{1}{2}\sum_{k=k_i}^{k_f-1}\left[x^T(k)Q(k)x(k) + u^T(k)R(k)u(k)\right] \tag{7.94}$$

where H and Q are real, symmetric, positive semi-definite $n \times n$ matrices and R is a real, symmetric, positive definite $p \times p$ matrix.

Here, k_i represents initial time instant and k_{f-1} represents the final time instant.

The auxiliary cost function is given by equation (7.95).

$$J_\lambda = \lambda^T(k+1)\left[A(k)x(k) + B(k)u(k)\right] \tag{7.95}$$

Assumptions

1. The control input $u(k)$ is an unconstrained input.
2. The final discrete instant is k_f is specified, i.e., k_f is fixed.

3. The initial condition $x(k = k_i) = x(k_i)$ is given. However, the final state $x(k_f)$ is not given, i.e., $x(k_f)$ is free.

Assumptions (2) and (3) are also called the boundary conditions.
From the Pontryagin's equation, we can write

$$H(x(k), u(k), \lambda(k+1)) = \frac{1}{2} x^T(k)Q(k)x(k) + \frac{1}{2} u^T(k)R(k)u(k)$$

$$+ \lambda^T(k+1)\left[A(k)x(k) + B(k)u(k)\right] \qquad (7.96)$$

Along the optimal trajectory, we have the following equations

$$\frac{\partial H}{\partial u(k)}(x(k), u(k), \lambda(k+1)) = 0 \qquad (7.97)$$

$$R(k)u(k) + B^T(k)\lambda(k+1) = 0$$

and

$$u*(k) = -R^{-1}(k)B^T(k)\lambda(k+1) \qquad (7.98)$$

Let us minimize the Pontryagin's function, $H(x(k), u(k), \lambda(k+1))$ to get optimal function $H*(x(k), \lambda(k+1))$.
By substituting equation (7.98) in equation (7.96) and simplifying, we get

$$H*(x(k), \lambda(k+1)) = \frac{1}{2} x^T(k)Q(k)x(k) + \lambda^T(k+1)A(k)x(k)$$

$$- \frac{1}{2} \lambda^T(k+1)B(k)R^{-1}(k)B^T(k)\lambda(k+1)$$

The state and co-state equations can be obtained as follows:
State equation:

$$x*(k+1) = \frac{\partial H}{\partial \lambda(k+1)} = A(k)x(k) - B(k)R^{-1}(k)B^T(k)\lambda(k+1) \qquad (7.99)$$

$$x*(k+1) = \frac{\partial H}{\partial \lambda(k+1)} = A(k)x(k) - E(k)\lambda(k+1) \qquad (7.100)$$

where $E(k) = B(k)R^{-1}(k)B^T(k)$
Co-state equation:

$$\lambda*(k) = \frac{\partial H}{\partial x(k)} = Q(k)x(k) + A^T(k)\lambda(k+1) \qquad (7.101)$$

$$\left[\begin{array}{c} x^*(k+1) \\ \lambda^*(k) \end{array} \right] = \left[\begin{array}{cc} A(k) & -E(k) \\ Q(k) & A^T(k) \end{array} \right] \left[\begin{array}{c} x(k) \\ \lambda(k+1) \end{array} \right] \tag{7.102}$$

From the boundary conditions, we have

$$\lambda(k_f) = H(k_f)x(k_f) \tag{7.103}$$

To obtain a closed loop optimal configuration, the relation between co-state and state equations can be written as given in equation (7.104)

$$\lambda^*(k) = P(k)x^*(k) \tag{7.104}$$

where $P(k)$ is an unknown value.

By substituting equation (7.104) in (7.100) and (7.101), we get

$$x^*(k+1) = A(k)x(k) - E(k)P(k+1)x^*(k+1) \tag{7.105}$$

$$P(k)x^*(k) = Q(k)x(k) + A^T(k)P(k+1)x^*(k+1) \tag{7.106}$$

From equation (7.105), we get

$$x^*(k+1) + E(k)P(k+1)x^*(k+1) = A(k)x(k)$$

$$x^*(k+1)[I + E(k)P(k+1)] = A(k)x(k) \tag{7.107}$$

Substitute equation (7.107) in (7.106), we get

$$P(k)x^*(k) = Q(k)x(k) + A^T(k)P(k+1)[I + E(k)P(k+1)]^{-1}A(k)x(k) \tag{7.108}$$

If equation (7.108) holds good for all values of $x(k)$, then $P(k)$ can be written as

$$P(k) = Q(k) + A^T(k)P(k+1)[I + E(k)P(k+1)]^{-1}A(k) \tag{7.109}$$

Equation (7.109) is called the matrix difference Riccati equation. To obtain the solution, the Riccati equation is solved backwards. Here, $P(k)$ is positive definite.

From the boundary conditions, it can also be assumed that

$$\lambda(k_f) = H(k_f)x(k_f)$$

$$\lambda(k_f) = P(k_f)x(k_f)$$

$$\therefore \quad P(k_f) = H(k_f)$$

Now, we can obtain the optimal control law, $u^*(k)$

From co-state equation (7.101), we know that

$$\lambda^*(k) = Q(k)X(k) + A^T(k)\lambda(k+1)$$

$$\lambda^*(k) - Q(k)X(k) = A^T(k)\lambda(k+1) \tag{7.110}$$

Substituting the value of $\lambda(k) = P(k)x(k)$ from equation (7.104) in equation (7.110), we get

$$P(k)X(k) - Q(k)X(k) = A^T(k)\lambda(k+1)$$

$$(P(k) - Q(k))X(k) = A^T(k)\lambda(k+1)$$

$$-A^T(k)(P(k) - Q(k))X(k) = \lambda(k+1) \tag{7.111}$$

By substituting equation (7.111) in equation (7.98), we get

$$u^*(k) = -R^{-1}(k)B^T(k)P(k+1)x(k+1) \tag{7.112}$$

By substituting equation (7.92) in equation (7.112), we get

$$u^*(k) = -R^{-1}(k)B^T(k)P(k+1)\left[A(k)x(k) + B(k)u(k)\right] \tag{7.113}$$

We know that

$$u^*(k) = -Kx^*(k) \tag{7.114}$$

$$\text{where } K = \left[R(k) + B^T(k)P(k+1)B(k)\right]^{-1} B^T(k)P(k+1)A(k) \tag{7.115}$$

Now, the optimal state equation (7.116) can be obtained by substituting equation (7.114) in equation (7.92)

$$x(k+1) = A(k)x(k) + B(k)u(k)$$

$$\dot{x}^*(k+1) = [A(k) - B(k)K(k)]x^*(k) \tag{7.116}$$

The block diagram for discrete LQR is shown in Figure 7.6.

Case 2: Infinite Horizon

For the infinite time linear state regulator, the objective function is as given in equation (7.117).

$$J = \frac{1}{2}\sum_{k=k_i}^{\infty}\left[x^T(k)Q(k)x(k) + u^T(k)R(k)u(k)\right] \tag{7.117}$$

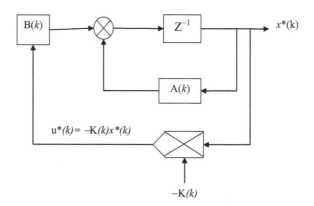

FIGURE 7.6
Block diagram for discrete linear quadratic controller design.

where the final time instant k_f tends to infinity ($k_f \to \infty$). Since $k_f \to \infty$ assumes the system to be time-invariant, this design produces a steady-state regulator design.

The design procedure is similar to that of finite horizon discrete LQR.

Procedural Steps for Solving Discrete LQR Optimal Control Problem

Step 1: Represent the plant by linear continuous time equation

$$x(k+1) = A(k)x(k) + B(k)u(k)$$

Step 2: Represent the performance index 'J' as

$$J = \frac{1}{2}x^T(k_f)H(k_f)x(k_f) + \frac{1}{2}\sum_{k=k_i}^{k_f-1}\left[x^T(k)Q(k)x(k) + u^T(k)R(k)u(k)\right]$$

Step 3: Form the Pontryagin's equation

$$H*(x(k), \lambda(k+1)) = \frac{1}{2}x^T(k)Q(k)x(k) + \lambda^T(k+1)(k)A(k)x(k)$$

$$-\frac{1}{2}\lambda^T(k+1)B(k)R^{-1}(k)B^T(k)\lambda(k+1)$$

Step 4: Minimize $H(x(k), u(k), \lambda(k+1))$

$$\text{i.e., } \frac{\partial H}{\partial u}(x(k), u(k), \lambda(k+1)) = 0$$

and hence, find $u*(k) = -\left[R^{-1}(k) + B^T(k)P(k+1)B(k)\right]^{-1}B^T(k)P(k+1)A(k)x(k)$

Step 5: Find the Pontryagin's function

$$H^*(x(k), \lambda(k+1)) = \min H(x(k), u(k), \lambda(k+1))$$

Step 6: Solve the state and co-state equations using the given boundary conditions

$$x^*(k+1) = \frac{\partial H}{\partial \lambda(k+1)} \quad \text{(State equation)}$$

$$\lambda^*(k) = \frac{\partial H}{\partial x(k)} \quad \text{(Co-state equation)}$$

Step 7: Substitute the results of Step 6 in $u^*(k)$, and hence, find the optimal control.

Linear Quadratic Tracking (LQT) Problem

As mentioned earlier, the main objective of the servo or tracking criteria is to make the system output track or follow a desired state $r(t)$ in a given time interval $[t_i, t_f]$. The general cost function associated with the tracking problem can be written as given in equation (7.25).

$$J = \frac{1}{2} e^T(t_f) H(t_f) e(t_f) + \frac{1}{2} \int_{t_i}^{t_f} (e^T(t)Q(t)e(t) + u^T(t)R(t)u(t)) dt$$

where $e(t)$ is an error value, i.e., $e(t) = x(t) - r(t)$, H and Q are positive, semi-definite, real, symmetric matrices. R is a real, symmetric, positive definite matrix.

Continuous Linear Quadratic Tracking (LQT) Controller Design

Consider the continuous LTV system as given in equations (7.26) and (7.27), respectively, which are reproduced here.

$$\dot{x} = A(t)x(t) + B(t)u(t) \tag{7.118}$$

$$y(t) = C(t)x(t) \tag{7.119}$$

where A is $n \times n$ state matrix, B is $n \times p$ control matrix, C is $q \times n$ output matrix, $x(t)$ is a $n \times 1$ state vector, $u(t)$ is a $p \times 1$ input vector, $y(t)$ is a $q \times 1$ output vector. For this continuous time-varying system, the control interval is given as $[t_i, t_f]$.

Objective

To obtain an optimal control law $u^*(t) = -K(t)x(t)$ such that the output tracks the desired state.

Case 1: Finite Horizon

The cost function for this optimal control problem is as given in equation (7.120).

$$J = \frac{1}{2}e^T(t_f)H(t_f)e(t_f) + \frac{1}{2}\int_{t_i}^{t_f}(e^T(t)Q(t)e(t) + u^T(t)R(t)u(t))dt \qquad (7.120)$$

where $e(t) = r(t) - y(t)$.
   From equation (7.119), it can be written as

$$e(t) = r(t) - C(t)x(t) \qquad (7.121)$$

Assumptions

1. The final time t_f is specified, i.e., t_f is fixed.
2. The initial condition $x(t_0) = x_0$ is given. However, the final state $x(t_f)$ is not given, i.e., $x(t_f)$ is free.

Assumption (2) is called the boundary conditions.
   From the Pontryagin's equation, we can write

$$H(x,u,\lambda,t) = \frac{1}{2}e^T(t)Q(t)e(t) + \frac{1}{2}u^T(t)R(t)u(t) + \lambda^T(t)[A(t)x(t) + B(t)u(t)] \qquad (7.122)$$

By substituting equation (7.121) in equation (7.122), we get

$$H(x,u,\lambda,t) = \frac{1}{2}[r(t) - C(t)x(t)]^T Q(t)[r(t) - C(t)x(t)]$$

$$+ \frac{1}{2}u^T(t)R(t)u(t) + \lambda^T(t)[A(t)x(t) + B(t)u(t)] \qquad (7.123)$$

Along the optimal trajectory, we have

$$\frac{\partial H}{\partial u}(x,u,\lambda,t)=0 \qquad\qquad (7.124)$$

$$R(t)u(t)+B^T(t)\lambda(t)=0$$

and

$$u*(t)=-R^{-1}(t)B^T(t)\lambda(t) \qquad\qquad (7.125)$$

Since, $\frac{\partial^2 H}{\partial u^2}(x,u,\lambda,t)=R(t)$ is positive definite, $u*(t)$ given in equation (7.125) minimizes the Pontryagin's function, $H(x,u,\lambda,t)$. By substituting equation (7.125) in equation (7.123) and after simplifying, we get

$$H*(x,\lambda,t)=\frac{1}{2}[r(t)-C(t)x(t)]^T Q(t)[r(t)-C(t)x(t)]+\lambda^T(t)A(t)x(t)$$

$$-\frac{1}{2}\lambda^T(t)B(t)R^{-1}(t)B^T(t)\lambda(t) \qquad\qquad (7.126)$$

The state and co-state equations can be obtained as follows:

$$\text{State equation } \dot{x}(t)=\frac{\partial H}{\partial\lambda}=A(t)x(t)+B(t)(-R^{-1}(t)B^T(t)\lambda(t)) \qquad (7.127)$$

$$\text{Co-state equation } \dot{\lambda}(t)=-\frac{\partial H}{\partial x}=-C^T(t)Q(t)C(t)x(t)-A^T(t)\lambda(t)+C^T(t)Q(t)r(t)$$

$$(7.128)$$

$$\begin{bmatrix} \dot{x}(t) \\ \dot{\lambda}(t) \end{bmatrix}=\begin{bmatrix} A(t) & -B(t)R^{-1}(t)B^T(t) \\ -C^T(t)Q(t)C(t) & -A^T(t) \end{bmatrix}\begin{bmatrix} x(t) \\ \lambda(t) \end{bmatrix}+\begin{bmatrix} 0 \\ C^T(t)Q(t) \end{bmatrix}r(t)$$

$$(7.129a)$$

Equation (7.129a) can be simplified as follows:

$$\begin{bmatrix} \dot{x}(t) \\ \dot{\lambda}(t) \end{bmatrix}=\begin{bmatrix} A(t) & -E(t) \\ -S(t) & -A^T(t) \end{bmatrix}\begin{bmatrix} x(t) \\ \lambda(t) \end{bmatrix}+\begin{bmatrix} 0 \\ w(t) \end{bmatrix}r(t) \qquad (7.129b)$$

where

$$S(t)=C^T(t)Q(t)C(t) \qquad\qquad (7.129c)$$

and

$$E(t) = B(t)R^{-1}(t)B^T(t) \qquad (7.129d)$$

From the boundary conditions, we have

$$\lambda(t) = \frac{\partial H}{\partial x}\bigg|_{t=t_f}$$

$$\lambda(t_f) = \frac{\partial H}{\partial x(t_f)}\left[\frac{1}{2}e^T(t_f)H(t_f)e(t_f)\right] \qquad (7.130)$$

$$\lambda(t_f) = \frac{\partial H}{\partial x(t_f)}\left[\frac{1}{2}\left[r(t_f)-C(t_f)x(t_f)\right]^T H(t_f)\left[r(t_f)-C(t_f)x(t_f)\right]\right]$$

(from equation (7.121))

By applying partial differential with respect to $x(t)$, we get

$$\lambda(t_f) = C^T(t_f)H(t_f)C(t_f)x(t_f) - C^T(t_f)H(t_f)r(t_f) \qquad (7.131)$$

To obtain a closed loop optimal configuration, the relation between co-state and state equations can be written as follows:

$$\lambda*(t) = P(t)x*(t) - f(t) \qquad (7.132)$$

where $P(t)$ is a $n \times n$ matrix and $f(t)$ is a n vector, are unknown values.
   Now, by differentiating equation (7.132), we get

$$\dot{\lambda}*(t) = \dot{P}(t)x*(t) + P(t)\dot{x}*(t) - \dot{f}(t) \qquad (7.133)$$

From equations (7.129a and 7.129c), we get

$$\dot{\lambda}(t) = -S(t)x(t) - A^T(t)\lambda(t) + w(t)r(t) \qquad (7.134)$$

where $w(t) = C^T(t)Q(t)$
   By substituting equations (7.132 and 7.133) in equation (7.134), we get

$$\dot{P}(t)x*(t) + P(t)\dot{x}*(t) - \dot{f}(t) = -S(t)x(t) - A^T(t)\left[P(t)x*(t) - f(t)\right] + w(t)r(t) \quad (7.135)$$

By substituting equations (7.127 and 7.129d) in equation (7.135), we get

$$\dot{P}(t)x*(t) + P(t)\left[A(t)x(t) + E(t)\lambda(t)\right] - \dot{f}(t)$$

$$= -S(t)x(t) - A^T(t)\left[P(t)x*(t) - f(t)\right] + w(t)r(t) \qquad (7.136)$$

By substituting equation (7.132) in equation (7.136), we get

$$\dot{P}(t)x^*(t) + P(t)\left[A(t)x(t) + E(t)\{P(t)x^*(t) - f(t)\}\right] - \dot{f}(t)$$

$$= -S(t)x(t) - A^T(t)\left[P(t)x^*(t) - f(t)\right] + w(t)r(t) \qquad (7.137)$$

Rearranging equation (7.137), we get

$$\left[\dot{P}(t) + P(t)A(t) + A^T(t)P(t) - P(t)E(t)P(t) + S(t)\right]x^*(t)$$

$$-\left[\dot{f}(t) + A^T(t)f(t) - P(t)E(t)f(t) + w(t)r(t)\right] = 0 \qquad (7.138)$$

If equation (7.138) holds good for all values of $x(t)$, then

$$\dot{P}(t) = -P(t)A(t) - A^T(t)P(t) + P(t)E(t)P(t) - S(t) \qquad (7.139)$$

Equation (7.139) is called the matrix difference Riccati equation. To obtain the solution, the Riccati equation is solved backwards. Here $P(k)$ is positive definite.

If equation (7.138) holds good for all values of $r(t)$ and t, then

$$\dot{f}(t) = \left[P(t)E(t) - A^T(t)\right]f(t) - w(t)r(t) \qquad (7.140)$$

By substituting equation (7.132) in control law $u(t)$, we get

$$\therefore \quad u^*(t) = -R^{-1}B^T(t)\left[P(t)x(t) - f(t)\right]$$

$$u^*(t) = -R^{-1}B^T(t)P(t)x(t) + R^{-1}B^T(t)f(t)$$

$$u^*(t) = -K(t)x(t) + R^{-1}(t)B^T(t)f(t) \qquad (7.141)$$

From the control law, the Kalman gain can be expressed as given in equation (7.142)

$$K(t) = R^{-1}B^T(t)P(t) \qquad (7.142)$$

To solve for the optimal state equation, substituting $u^*(t)$ in equation (7.118), we get

$$\dot{x}^* = [A(t) - B(t)R^{-1}B^T(t)P(t)]x(t) + B(t)R^{-1}B^T(t)f(t) \qquad (7.143)$$

$$\dot{x}^* = [A(t) - E(t)P(t)]x(t) + E(t)f(t) \qquad (7.144)$$

The block diagram for continuous time LQT controller design is shown in Figure 7.7.

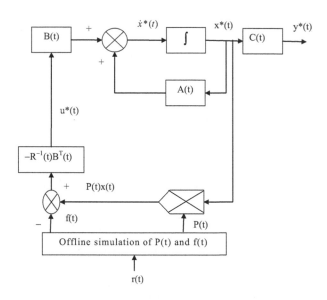

FIGURE 7.7
Block diagram of continuous time LQT controller design.

Case 2: Infinite Horizon

For the infinite time linear state regulator, the objective function is as in equation (7.145)

$$J = \frac{1}{2}\int_{t_i}^{\infty} (e^T(t)Q(t)e(t) + u^T(t)R(t)u(t))dt \qquad (7.145)$$

where the final time t_f tends to infinity ($t_f \to \infty$).

The design procedure is similar to that of finite horizon LQR.

Procedural Steps for Solving LQT Optimal Control Problem

Step 1: Represent the plant by linear continuous time equation

$$\dot{x} = A(t)x(t) + B(t)u(t)$$

$$y(t) = C(t)x(t)$$

Step 2: Represent the performance index 'J' as

$$J = \frac{1}{2}e^T(t_f)H(t_f)e(t_f) + \frac{1}{2}\int_{t_i}^{t_f} (e^T(t)Q(t)e(t) + u^T(t)R(t)u(t))dt$$

Step 3: Form the Pontryagin's equation

$$H(x,u,\lambda,t) = \frac{1}{2}e^T(t)Q(t)e(t) + \frac{1}{2}u^T(t)R(t)u(t) + \lambda^T(t)[A(t)x(t) + B(t)u(t)]$$

$$H(x,u,\lambda,t) = \frac{1}{2}[r(t) - C(t)x(t)]^T Q(t)[r(t) - C(t)x(t)] + \frac{1}{2}u^T(t)R(t)u(t)$$

$$+ \lambda^T(t)[A(t)x(t) + B(t)u(t)]$$

Step 4: Minimize $H(x,u,\lambda,t)$

$$\text{i.e., } \frac{\partial H}{\partial u}(x,u,\lambda,t) = 0$$

and hence, find $u(t) = -R^{-1}(t)B^T(t)\lambda(t)$
   Step 5: Find the Pontryagin's function

$$H^*(x,\lambda,t) = \min H(x,u,\lambda,t)$$

Step 6: Solve the state and co-state equations using the given boundary conditions

$$\dot{x} = \frac{\partial H}{\partial \lambda} \quad \text{(State equation)}$$

$$\dot{\lambda} = -\frac{\partial H}{\partial x} \quad \text{(Co-state equation)}$$

Step 7: Substitute the results of Step 6 in $u^*(t)$, and hence, find the optimal control.

Discrete LQT Optimal Regulator Design

The LQT design for a discrete, linear, time-varying system is extended here. Consider the discrete LTV system as given in equations (7.146) and (7.147).

$$x(k+1) = A(k)x(k) + B(k)u(k) \tag{7.146}$$

$$y(k) = C(k)x(k) \tag{7.147}$$

where x is a $n \times 1$ state vector, u is the $p \times 1$ input vector, y is a $q \times 1$ output vector. A, B and C are $n \times n$ state matrix, $n \times p$ input matrix and $q \times n$ output matrix, respectively. All the three matrices are time-varying real matrices.

Objective

To obtain an optimal control law $u^*(k) = -K(k)x(k)$ such that the output tracks the desired state.

Case 1: Finite Horizon

The cost function for this optimal control problem is as given in equation (7.148).

$$J = \frac{1}{2}e^T(k_f)H(k_f)e(k_f) + \frac{1}{2}\sum_{k=k_i}^{k_f-1}\int_{t_i}^{t_f}(e^T(k)Q(k)e(k) + u^T(k)R(k)u(k)) \quad (7.148)$$

where $e(k) = y(k)-r(k) = C(k)x(k)-r(k)$. H and Q are real, symmetric, positive semi-definite $n \times n$ matrices and R is a real, symmetric, positive definite $p \times p$ matrix. Here, k_i represents initial time instant and k_{f-1} represents the final time instant.

The auxiliary cost function is given by equation (7.149).

$$J_\lambda = \lambda^T(k+1)[A(k)x(k) + B(k)u(k)] \quad (7.149)$$

Assumptions

1. The control input $u(k)$ is an unconstrained input.
2. The final discrete instant is k_f is specified, i.e., k_f is fixed.
3. The initial condition $x(k = k_i) = x(k_i)$ is given. However, the final state $x(k_f)$ is not given, i.e., $x(k_f)$ is free.

Assumptions (2) and (3) are called the boundary conditions.

From the Pontryagin's equation, we can write

$$H(x(k), u(k), \lambda(k+1)) = \frac{1}{2}e^T(k)Q(k)e(k) + \frac{1}{2}u^T(k)R(k)u(k)$$

$$+ \lambda^T(k+1)[A(k)x(k) + B(k)u(k)] \quad (7.150)$$

Substituting $e(k) = C(k)x(k)-r(k)$ in equation (7.150), we get

$$H(x(k), u(k), \lambda(k+1)) = \frac{1}{2}[C(k)x(k) - r(k)]^T Q(t)[C(k)x(k) - r(k)]$$

$$+ \frac{1}{2}u^T(k)R(k)u(k) + \lambda^T(k+1)[A(k)x(k) + B(k)u(k)]$$

Along the optimal trajectory, we have

$$\frac{\partial H}{\partial u(k)}(x(k), u(k), \lambda(k+1)) = 0 \qquad (7.151)$$

$$R(k)u(k) + B^T(k)\lambda(k+1) = 0$$

and

$$u^*(k) = -R^{-1}(k)B^T(k)\lambda(k+1) \qquad (7.152)$$

Let us minimize the Pontryagin's function, $H(x(k), u(k), \lambda(k+1))$ to get optimal function $H^*(x(k), \lambda(k+1))$.

By substituting equation (7.152) in equation (7.150) and simplifying, we get

$$H^*(x(k), \lambda(k+1)) = \frac{1}{2}e^T(k)Q(k)e(k) + \lambda^T(k+1)(k)A(k)x(k)$$

$$-\frac{1}{2}\lambda^T(k+1)B(k)R^{-1}(k)B^T(k)\lambda(k+1)$$

The state and co-state equations can be obtained as follows:
State equation:

$$x^*(k+1) = \frac{\partial H}{\partial \lambda(k+1)} = A(k)x(k) - B(k)R^{-1}(k)B^T(k)\lambda(k+1) \qquad (7.153)$$

$$x^*(k+1) = \frac{\partial H}{\partial \lambda(k+1)} = A(k)x(k) - E(k)\lambda(k+1) \qquad (7.154)$$

where $E(k) = B(k)R^{-1}(k)B^T(k)$
Co-state equation:

$$\lambda^*(k) = \frac{\partial H}{\partial x(k)} = C^T(k)Q(k)C(k)x(k) + A^T(k)\lambda(k+1) - C^T(k)Q(k)r(k) \qquad (7.155)$$

$$\begin{bmatrix} x^*(k+1) \\ \lambda^*(K) \end{bmatrix} = \begin{bmatrix} A(k) & -B(k)R^{-1}(k)B^T(k) \\ C^T(k)Q(k)C(k) & A^T(k) \end{bmatrix}$$

$$\times \begin{bmatrix} x^*(k) \\ \lambda^*(k+1) \end{bmatrix} + \begin{bmatrix} 0 \\ -C^T(k)Q(k) \end{bmatrix}r(k)$$

$$\begin{bmatrix} x^*(k+1) \\ \lambda^*(K) \end{bmatrix} = \begin{bmatrix} A(k) & -E(k) \\ S(k) & A^T(k) \end{bmatrix}\begin{bmatrix} x^*(k) \\ \lambda^*(k+1) \end{bmatrix} + \begin{bmatrix} 0 \\ -w(k) \end{bmatrix}r(k) \qquad (7.156)$$

where
$$E(k) = B(k)R^{-1}(k)B^T(k)$$
$$S(k) = C^T(k)Q(k)C(k)$$
$$w(k) = C^T(k)Q(k)$$

From the boundary conditions, we have

$$\lambda(k_f) = C^T(k_f)H(k_f)C(k_f)x(k_f) - C^T(k_f)H(k_f)z(k_f) \tag{7.157}$$

To obtain a closed loop optimal configuration, the relation between co-state and state equations can be written as follows:

$$\lambda^*(k) = P(k)x^*(k) - f(k) \tag{7.158}$$

where $P(k)$ and $f(k)$ are unknown values.

By substituting equation (7.158) in (7.154) and (7.155), we get

$$x^*(k+1) = A(k)x(k) - E(k)P(k+1)x^*(k+1) + E(k)f(k+1) \tag{7.159}$$

$$P(k)x^*(k) - f(k) = C^T(k)Q(k)C(k)x(k) + \Phi^T(k)\lambda(k+1) - C^T(k)Q(k)r(k) \tag{7.160}$$

From equation (7.159), we have

$$x^*(k+1) + E(k)P(k+1)x^*(k+1) = A(k)x(k) + E(k)f(k+1)$$

$$x^*(k+1)[I + E(k)P(k+1)] = A(k)x(k) + E(k)f(k+1)$$

$$x^*(k+1) = [I + E(k)P(k+1)]^{-1}\left[A(k)x(k) + E(k)f(k+1)\right] \tag{7.161}$$

From equation (7.160), we have

$$P(k)x^*(k) - f(k) = C^T(k)Q(k)C(k)x(k) + A^T(k)\lambda(k+1) - C^T(k)Q(k)r(k)$$

By substituting equation (7.158) in (7.160), we get

$$P(k)x^*(k) - f(k) = C^T(k)Q(k)C(k)x(k) + A^T(k)\left[P(k+1)x^*(k+1) - f(k)\right]$$
$$- C^T(k)Q(k)r(k) \tag{7.162}$$

By substituting equation (7.161) in (7.162), we get

$$P(k)x^*(k) - f(k) = C^T(k)Q(k)C(k)x(k)$$
$$+ A^T(k)\Big[P(k+1)\big\{[I + E(k)P(k+1)]^{-1}$$
$$\times[A(k)x(k) + E(k)f(k+1)]\big\} - f(k)\Big] - C^T(k)Q(k)r(k) \tag{7.163}$$

Rearranging equation (7.163), we get

$$\left[-P(k)+A^T P(k+1)[I+E(k)P(k+1)]^{-1}A(k)+C^T(k)Q(k)C(k)\right]x(k)$$

$$+\left[f(k)+A^T P(k+1)[I+E(k)P(k+1)]^{-1}E(k)f(k+1)-A^T f(k+1)\right.$$

$$\left.-C^T(k)Q(k)r(k)\right]=0$$

$$\left[-P(k)+A^T P(k+1)[I+E(k)P(k+1)]^{-1}A(k)+S(k)\right]x(k)$$

$$+\left[f(k)+A^T P(k+1)[I+E(k)P(k+1)]^{-1}E(k)f(k+1)-A^T f(k+1)-w(k)r(k)\right]$$

$$=0 \tag{7.164}$$

If equation (7.164) holds good for all values of $x(k)$, then from equation (7.164), we get

$$-P(k)+A^T P(k+1)[I+E(k)P(k+1)]^{-1}A(k)+S(k)=0$$

$$P(k)=A^T P(k+1)[I+E(k)P(k+1)]^{-1}A(k)+S(k) \tag{7.165}$$

Equation (7.165) is called the matrix difference Riccati equation. To obtain the solution, the Riccati equation is solved backwards. Here $P(k)$ is positive definite.

If equation (7.164) holds good for all values of $r(k)$ and k, then from equation (7.164), we get

$$f(k)+A^T(k)P(k+1)[I+E(k)P(k+1)]^{-1}E(k)f(k+1)$$

$$-A^T(k)f(k+1)-w(k)r(k)=0$$

$$f(k)=-A^T(k)P(k+1)[I+E(k)P(k+1)]^{-1}E(k)f(k+1)+A^T(k)f(k+1)+w(k)r(k)$$

$$f(k)=A^T(k)\left\{I-[P^{-1}(k+1)+E(k)]^{-1}E(k)\right\}f(k+1)+w(k)r(k) \tag{7.166}$$

Now, to get optimal control law, substitute equation (7.158) in equation (7.152)

$$\therefore \quad u^*(k)=-R^{-1}(k)B^T(k)\left[P(k+1)x(k+1)-f(k+1)\right] \tag{7.167}$$

$$u^*(k)=-R^{-1}(k)B^T(k)P(k+1)x(k+1)+R^{-1}(k)B^T(k)f(k+1)$$

$$u^*(k)=-R^{-1}(k)B^T(k)P(k+1)\left[Ax(k)+Bu(k)\right]+R^{-1}(k)B^T(k)f(k+1)$$

$$u^*(k)=-R^{-1}(k)B^T(k)P(k+1)A(k)x(k)-R^{-1}(k)B^T(k)P(k+1)B(k)u(k)$$

$$+R^{-1}(k)B^T(k)f(k+1) \tag{7.168}$$

Equation (7.168) can also be represented as follows:

$$u^*(k) = -K(k)x(k) + K_f(k)f(k+1) \tag{7.169}$$

where

$K(k) = \left[R(k) + B^T(k)P(k+1)B(k) \right]^{-1} B^T P(k+1)A(k)$ is the feedback gain or the Kalman gain.

$K_f(k) = \left[R(k) + B^T(k)P(k+1)B(k) \right]^{-1} B^T$ is the feed forward gain.

By substituting equation (7.169) in equation (7.146), we can obtain the state equation as given in equation (7.170)

$$x(k+1) = \Phi(k)x(k) + \Gamma(k)u(k)$$

$$x(k+1) = \left[\Phi(k) - \Gamma(k)K(k) \right] x(k) + \Gamma(k)K_f(k)f(k+1) \tag{7.170}$$

The block diagram for discrete LQT is shown in Figure 7.8.

Procedural Steps for Solving Discrete LQT Optimal Control Problem

Step 1: Represent the plant by linear continuous time equation

$$x(k+1) = A(k)x(k) + B(k)u(k)$$

$$y(k) = C(k)x(k)$$

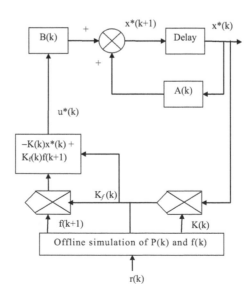

FIGURE 7.8
Block diagram for discrete LQT control.

Step 2: Represent the performance index 'J' as

$$J = \frac{1}{2}e^T(k_f)H(k_f)e(k_f) + \frac{1}{2}\sum_{k=k_i}^{k_f-1}(e^T(k)Q(k)e(k) + u^T(k)R(k)u(k))$$

Step 3: Form the Pontryagin's equation

$$H*(x(k),\lambda(k+1)) = \frac{1}{2}e^T(k)Q(k)e(k) + \lambda^T(k+1)(k)A(k)x(k)$$

$$-\frac{1}{2}\lambda^T(k+1)B(k)R^{-1}(k)B^T(k)\lambda(k+1)$$

Step 4: Minimize $H(x(k),u(k),\lambda(k+1))$

$$\text{i.e., } \frac{\partial H}{\partial u}(x(k),u(k),\lambda(k+1)) = 0$$

and hence, find $u*(k) = -R^{-1}(k)B^T(k)\lambda(k+1)$

Step 5: Find the Pontryagin's function

$$H*(x(k),\lambda(k+1)) = \min H(x(k),u(k),\lambda(k+1))$$

Step 6: Solve the state and co-state equations using the given boundary conditions

$$x*(k+1) = \frac{\partial H}{\partial\lambda(k+1)} \quad \text{(State equation)}$$

$$\lambda*(k) = \frac{\partial H}{\partial x(k)} \quad \text{(Co-state equation)}$$

Step 7: Substitute the results of Step 6 in $u*(k)$, and hence, find the optimal control.

Linear Quadratic Gaussian (LQG) Control

The LQG problem deals with linear system corrupted by Gaussian white noise which is additive in nature. LQG is a dynamic problem. The LQG problem is a combination of LQR with Linear Quadratic Estimator (LQE). Here, both LQR and LQE can be separately implemented. This is based on the separation principle. LQG can be designed for both time-variant and time-invariant systems. It can be continuous or discrete in time.

Continuous Time Linear Quadratic Gaussian (LQG) Control

Consider the continuous LTV system as given in equations (7.171) and (7.172), respectively

$$\dot{x}(t) = A(t)x(t) + B(t)u(t) + w(t) \tag{7.171}$$

$$y(t) = C(t)x(t) + v(t) \tag{7.172}$$

where A is $n \times n$ state matrix, B is $n \times p$ control matrix, C is $q \times n$ output matrix, $x(t)$ is a $n \times 1$ state vector, $u(t)$ is a $p \times 1$ input vector, $y(t)$ is a $q \times 1$ output vector. Additionally, $w(t)$ is a $n \times 1$ input noise vector and $v(t)$ is $q \times 1$ measurement noise vector. Both $w(t)$ and $v(t)$ are Gaussian white noise.

Objective

To obtain the optimal control law $u^*(t) = -K(t)x(t)$ that can minimize the cost function.

Assumptions

1. The final time t_f is specified, i.e., t_f is fixed.
2. The initial condition $x(t_i) = x_0$ is given. However, the final state $x(t_f)$ is not given, i.e., $x(t_f)$ is free.
3. $w(t)$ and $v(t)$ are Gaussian white noises, i.e., zero mean white noise.
4. $E[w(t)] = E[v(t)] = 0$ and also $E[w(t)w^T(\tau)] = Q\delta(t)$ where Q is known as input noise covariance matrix. Also $E[v(t)v^T(\tau)] = R\delta(t)$ where R is known as measurement noise covariance matrix.
5. The noises are uncorrelated, i.e., $E[w(t)v^T(t)] = 0$.
6. Initial state $x(t_i) = x_0$ also has zero mean. $E[x_0] = 0$. Also $E[x_0 x_0^T] = S\delta(t)$ where S is covariance matrix.
7. The initial state is uncorrelated with noise matrices.
   $E[x_0 w^T(t)] = E[x_0 v^T(t)] = 0$ for all values of t.

Also the following assumptions are made

1. The system is controllable.
2. The system is observable.
3. All the state variables are not measurable.

Case 1: Finite Horizon

The cost function (J) for this optimal control problem is as given in equation (7.173).

$$J = \frac{1}{2} E \left[x^T(t_f) H(t) x(t_f) + \int_{t_i}^{t_f} (x^T(t)Q(t)x(t) + u^T(t)R(t)u(t))dt \right] \quad (7.173)$$

where E is the expected value. The control interval is given as $[t_i, t_f]$. Here, H and Q are positive semi-definite, real, symmetric matrices. R is a real, symmetric, positive definite matrix.

For an LQG problem, both LQ regulator and LQ estimator can be designed separately. Thus, LQG follows the property of separation principle.

LQR Design

The continuous time LQR design is same as that discussed previously. Therefore, only some important conclusions from the LQR derivation are repeated here.

From equation (7.38), the control law $u(t)$ for the regulator design can be given as follows:

$$u(t) = -R^{-1}(t)B^T(t)P(t)x(t) \quad (7.174)$$

$$u(t) = -Kx(t) \quad (7.175)$$

where K is the feedback gain given by $K = R^{-1}(t)B^T(t)P(t)$ and $P(t)$ is the solution of the MRE as given in equation (7.41). It can be rewritten as follows:

$$A^T(t)P(t) + P(t)A(t) - P(t)B(t)R^{-1}(t)B^T(t)P(t) + Q(t) = 0 \quad (7.176)$$

where $\dot{P}(t)$ is zero.

From Assumption (3), it can be noted that, in LQG problem, it is assumed that all the state variables are not measurable. Therefore, it can be noted from equation (7.175), the optimal control law cannot be obtained unless all the state variables $x(t)$ are estimated. Hence, an LQE must be designed.

LQE Design

Here, we design a linear estimator like a Kalman filter for estimation.

Let $\hat{x}(t)$ be the estimated state. The cost function for the estimator can be given as in equation (7.177)

$$J = E\left[(x(t) - \hat{x}(t))^T (x(t) - \hat{x}(t)) \right] \quad (7.177)$$

For simplicity, the Kalman estimator equation is given in equation (7.178) directly. However, it is dealt in detail in the Chapter 8. Moreover in this chapter, to differentiate Kalman gain with feedback gain (K), the Kalman gain is denoted as 'L' whereas in Chapter 8, it is denoted as 'K'.

$$\dot{\hat{x}}(t) = A(t)\hat{x}(t) + B(t)u(t) + L\big(y(t) - C(t)\hat{x}(t)\big) \tag{7.178}$$

where L is a $n \times 1$ Kalman gain and the term $y(t) - C(t)\hat{x}(t)$ is called the error term.

Equation (7.178) can also be written as in equation (7.179)

$$\dot{\hat{x}}(t) = \big(A(t) - L(t)C(t)\big)\hat{x}(t) + B(t)u(t) + Ly(t) \tag{7.179}$$

To find the value of $L(k)$, equations from LQR design can be used. Since, regulator problem is a dual of estimator problem, due to the duality property, LQR equations can be rewritten as LQE equations by replacing A as A^T, B as C^T, Q as W, R as V and L as K^T.

NOTE

1. W is the process noise covariance matrix.
2. V is the measurement noise covariance matrix.
3. In Chapter 8, the process noise covariance matrix and measurement noise covariance matrices are denoted by Q and R, respectively.
4. Whereas in this chapter to differentiate weighing matrices Q and R, the process noise covariance matrix and measurement noise covariance matrix are denoted by W and C, respectively.

Therefore, equation $K = R^{-1}(t)B^T(t)P(t)$ can be modified as follows:

$$L = P_e(t)C^T(t)V^{-1}(t) \tag{7.180}$$

In equation (7.180), C and V are known. P_e is unknown.

To find the value of P_e, consider equation (7.176) and modify it using the duality property as given in equation (7.181)

$$A(t)P_e(t) + P_e(t)A^T(t) - P_e(t)C^T(t)V^{-1}(t)C(t)P_e(t) + W(t) = 0 \tag{7.181}$$

$$A(t)P_e(t) + P_e(t)A^T(t) - P_e(t)C^T(t)V^{-1}(t)C(t)P_e(t) = 0 \quad (\because W(t) = 0)$$

The solution can be solved to get the value of $P_e(t)$.

Thus, the LQG design can be summarized using equations (7.175), (7.178) and (7.180) as given below.

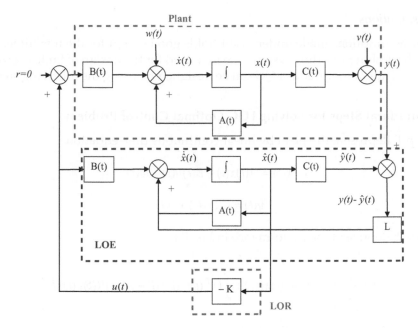

FIGURE 7.9
Block diagram for LQG design.

State Equation (Estimator)

$$\dot{\hat{x}}(t) = A(t)\hat{x}(t) + B(t)u(t) + L\left(y(t) - C\hat{x}(t)\right)$$

where 'L' is the Kalman gain given by $L = P_e(t)C^T(t)V^{-1}(t)$

Control Law (Regulator)

$$u(t) = -K\hat{x}(t)$$

where 'K' is the feedback gain given by $K = R^{-1}(t)B^T(t)P(t)$
The block diagram for LQG design is shown in Figure 7.9.

Case 2: Infinite Horizon

For the infinite time LQG, the objective function is as in equation (7.182)

$$J = \frac{1}{2}E\left[\int_{t_i}^{t_f}(x^T(t)Q(t)x(t) + u^T(t)R(t)u(t))dt\right] \qquad (7.182)$$

where the final time t_f tends to infinity $(t_f \rightarrow \infty)$.

Assumptions

The assumptions made under case 1 holds good except for the final time.
   The design procedure is similar to that used for the design of finite horizon LQG. Thus, a continuous LQG can be designed by using the following steps.

Procedural Steps for Solving LQG Optimal Control Problem

Step 1: Represent the plant by linear continuous time equation

$$\dot{x}(t) = A(t)x(t) + B(t)u(t) + w(t)$$

$$y(t) = C(t)x(t) + v(t)$$

Step 2: Represent the performance index 'J' as

$$J = \frac{1}{2}E\left[x^T(t_f)H(t)x(t_f) + \int_{t_i}^{t_f}(x^T(t)Q(t)x(t) + u^T(t)R(t)u(t))dt \right]$$

Step 3: Design LQE to estimate the states

$$\dot{\hat{x}}(t) = A(t)\hat{x}(t) + B(t)u(t) + L\big(y(t) - C\hat{x}(t)\big)$$

where $L = P_e(t)C^T(t)V^{-1}(t)$
   Step 4: Design the LQR and determine the optimal control law

$$u(t) = -K\hat{x}(t)$$

where $K = R^{-1}(t)B^T(t)P(t)$
   Step 5: Combine Steps 3 and 4, to complete LQG design problem.

8

Optimal Estimation

This chapter introduces Kalman filter algorithm for optimal state estimation. The review of statistical tools needed in the design of Kalman filter namely: Sample mean, sample standard deviation, sample variance, sample covariance and variance, covariance and cross-covariance of stochastic process are discussed. The introduction to Kalman filter design, Gaussian noise, advantages, disadvantages, applications of Kalman filter and assumptions used in Kalman filter design are initially explained. The Kalman filter design for continuous-time systems and time-invariant systems and discrete-time systems are also discussed in detail with the help of flowcharts and block diagrams. The estimation procedures are illustrated with the help of numerical examples. The Extended Kalman Filter (EKF) for multi-dimensional system is also explained briefly.

Statistical Tools

The statistical tools play a major role in analysing the process data. The tools such as mean, median, mode, geometric mean and harmonic mean help in measuring the central tendency. The standard deviation and variance measure the dispersion. The asymmetry is measured using skewness and kurtosis. In the context of the Kalman filter design, the statistical tools, namely mean, variance and standard deviation are briefly reviewed.

Mean ($\bar{X}$)

It is the most common tool used for measuring the central tendency.

$$\text{Mean}(\bar{X}) = \frac{\sum\limits_{i=1}^{n} x_i}{n} = \frac{x_1 + x_2 + \cdots + x_n}{n} \tag{8.1}$$

where

    $'n'$ is the number of data samples

    $'x_i'$ is the value of ith data sample

    $\bar{X}$ is the mean or average value

Standard Deviation (σ)

The mean or average cannot reveal the entire dynamics. In order to measure the 'scatter' of values of the process data, the statistical tool, namely, standard deviation (σ) is used. It is given by square root of the average of squares of deviation.

$$\sigma = \sqrt{\dfrac{\displaystyle\sum_{i=1}^{n} (\bar{X} - x_i)^2}{(n-1)}} \tag{8.2}$$

where

    'n' is the number of data samples

    'x_i' is the value of ith data sample

    '$\bar{X}$' is the mean or average value

    'σ' is the standard deviation

Variance (σ^2)

It is the average of squared differences from the mean. In other words, it is square of standard deviation. It measures how far a set of random numbers are spread out from their average value.

$$\sigma^2 = \frac{\sum_{i=1}^{n}(\bar{X} - x_i)^2}{(n-1)} \qquad (8.3)$$

NOTE

i. Variance in two dimensions $= \begin{bmatrix} \sigma_x^2 & \sigma_x\sigma_y \\ \sigma_y\sigma_x & \sigma_y^2 \end{bmatrix}$ $\qquad (8.4)$

ii. Variance in three dimensions $= \begin{bmatrix} \sigma_x^2 & \sigma_x\sigma_y & \sigma_x\sigma_z \\ \sigma_y\sigma_x & \sigma_y^2 & \sigma_y\sigma_z \\ \sigma_z\sigma_x & \sigma_z\sigma_y & \sigma_z^2 \end{bmatrix}$ $\qquad (8.5)$

Covariance

It is a measure of joint variability of two random variables. Though it does not assess the dependency, it evaluates to what extent the two variables change together.

$$\text{If } \bar{X} = \frac{\sum_{i=1}^{n} x_i}{n} \text{ and } \bar{Y} = \frac{\sum_{i=1}^{n} y_i}{n}$$

$$\text{Then, } Cov(X,Y) = \frac{\sum_{i=1}^{n}(\bar{X} - x_i)(\bar{Y} - y_i)}{(n-1)} \qquad (8.6)$$

Mean, Variance and Covariance and Cross-Covariance of Stochastic Process

In the earlier session, we have discussed the sample mean, sample variance and sample covariance. Here, the mean, variance and covariance and cross-covariance of random vectors, continuous-time vector stochastic process and discrete-time vector stochastic process are discussed briefly.

Case 1: Sample Random Variable (*x*)

In the case of sample random variable (*x*),

$$\text{Mean or Average or expected value, } \bar{X} = E\{x\} \tag{8.7}$$

where '*E*' is called the expectation operator

$$\text{Variance, } V_x = E\{(x - \bar{x})^2\} \tag{8.8}$$

$$\text{Standard deviation} = \sqrt{V_x} \tag{8.9}$$

Covariance between two random variables *x* and $y = V_{x,y} = \text{Cov}(x, y)$

$$= E\{(x - \bar{x})(y - \bar{y})\} \tag{8.10}$$

Case 2: Random Vector (*X*)

If $x_1, x_2, \ldots, x_n$ are the random variables, then the random vector is given by

$$\text{Random vector } (X) = \begin{bmatrix} x_1 & x_2 & \cdots & x_n \end{bmatrix}^T \tag{8.11}$$

$$\text{Mean value vector } (\bar{X}) = E\{x\} = \begin{bmatrix} \bar{x}_1 & \bar{x}_2 & \cdots & \bar{x}_n \end{bmatrix}^T \tag{8.12}$$

$$\text{Covariance matrix } (V_x) = E\left\{\left(X - \bar{X}\right)\left(X - \bar{X}\right)^T\right\} \tag{8.13}$$

NOTE

i. V_x is a symmetric and non-negative definite matrix.
ii. The *i*th diagonal element will be the variance (x_i).
iii. The *ij*th element will be the *Cov* (x_i, x_j).

Case 3: Continuous-Time Vector Stochastic Process

It is a collection of random vectors $\{x(t) : t_0 \leq t \leq t_1\}$

$$\text{Mean value, } \bar{X}(t) = E\{x(t)\}; t_0 \leq t \leq t_1 \tag{8.14}$$

$$\text{Variance matrix } (Q(t)) = V_{X,X}(t, \tau) = E\left\{\left(X(t) - \bar{X}(\tau)\right)\left(X(t) - \bar{X}(\tau)\right)^T\right\} \tag{8.15}$$

$$\text{Covariance matrix } V_{X,X}(t, \tau) = \text{Cov}(X(t), X(\tau)) \tag{8.16}$$

$$V_{X,X}(t, \tau) = E\left\{(X(t) - \bar{X}(t))(X(\tau) - \bar{X}(\tau))^T\right\} \tag{8.17}$$

NOTE

1. The cross-covariance matrix (between $X(t)$ and $Y(t)$ is given by,
$$V_{X,Y}(t, \tau) = Cov(X(t), Y(\tau))$$

or

$$V_{X,Y}(t, \tau) = E\left\{(X(t) - \bar{X}(t))(Y(\tau) - \bar{Y}(\tau))^T\right\} \tag{8.18}$$

2. In the case of wide-sense white noise stochastic process (w), where $w(t)$ and $w(\tau)$ are uncorrelated, the covariance matrix is given by

$$V_{w,w}(t, \tau) = w(t)\delta(t - \tau) \tag{8.19}$$

where $\delta(t - \tau)$ is a delta function.

Case 4: Discrete-Time Vector Stochastic Process

In this case, mean, variance and covariance can be determined as follows:

$$\text{Mean value,} \left(\bar{X}(i)\right) = E\{X(i)\} \tag{8.20}$$

$$\text{Variance matrix } V_{X,X}(i, i) = Q(i)$$

$$V_{X,X}(i, i) = E\left\{\left(X(i) - \bar{X}(i)\right)\left(X(i) - \bar{X}(i)\right)^T\right\} \tag{8.21}$$

NOTE

In the case of uncorrelated vector-valued stochastic variable, the covariance matrix is given by

$$V_{X,X}(i, j) = W(i)\mu(i - j) \tag{8.22}$$

where $\mu(i - j)$ is the unit pulse signal.

Illustration 1

The values of two variables (x and y) in a system vary as follows:

x	2	4	6	8	10
y	3	5	7	9	11

Compute the following:

    i. Mean $(\bar{X})$ and $(\bar{Y})$
    ii. Standard deviation $(\sigma_x$ and $\sigma_y)$
    iii. Variance $(\sigma_x^2$ and $\sigma_y^2)$
    iv. Covariance $(Cov(X,Y))$

Solution

Given that there are two variables x and y whose values are as follows.

x	2	4	6	8	10
y	3	5	7	9	11

i. To find the mean

$$\bar{X} = \frac{\sum_{i=1}^{n} x_i}{n} = \frac{\sum_{i=1}^{5} x_i}{5} \quad (\because n = 5)$$

$$\bar{X} = \frac{2+4+6+8+10}{5} = 6$$

$$\bar{Y} = \frac{\sum_{i=1}^{n} y_i}{n} = \frac{\sum_{i=1}^{5} y_i}{5} \quad (\because n = 5)$$

$$\bar{Y} = \frac{3+5+7+9+11}{5} = 7$$

ii. To find the standard deviation

$$\text{Standard deviation } (\sigma_x) = \sqrt{\frac{\sum_{i=1}^{n} (\bar{X} - x_i)^2}{(n-1)}}$$

$$\sigma_x = \sqrt{\frac{(6-2)^2 + (6-4)^2 + (6-6)^2 + (6-8)^2 + (6-10)^2}{4}}$$

$$\sigma_x = \sqrt{\frac{40}{4}} = \sqrt{10} = 3.16$$

$$\text{Standard deviation } (\sigma_y) = \sqrt{\frac{\sum\limits_{i=1}^{n}(\bar{Y}-y_i)^2}{(n-1)}}$$

$$\sigma_y = \sqrt{\frac{(5-3)^2 + (5-5)^2 + (5-7)^2 + (5-9)^2 + (5-11)^2}{4}}$$

$$\sigma_y = \sqrt{\frac{60}{4}} = \sqrt{15} = 3.87$$

iii. Variance $(\sigma_x^2) = (3.16)^2 = 10$

    Variance $(\sigma_y^2) = (3.87)^2 = 15$

iv. Covariance $(Cov(X,Y)) = \dfrac{\sum\limits_{i=1}^{n}(\bar{X}-x_i)(\bar{Y}-y_i)}{(n-1)}$

$$\left(\bar{X}-x_i\right)_{i=1\,to\,5} = (6-2),(6-4),(6-6),(6-8),(6-10)$$

$$\left(\bar{X}-x_i\right)_{i=1\,to\,5} = 4,\,2,\,0,\,-2,\,-4$$

$$\left(\bar{Y}-y_i\right)_{i=1\,to\,5} = (7-3),(7-5),(7-7),(7-9),(7-11)$$

$$\left(\bar{Y}-y_i\right)_{i=1\,to\,5} = 4,2,0,-2,-4$$

$$\frac{\sum\limits_{i=1}^{n}\left(\bar{X}-x_i\right)\left(\bar{Y}-y_i\right)}{(n-1)} = \frac{(4\times4)+(2\times2)+(0\times0)+(-2\times-2)+(-4\times-4)}{(5-1)}$$

$$Cov(X,Y) = \frac{16+4+0+4+16}{4} = \frac{40}{4} = 10$$

It is a positive covariance. It means that the relationship between 'X' and 'Y' is a linear relationship. It is also observed that, when X is increasing, Y is also increasing.

NOTE

a. Positive covariance indicates that the two variables move in the same direction.

b. Negative covariance indicates that the two variables move in opposite direction.

Illustration 2

The marks obtained by three students (A, B, C) in three subjects offered in an undergraduate programme are given below.

		Marks Secured	
Student	Process Control (P)	System Identification (S)	Transducers (T)
A	90	60	30
B	60	60	60
C	30	30	60

Compute the covariance matrix and hence comment on the results.

Solution

The marks secured by the three students in three subjects, namely, Process Control (P), System Identification (S), Transducers (T) can be arranged in the form of a matrix (M).

$$
M = \begin{bmatrix} 90 & 60 & 30 \\ 60 & 60 & 60 \\ 30 & 30 & 60 \end{bmatrix}
$$

Step 1: Find the mean ($\bar{P}, \bar{S}$ and $\bar{T}$ for Process Control, System Identification and Transducers subjects)

$$
\text{Class average in Process Control } (\bar{P}) = \frac{90+60+30}{3} = 60
$$

$$
\text{Class average in System Identification } (\bar{S}) = \frac{60+60+30}{3} = 50
$$

$$
\text{Class average in Transducers } (\bar{T}) = \frac{30+60+60}{3} = 50
$$

Step 2: Find the variance ($\sigma_P{}^2, \sigma_S{}^2, \sigma_T{}^2$)

$$
\sigma_P{}^2 = \frac{(60-90)^2 + (60-60)^2 + (60-30)^2}{(3-1)} = 900
$$

$$
\sigma_S{}^2 = \frac{(50-60)^2 + (50-60)^2 + (50-30)^2}{(3-1)} = 300
$$

$$
\sigma_T{}^2 = \frac{(50-30)^2 + (50-60)^2 + (50-60)^2}{(3-1)} = 300
$$

Step 3: Find the covariance ($Cov(P,S)$, $Cov(S,T)$ and $Cov(T,P)$)

$$Cov\,(P,S) = \frac{\big((60-90)(50-60)\big) + \big((60-60)(50-60)\big) + \big((60-30)(50-30)\big)}{}$$

$$Cov(P,S) = 450$$

$$Cov\,(S,T) = \frac{\big((50-60)(50-30)\big) + \big((50-60)(50-60)\big) + \big((50-30)(50-60)\big)}{(3-1)}$$

$$Cov(S,T) = -150$$

$$Cov(T,P) = \frac{\big((50-30)(60-90)\big) + \big((50-60)(60-60)\big) + \big((50-60)(60-30)\big)}{(3-1)}$$

$$Cov(T,P) = -450$$

Thus, the covariance matrix can be written as follows:

$$\text{Overall covariance matrix} = \begin{bmatrix} 900 & 450 & -450 \\ 450 & 300 & -150 \\ -450 & -150 & 300 \end{bmatrix} = \begin{bmatrix} PP & PS & PT \\ SP & SS & ST \\ TP & TS & TT \end{bmatrix}$$

Comments

1. In the above matrix, the value '+450' corresponding to 'PS' indicates that the marks scored by A, B and C in both Process Control and System Identification subjects are in descending order (same order).
2. The value '−450' corresponding to 'PT' indicates that the marks scored by A, B and C in Process Control is in descending order whereas the marks scored in Transducers is in ascending order (opposite order).

Gaussian Noise

It is a statistical noise with a probability density function equal to that of Gaussian distribution. It derives the name from Carl Friedrich Gauss, a German mathematician and physicist.

Kalman Filter

Kalman filter is named after Rudolf Emil Kalman a Hungarian-American electrical engineer and mathematician. It is an optimal recursive estimator which minimizes mean square error of the estimated parameters. The name filter refers to filtering the error or uncertainty or noise from noisy measurements to give the values of desired states.

Advantages of Kalman Filter

- It is easily implementable on a digital computer
- Compatible with state space formulation of optimal controllers
- Computation increases only linearly with the number of detectors
- Easy to formulate and implement
- No need to invert measurement equations
- More robust.

Disadvantages of Kalman Filter

- Assumptions are too restrictive
- It can represent only Gaussian distributions.

Applications of Kalman Filter

Kalman filter is used in many applications; some of them include the following:

- Multi-sensor fusion (using data from radar, scanner and camera for depth and velocity measurements)
- Tracking in interactive computer graphics
- Feature tracking/cluster tracking
- Tracking objects (missiles, heads, hands)
- Navigation
- Economics.

Continuous-Time Kalman Filter Algorithm

Let us consider a linear continuous-time dynamic system represented by the following equations (8.23 and 8.24)

$$\dot{x}(t) = A(t)x(t) + B(t)u(t) + w(t) \tag{8.23}$$

$$y(t) = C(t)x(t) + v(t) \tag{8.24}$$

where

 $x(t)$ is a state vector $(n \times 1)$

 $u(t)$ is a control vector $(p \times 1)$

 $y(t)$ is a measurement vector $(q \times 1)$

 $A(t)$ is a system matrix containing coefficients of state terms $(n \times n)$

 $B(t)$ is a system matrix containing coefficients of input terms $(n \times p)$

 $C(t)$ is a measurement or observation matrix $(q \times n)$

 $w(t)$ is a state excitation noise vector $(n \times 1)$

 $v(t)$ is a measurement (observation) noise vector $(q \times 1)$

The block diagram of continuous-time dynamic system with process noise and measurement noise is as shown in Figure 8.1.

Steps Involved in the Continuous-Time Kalman Filter Design

Let us first define the notation to be used in the flowchart for the design of continuous-time Kalman filter.

 $x(t_0)$ – Initial state vector

 $E(x(t_0))$ – Mean of initial state vector = $\overline{X}_0$

 $E(w(t))$ – Mean of state noise vector

 $E(v(t))$ – Mean of measurement noise vector

 $Q(t)$ – Process noise covariance matrix

 $R(t)$ – Measurement noise covariance matrix

 P_0 – Variance matrix (of initial state vector)

 $\delta(t - \tau)$ – Delta function

 $E(w(t)w^T(\tau))$ – Process noise covariance matrix

 $E(v(t)\,v^T(\tau))$ – Measurement noise covariance matrix

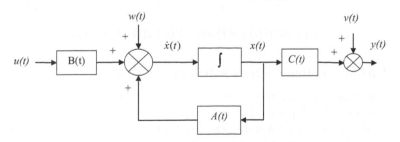

FIGURE 8.1
Block diagram of continuous-time dynamic system with process noise and measurement noise.

$E(w(t)\, v^T(\tau))$ – Cross-covariance matrix (between w and v)

$E(x(t_0)\, v^T(t))$ – Cross-covariance matrix (between x and v)

$E(x(t_0)w^T(t))$ – Cross-covariance matrix (between x and w)

$\hat{x}(t)$ – Current estimated state

$P(t)$ – Variance matrix

Step 1: Initialize the following

$E(x(t_0)) = \bar{x}_0$

$E(w(t)) = E(v(t)) = 0$

$E(w(t)w^T(\tau)) = Q(t)\delta(t-\tau)$

$E(v(t)v^T(\tau)) = R(t)\delta(t-\tau)$

$E(w(t)\, v^T(\tau)) = 0$

$E(x(t_0)\, v^T(t)) = 0$

$E(x(t_0)w^T(t)) = 0$

$E\left\{(x(t_0)-\bar{x}_0)(x(t_0)-\bar{x}_0)^T\right\} = P_0$

$\hat{X}(0) = \bar{X}_0$

Step 2: Solve the matrix Riccati equation for $P(t)$

$$\dot{P}(t) = Q(t) - P(t)C^T(t)R^{-1}(t)C(t)P(t) + P(t)A^T(t) + A(t)P(t) \tag{8.25}$$

$$P(t_0) = P_0 \tag{8.26}$$

Step 3: Compute the Kalman gain $K(t)$

$$K(t) = P(t)C^T(t)R^{-1}(t) \tag{8.27}$$

Step 4: Find the values of $\dot{\hat{x}}(t)$

$$\dot{\hat{x}}(t) = A(t)\hat{x}(t) + B(t)u(t) + K(t)\left[y(t) - C(t)\hat{x}(t)\right] \tag{8.28}$$

Flowchart for Continuous-Time Kalman Filter Design

Thus, the flowchart for the continuous-time Kalman filter design can be drawn as shown in Figure 8.2.

Thus, the block diagram for continuous-time Kalman filter with the dynamic system can be obtained as follows:

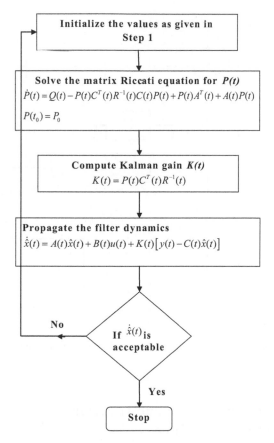

FIGURE 8.2
Flowchart for continuous-time Kalman filter algorithm.

From the continuous-time Kalman filter algorithm, the estimated value of the state is given by

$$\dot{\hat{x}}(t) = \left[A(t)\hat{x}(t) + B(t)u(t) \right] + K(t)\left[y(t) - C(t)\hat{x}(t) \right] \tag{i}$$

Also the plant is represented by the following equations:

$$\dot{x}(t) = A(t)x(t) + B(t)u(t) + w(t) \tag{ii}$$

$$y(t) = C(t)x(t) + v(t) \tag{iii}$$

From equations (i), (ii) and (iii), we can draw the block diagram of continuous-time Kalman filter with the dynamic system as shown in Figure 8.3.

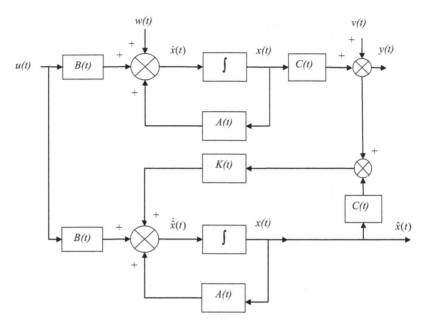

FIGURE 8.3
Block diagram of continuous-time Kalman filter with dynamic systems.

State Estimation in Linear Time-Invariant Systems Using Kalman Filter

In the case of time-invariant system, the matrices A, B and C are constant. Hence, the dynamic system can be represented as follows:

$$\dot{x}(t) = Ax(t) + Bu(t) + w(t)$$

$$y(t) = Cx(t) + v(t)$$

Since, both the state excitation noise $w(t)$ and measured noise $v(t)$ are stationary, the matrices Q and R also become constant matrices.

Moreover, if the observation time is long compared to the dominant time constant of the process, then $\hat{x}(t)$, and hence, $e(t) = x(t) - \hat{x}(t)$ becomes stationary.

In the light of above discussion, we have

$$P(t) = P(t_0) = P^C \quad \text{(Constant matrix)} \tag{8.29}$$

and

$$\dot{P}(t) = 0 \tag{8.30}$$

$$0 = Q - P^c C^T R^{-1} C P^c + P^c A^T + A P^c \qquad (8.31)$$

Solve the above equation for P^c with appropriate values for A, B, C, Q and R.

Thus, the expression for Kalman gain (K) and for the estimated state for a linear time-invariant system can be given as follows.

$$K = P^c C^T R^{-1} \qquad (8.32)$$

and

$$\dot{\hat{x}}(t) = A\hat{x}(t) + Bu(t) + K(t)\left[y(t) - C\hat{x}(t)\right] \qquad (8.33)$$

NOTE

1. The Kalman filter for time-invariant system is easy for implementation.
2. The covariance of the error in the initial estimate P_0 has no effects on Kalman gain (K).

Illustration 3

The state and observation equations of a linear time-invariant system are as follows:

$$\dot{x}_1(t) = x_2(t)$$

$$\dot{x}_2(t) = u(t) + w(t)$$

$$y(t) = x(t) + v(t)$$

Also given that $Q = 0.5$ and $R = \begin{bmatrix} 8 & 0 \\ 0 & 0.5 \end{bmatrix}$. Compute the Kalman gain for the above system.

Solution

Given

$$\dot{x}_1(t) = x_2(t) = 0 \cdot x_1(t) + 1 \cdot x_2(t) + 0 \cdot u(t) + 0 \cdot w(t)$$

$$\dot{x}_2(t) = u(t) + w(t) = 0 \cdot x_1(t) + 0 \cdot x_2(t) + 1 \cdot u(t) + 1 \cdot w(t)$$

$$y(t) = 1 \cdot x(t) + 1 \cdot v(t)$$

Thus, the state space model can be written as follows:

$$\begin{bmatrix} \dot{x}_1(t) \\ \dot{x}_2(t) \end{bmatrix} = \begin{bmatrix} 0 & 1 \\ 0 & 0 \end{bmatrix} \begin{bmatrix} x_1(t) \\ x_2(t) \end{bmatrix} + \begin{bmatrix} 0 \\ 1 \end{bmatrix} u(t) + \begin{bmatrix} 0 \\ 1 \end{bmatrix} w(t)$$

$$y(t) = \begin{bmatrix} 1 & 0 \\ 0 & 1 \end{bmatrix} x(t) + \begin{bmatrix} 1 \\ 1 \end{bmatrix} v(t)$$

Thus, we have

$$A = \begin{bmatrix} 0 & 1 \\ 0 & 0 \end{bmatrix}; B = \begin{bmatrix} 0 \\ 1 \end{bmatrix}; C = \begin{bmatrix} 1 & 0 \\ 0 & 1 \end{bmatrix}; Q = 0.5; G = \begin{bmatrix} 0 \\ 1 \end{bmatrix}$$

$$\text{Also } R = \begin{bmatrix} 8 & 0 \\ 0 & 0.5 \end{bmatrix} \Rightarrow R^{-1} = \begin{bmatrix} 0.125 & 0 \\ 0 & 2 \end{bmatrix}$$

We have the Algebraic Riccati equation as

$$AP + PA^T - PC^T R^{-1} CP + GQG^T = 0$$

Let us choose $P = \begin{bmatrix} P_{11} & P_{12} \\ P_{21} & P_{22} \end{bmatrix}$

$$\begin{bmatrix} 0 & 1 \\ 0 & 0 \end{bmatrix} \begin{bmatrix} P_{11} & P_{12} \\ P_{21} & P_{22} \end{bmatrix} + \begin{bmatrix} P_{11} & P_{12} \\ P_{21} & P_{22} \end{bmatrix} \begin{bmatrix} 0 & 0 \\ 1 & 0 \end{bmatrix}$$

$$- \begin{bmatrix} P_{11} & P_{12} \\ P_{21} & P_{22} \end{bmatrix} \begin{bmatrix} 1 & 0 \\ 0 & 1 \end{bmatrix} \begin{bmatrix} 0.125 & 0 \\ 0 & 2 \end{bmatrix} \begin{bmatrix} 1 & 0 \\ 0 & 1 \end{bmatrix}$$

$$\begin{bmatrix} P_{11} & P_{12} \\ P_{21} & P_{22} \end{bmatrix} + \begin{bmatrix} 0 \\ 1 \end{bmatrix} [0.5] \begin{bmatrix} 0 & 1 \end{bmatrix} = 0$$

$$\begin{bmatrix} P_{12} & P_{22} \\ 0 & 0 \end{bmatrix} + \begin{bmatrix} P_{12} & 0 \\ P_{22} & 0 \end{bmatrix} - \begin{bmatrix} 0.125 P_{11} & 2 P_{12} \\ 0.125 P_{12} & 2 P_{22} \end{bmatrix} \begin{bmatrix} P_{11} & P_{12} \\ P_{21} & P_{22} \end{bmatrix}$$

$$+ \begin{bmatrix} 0 & 0 \\ 0 & 0.5 \end{bmatrix} = \begin{bmatrix} 0 & 0 \\ 0 & 0 \end{bmatrix}$$

$$\begin{bmatrix} P_{11} & P_{22} \\ 0 & 0 \end{bmatrix} + \begin{bmatrix} P_{12} & 0 \\ P_{22} & 0 \end{bmatrix}$$

$$- \begin{bmatrix} (0.125P_{11}^2 + 2P_{12}^2) & (0.125P_{11}P_{12} + 2P_{12}P_{22}) \\ (0.125P_{11}P_{12} + 2P_{12}P_{22}) & (0.125P_{12}^2 + 2P_{22}^2) \end{bmatrix}$$

$$+ \begin{bmatrix} 0 & 0 \\ 0 & 0.5 \end{bmatrix} = \begin{bmatrix} 0 & 0 \\ 0 & 0 \end{bmatrix}$$

After further simplification, we get

$$P_{12} + P_{12} - 0.125P_{11}^2 - 2P_{12}^2 = 0 \tag{i}$$

$$P_{22} - 0.125P_{11}P_{12} - 2P_{12}P_{22} = 0 \tag{ii}$$

$$-(0.125P_{12}^2 + 2P_{22}^2) + 0.5 = 0 \tag{iii}$$

Let us substitute $P_{11} = 4P_{22}$ in equation (ii), we get

$$P_{22} - 0.25 \times 4P_{22} \times P_{12} - 2P_{12}P_{22} = 0$$

$$P_{22}(1 - 2.5P_{12}) = 0$$

$$P_{12} = \frac{1}{2.5} = \frac{2}{5}$$

Substituting $P_{12} = \dfrac{2}{5}$ in equation (iii), we get

$$-0.125P_{12}^2 + 2P_{22}^2 = 0.5$$

$$-0.125\left(\frac{2}{5}\right)^2 + 2P_{22}^2 = 0.5$$

$$P_{22} = \frac{\sqrt{6}}{5}$$

$$\therefore \quad P_{11} = 4P_{22} = \frac{4\sqrt{6}}{5}$$

$$P = \begin{bmatrix} P_{11} & P_{12} \\ P_{21} & P_{22} \end{bmatrix} = \begin{bmatrix} \dfrac{4\sqrt{6}}{5} & \dfrac{2}{5} \\ \dfrac{2}{5} & \dfrac{\sqrt{6}}{5} \end{bmatrix}$$

∴ Kalman gain $(K) = PC^T R^{-1}$

$$= \begin{bmatrix} \dfrac{4\sqrt{6}}{5} & \dfrac{2}{5} \\[2ex] \dfrac{2}{5} & \dfrac{\sqrt{6}}{5} \end{bmatrix} \begin{bmatrix} 1 & 0 \\ 0 & 1 \end{bmatrix} \begin{bmatrix} 0.125 & 0 \\ 0 & 2 \end{bmatrix}$$

$$K = \begin{bmatrix} 0.2449 & 0.8 \\ 0.05 & 0.9798 \end{bmatrix}$$

Discrete-Time Kalman Filter Design for a Multi-Dimension Model

Let us consider a discrete-time dynamic system represented by equations (8.34) and (8.35)

$$\dot{x}_K = A_{K-1} x_{K-1} + B_{K-1} u_{K-1} + w_{K-1} \tag{8.34}$$

$$y_K = C_K x_K + v_K \tag{8.35}$$

where
   x is state vector
   y is output vector
   u is input vector
   A, B, C are the system matrices
   w is the process noise vector
   v is the measurement noise vector

The block diagram of the discrete dynamic system with process noise and measurement noise is as shown in Figure 8.4.

Equation (8.34) describes the state dynamics, whereas equation (8.35) represents the output. The input vector u contains information about the system dynamics obtained through sensor measurements. The output vector y contains values that can be obtained through independent measurements as well as can be computed from states. The values of state vector x have to be estimated by the filter. The system matrices A, B and C contain coefficients of state, input and output terms, respectively. The w and v terms correspond to process and measurement noise vectors accounting for the errors due to sensor measurements. These two terms are used to determine the values

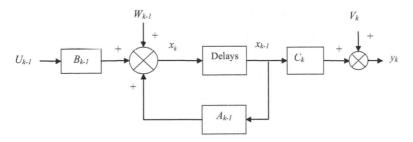

FIGURE 8.4
Block diagram of discrete-time dynamic system with process noise and measurement noise.

of process noise covariance matrix (Q) and measurement noise covariance matrix (R) in the implementation of Kalman filter algorithm.

Assumptions Used in the Design of Kalman Filter

The following are the assumptions used in the design of Kalman filter algorithm:

- Equations (8.34) and (8.35) represent the given plant.
- Inputs, outputs and states are independent, and hence, the matrices B and C can be non-square matrices whereas A matrix is a square matrix.
- Initial state is a Gaussian random variable independent of w and v and has known mean and covariance.
- Both the process noise (w) and measurement noise (v) are independent, Gaussian and of zero mean with known covariances.
- The covariance matrices Q an R are symmetric and non-negative definite. Also R matrix is non-singular.

Steps Involved in the Kalman Filter Algorithm for a Multi-Dimensional Model

Let us first define the notations to be used in the flowchart for the design of discrete-time Kalman filter.

X_0 – Initial state matrix

P_0 – Initial process covariance matrix

X_{K-1} – Previous state matrix

P_{K-1} – Previous process covariance matrix

$\hat{X}_{K_p}$ – Predicted new state matrix

P_{K_p} – Predicted process covariance matrix

A – System matrix containing coefficients of state terms
B – System matrix containing coefficients of input terms
u_K – Control variable matrix
w_K – Predicted state noise matrix
Q_{K-1} – Process noise covariance matrix
K – Kalman gain
C – Observation matrix
R_K – Sensor (measurement) noise covariance matrix
y_K – Output matrix
v_K – Measurement noise matrix
I – Identity matrix
$\hat{x}_K$ – Current estimated state matrix
P_K – Updated process covariance matrix
k – Discrete time index

Step 1: Initialize the following

Let X_0 be the initial state.
Let P_0 be the initial process covariance matrix.
Let X_{K-1} be the previous state.
Let P_{K-1} be the previous process covariance matrix.

Step 2: Predict the new state

$$\hat{x}_{k_p} = A_{k-1}x_{k-1} + B_{k-1}u_{k-1} + w_{k-1} \tag{8.36}$$

Step 3: Predict the process covariance matrix

$$P_{k_p} = A_{k-1}P_{k-1}A_{k-1}{}^T + Q_{k-1} \tag{8.37}$$

Step 4: Compute the Kalman gain (K)

$$K = P_{k_p}C_k{}^T(C_kP_{k_p}C_k{}^T + R)^{-1} \tag{8.38}$$

Step 5: Find $y_k = C_kx_{k_m} + v_k$ $\hspace{3cm}$ (8.39)
Step 6: Find the current estimated state ($\hat{x}_k$)

$$\hat{x}_k = \hat{x}_{k_p} + K[y_k - C_k\hat{x}_{k_p}] \tag{8.40}$$

Step 7: Update the process covariance matrix (P_k)

$$P_k = (I - KC_k)P_{k_p} \tag{8.41}$$

Step 8: If $\hat{x}_k$ and P_k obtained from steps 6 and 7, respectively, are acceptable, then stop. Otherwise go to step 1 and repeat the above steps till we get the acceptable values. In that case, 'k' becomes '$k-1$', i.e., current updated values become previous values for the next iteration.

Flowchart for Discrete-Time Kalman Filter Design (for Multi-Dimensional System)

Thus, the detailed flowchart for the design of discrete-time Kalman filter to determine the Kalman gain, current estimated state and updated process covariance matrix is shown in Figure 8.5.

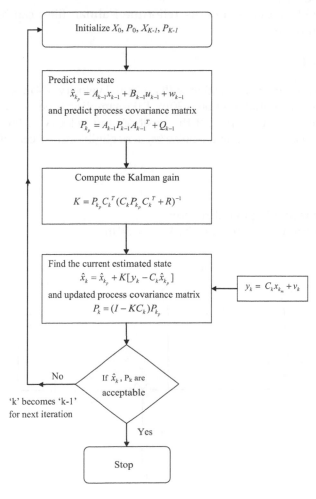

FIGURE 8.5
Flowchart for discrete-time Kalman filter algorithm (multi-dimensional system).

Thus, the block diagram of discrete-time Kalman filter can be obtained as follows:

From the discrete Kalman filter algorithm, we have the following equations for the predicted value and estimated value of the state.

Predicted new value of the state $(\hat{x}_{k_p})$ is given by

$$\hat{x}_{k_p} = A_{k-1}x_{k-1} + B_{k-1}u_{k-1} + w_{k-1} \tag{i}$$

Also the estimated value of the state $(\hat{x}_k)$ is given by

$$\hat{x}_k = \hat{x}_{k_p} + K[y_k - C_k\hat{x}_{k_p}] \tag{ii}$$

where $y_k = C_k x_k + v_k$

Thus, the block diagram of discrete-time Kalman filter can be drawn as shown in Figure 8.6.

Illustration 4

An object is moving in y-direction with an initial velocity of 4 m/s and acceleration of 2 m/s². The initial position is 30 m. Take sampling time as 1 second. Using Kalman filter algorithm, predict the position and velocity of the object after three iterations. Assume that the noise (w) is negligible and acceleration is fixed.

Solution

Given
    An object is moving in y-direction
    Initial position in y-direction (X_{o_y}) = 30 m

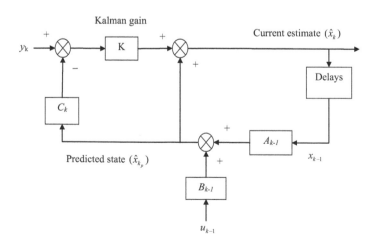

FIGURE 8.6
Block diagram of discrete Kalman filter.

Initial velocity in y-direction $(\dot{X}_{o_y})$ $= 4\,\text{m/s}$
Acceleration in y-direction $(\ddot{X}_y)$ $= 2\,\text{m/s}^2$
Sampling time (Δt) $= 1\,\text{s}$
State noise (w) $= 0$

The kinematic equations governing the system can be written as follows:

$$X_y = X_{o_y} + \dot{X}_{o_y}\Delta T + \frac{1}{2}\ddot{X}_y\Delta T^2 \tag{i}$$

$$\dot{X}_y = \dot{X}_{o_y} + \ddot{X}_y\Delta T \tag{ii}$$

We have from Kalman filter algorithm; the expression for predicted new state is given by

$$\hat{x}_{k_p} = A_{k-1}x_{k-1} + B_{k-1}u_{k-1} + w_{k-1} \tag{iii}$$

From equations (i), (ii) and (iii), we have

$$A = \begin{bmatrix} 1 & \Delta T \\ 0 & 1 \end{bmatrix} \text{ and } B = \begin{bmatrix} \frac{1}{2}\Delta T^2 \\ \Delta T \end{bmatrix}$$

$$\therefore \hat{X}_{kp} = \begin{bmatrix} 1 & \Delta T \\ 0 & 1 \end{bmatrix}\begin{bmatrix} X_{yk-1} \\ \dot{X}_{yk-1} \end{bmatrix} + \begin{bmatrix} \frac{1}{2}\Delta T^2 \\ \Delta T \end{bmatrix}[\ddot{X}_y] + [0]$$

$$\hat{X}_{kp} = \begin{bmatrix} \left(X_{yk-1} + \dot{X}_{yk-1}\Delta T\right) \\ \dot{X}_{yk-1} \end{bmatrix} + \begin{bmatrix} (\ddot{X}_y)\left(\frac{1}{2}\Delta T^2\right) \\ (\ddot{X}_y)(\Delta T) \end{bmatrix}$$

$$\therefore \hat{X}_{kp} = \begin{bmatrix} \left(X_{yk-1} + \dot{X}_{yk-1}\Delta T + \frac{1}{2}\ddot{X}_y\Delta T^2\right) \\ (\dot{X}_{yk-1} + \ddot{X}_y\Delta T) \end{bmatrix}$$

When $\Delta T = 0$,

$$\hat{X}_{kp} = \begin{bmatrix} 30 \\ 4 \end{bmatrix}$$

When $\Delta T = 1\,\text{s}$,

$$\hat{X}_{kp} = \begin{bmatrix} \left(30 + (4 \times 1) + \frac{1}{2}(2 \times 1^2)\right) \\ (4 + (2 \times 1)) \end{bmatrix} = \begin{bmatrix} 35 \\ 6 \end{bmatrix}$$

When $\Delta T = 2\,\text{s}$,

$$\hat{X}_{kp} = \begin{bmatrix} \left(35 + (6 \times 1) + \dfrac{1}{2}(2 \times 1^2)\right) \\ (6 + (2 \times 1)) \end{bmatrix} = \begin{bmatrix} 42 \\ 8 \end{bmatrix}$$

When $\Delta T = 3\,\text{s}$,

$$\hat{X}_{kp} = \begin{bmatrix} \left(42 + (8 \times 1) + \dfrac{1}{2}(2 \times 1^2)\right) \\ (8 + (2 \times 1)) \end{bmatrix} = \begin{bmatrix} 51 \\ 10 \end{bmatrix}$$

The same can be verified using the kinematics equations (i) and (ii)
When $\Delta T = 1\,\text{s}$,

$$X_y = \left(30 + (4 \times 1) + \frac{1}{2}(2 \times 1^2)\right) = 35$$

$$\dot{X}_y = 4 + (2 \times 1) = 6$$

When $\Delta T = 2\,\text{s}$,

$$X_y = \left(30 + (4 \times 2) + \frac{1}{2}(2 \times 2^2)\right) = 42$$

$$\dot{X}_y = 4 + (2 \times 2) = 8$$

When $\Delta T = 3\,\text{s}$,

$$X_y = \left(30 + (4 \times 3) + \frac{1}{2}(2 \times 3^2)\right) = 51$$

$$\dot{X}_y = 4 + (2 \times 3) = 10$$

Illustration 5

A discrete-time dynamic system is represented as follows:

$$x(k) = 2x(k-1) + w(k-1)$$

$$y(k) = x(k) + v(k)$$

Take $Q(k-1) = 1$, $R(k) = 0.5$, $X_0 = 1$, $P_0 = 2$, $y(1) = 1$, and hence, compute the value of $\hat{x}_k$ for $K = 1$.

Solution

Given that
$A(k-1) = 2, B(k-1) = 0, C(k) = 1, Q(k-1) = 1, R(k) = 0.5, X_0 = 1, P_0 = 2, y(1) = 1,$
$w(k-1) = 0$.

Step 1: Find $\hat{x}_{k_p}$

$$\hat{x}_{k_p} = A_{k-1}x_{k-1} + B_{k-1}u_{k-1} + w_{k-1}$$

$$= (2 \times 1) + (0) + (0) = 2$$

Step 2: Find P_{k_p}

$$P_{k_p} = A_{k-1}P_0A_{k-1}^T + Q_{k-1}$$

$$P_{k_p} = (2 \times 2 \times 2) + 1 = 7$$

Step 3: Compute the Kalman gain (K)

$$K = \frac{P_{k_p}C^T}{CP_{k_p}C^T + R} = \frac{7 \times 1}{[(1 \times 7 \times 1) + 0.5]} = 0.93$$

Step 4: Find $\hat{x}_k$

$$\hat{x}_k = \hat{x}_{k_p} + K[y_k - C_k\hat{x}_{k_p}]$$

$$\hat{x}_k = 2 + 0.93[1 - (1 \times 2)] = 1.07$$

Illustration 6

A discrete-time dynamic system is represented as follows:

$$x(k) = \begin{bmatrix} 0 & 0 \\ 0 & 0.5 \end{bmatrix} x(k-1) + w(k-1)$$

$$y(k) = x(k) + v(k)$$

Take $Q(k-1) = \begin{bmatrix} 0 & 0 \\ 0 & 0.5 \end{bmatrix}$, $P_0 = \begin{bmatrix} 5 & 0 \\ 0 & 5 \end{bmatrix} = 2$, $R(k) = 1$. Compute the Kalman gain for $k = 1$.

Solution

Given that

$$A_{k-1} = \begin{bmatrix} 1 & 1 \\ 0 & 1 \end{bmatrix}; B_{k-1} = [0]; C_k = \begin{bmatrix} 1 & 0 \end{bmatrix}; Q_{k-1} = \begin{bmatrix} 0 & 0 \\ 0 & 0.5 \end{bmatrix};$$

$$P_0 = P_{k-1} = \begin{bmatrix} 5 & 0 \\ 0 & 5 \end{bmatrix}; R(k) = 1,$$

To compute the Kalman gain for $k = 1$
  Step 1: Find P_{k_p} where

$$P_{k_p} = A_{k-1}P_0A_{k-1}{}^T + Q_{k-1}$$

$$P_{k_p} = \begin{bmatrix} 1 & 1 \\ 0 & 1 \end{bmatrix}\begin{bmatrix} 5 & 0 \\ 0 & 5 \end{bmatrix}\begin{bmatrix} 1 & 0 \\ 1 & 1 \end{bmatrix} + \begin{bmatrix} 0 & 0 \\ 0 & 0.5 \end{bmatrix}$$

$$P_{k_p} = \begin{bmatrix} 10 & 5 \\ 5 & 5 \end{bmatrix} + \begin{bmatrix} 0 & 0 \\ 0 & 0.5 \end{bmatrix} = \begin{bmatrix} 10 & 5 \\ 5 & 5.5 \end{bmatrix}$$

Step 2: Compute the Kalman gain (K)

$$K = P_{k_p}C_k{}^T(C_kP_{k_p}C_k{}^T + R)^{-1}$$

$$K = \begin{bmatrix} 10 & 5 \\ 5 & 5.5 \end{bmatrix}\begin{bmatrix} 1 \\ 0 \end{bmatrix}\left\{ \begin{bmatrix} 1 & 0 \end{bmatrix}\begin{bmatrix} 10 & 5 \\ 5 & 5.5 \end{bmatrix}\begin{bmatrix} 1 \\ 0 \end{bmatrix} + 1 \right\}^{-1}$$

$$K = \begin{bmatrix} 10 \\ 5 \end{bmatrix}\left\{ \begin{bmatrix} 10 & 5 \end{bmatrix}\begin{bmatrix} 1 \\ 0 \end{bmatrix} + 1 \right\}^{-1}$$

$$K = \begin{bmatrix} 10 \\ 5 \end{bmatrix}[10+1] = \begin{bmatrix} \dfrac{10}{11} \\ \dfrac{5}{11} \end{bmatrix}$$

Illustration 7

An aeroplane travelling in two-dimensional mode (position and velocity) is tracked. The following are the recorded data.
  Initial position in x-direction (X_{o_x}) $= 5000\,m$
  Initial velocity in x-direction ($\dot{X}_{o_x}$) $= 350\,m/s$

Initial process errors in process
covariance matrix (for position) (ΔP_x) = 15 m
Initial process errors in process
covariance matrix (for velocity) $(\Delta P_{\dot{x}})$ = 3 m/s
Observation error in position in x-direction (ΔX_x) = 20 m
Observation error in velocity in x-direction $(\Delta \dot{X}_x)$ = 5 m/s
Acceleration $(\ddot{X}_x)$ = 6 m/s^2
Position in x-direction of $\Delta T = 1$ s, (X_{1_x}) = 5300 m
Velocity in x-direction of $\Delta T = 1$ s, $(\dot{X}_{1_x})$ = 360 m/s
Position in x-direction of $\Delta T = 2$ s, (X_{2_x}) = 5500 m
Velocity in x-direction of $\Delta T = 2$ s, $(\dot{X}_{2_x})$ = 375 m/s
Assume that $w_k = 0$, $v_k = 0$ and Q_{k-1} = 0

Using Kalman filter algorithm compute the values of x_k and P_k when $\Delta T = 1$ s and $\Delta T = 2$ s.

Solution

Iteration – 1: ($\Delta T = 1$ s)

Step 1: Initialize the states (x_{k-1}) and process covariance matrix (P_{k-1})
From the given data

$$X_{k-1} = \begin{bmatrix} X_{0x} \\ \dot{X}_{0x} \end{bmatrix} = \begin{bmatrix} 5000 \\ 350 \end{bmatrix}$$

$$P_{k-1} = \begin{bmatrix} (\Delta P_x^2) & (\Delta P_x)(\Delta P_{\dot{x}}) \\ (\Delta P_x)(\Delta P_{\dot{x}}) & (\Delta P_{\dot{x}}^2) \end{bmatrix} = \begin{bmatrix} 15^2 & (15 \times 3) \\ (15 \times 3) & 3^2 \end{bmatrix}$$

$$P_{k-1} = \begin{bmatrix} 225 & 45 \\ 45 & 9 \end{bmatrix} \approx \begin{bmatrix} 225 & 0 \\ 0 & 9 \end{bmatrix}$$

Step 2: Predict the new state $(\hat{x}_{k_p})$

$$\hat{x}_{k_p} = A_{k-1} x_{k-1} + B_{k-1} u_{k-1} + w_{k-1}$$

$$\hat{X}_{k_p} = \begin{bmatrix} 1 & \Delta T \\ 0 & 1 \end{bmatrix} \begin{bmatrix} X_{0x} \\ \dot{X}_{0x} \end{bmatrix} + \begin{bmatrix} \frac{1}{2}\Delta T^2 \\ \Delta T \end{bmatrix} [\ddot{X}] + w_k$$

$$\hat{X}_{k_p} = \begin{bmatrix} 1 & 1 \\ 0 & 1 \end{bmatrix} \begin{bmatrix} 5000 \\ 350 \end{bmatrix} + \begin{bmatrix} \frac{1}{2} \\ 1 \end{bmatrix} [6] + 0$$

$$\hat{X}_{k_p} = \begin{bmatrix} 5350 \\ 350 \end{bmatrix} + \begin{bmatrix} 3 \\ 6 \end{bmatrix} = \begin{bmatrix} 5353 \\ 356 \end{bmatrix}$$

Step 3: Predict the process covariance matrix (P_{k_p})

$$P_{k_p} = A_{k-1}P_{k-1}A_{k-1}{}^T + Q_{k-1}$$

$$P_{k_p} = \begin{bmatrix} 1 & 1 \\ 0 & 1 \end{bmatrix} \begin{bmatrix} 225 & 0 \\ 0 & 9 \end{bmatrix} \begin{bmatrix} 1 & 0 \\ 1 & 1 \end{bmatrix} + 0$$

$$P_{k_p} = \begin{bmatrix} 225 & 9 \\ 0 & 9 \end{bmatrix} \begin{bmatrix} 1 & 0 \\ 1 & 1 \end{bmatrix} = \begin{bmatrix} 234 & 9 \\ 9 & 9 \end{bmatrix}$$

$$P_{k_p} \cong \begin{bmatrix} 234 & 0 \\ 0 & 9 \end{bmatrix}$$

Step 4: Compute the Kalman gain K

$$K = P_{k_p}C_k{}^T (C_k P_{k_p} C_k{}^T + R)^{-1}$$

$$K = \begin{bmatrix} 234 & 0 \\ 0 & 9 \end{bmatrix} \begin{bmatrix} 1 & 0 \\ 0 & 1 \end{bmatrix}$$

$$\left(\begin{bmatrix} 1 & 0 \\ 0 & 1 \end{bmatrix} \begin{bmatrix} 234 & 0 \\ 0 & 9 \end{bmatrix} \begin{bmatrix} 1 & 0 \\ 0 & 1 \end{bmatrix} + \begin{bmatrix} \Delta x_x^2 & 0 \\ 0 & \Delta \dot{x}_x^2 \end{bmatrix} \right)^{-1}$$

$$K = \begin{bmatrix} 234 & 0 \\ 0 & 9 \end{bmatrix} \left(\begin{bmatrix} 234 & 0 \\ 0 & 9 \end{bmatrix} + \begin{bmatrix} 400 & 0 \\ 0 & 25 \end{bmatrix} \right)^{-1}$$

$$K = \begin{bmatrix} 234 & 0 \\ 0 & 9 \end{bmatrix} \begin{bmatrix} 634 & 0 \\ 0 & 34 \end{bmatrix}^{-1} = \begin{bmatrix} 0.37 & 0 \\ 0 & 0.26 \end{bmatrix}$$

Step 5: Find $y_k = Cx_{k_m} + v_k$

$$y_k = \begin{bmatrix} 1 & 0 \\ 0 & 1 \end{bmatrix} \begin{bmatrix} X_{1x} \\ \dot{X}_{1x} \end{bmatrix} + [0]$$

$$y_k = \begin{bmatrix} 1 & 0 \\ 0 & 1 \end{bmatrix} \begin{bmatrix} 5300 \\ 360 \end{bmatrix} + [0] = \begin{bmatrix} 5300 \\ 360 \end{bmatrix}$$

Step 6: Find the current estimated state $(\hat{x}_k)$

$$\hat{x}_k = \hat{x}_{k_p} + K[y_k - Cx_{k_p}]$$

$$\hat{x}_k = \begin{bmatrix} 5353 \\ 356 \end{bmatrix} + \begin{bmatrix} 0.37 & 0 \\ 0 & 0.26 \end{bmatrix}\left\{ \begin{bmatrix} 5300 \\ 360 \end{bmatrix} - \begin{bmatrix} 1 & 0 \\ 0 & 1 \end{bmatrix}\begin{bmatrix} 5353 \\ 356 \end{bmatrix}\right\}$$

$$\hat{x}_k = \begin{bmatrix} 5336 \\ 357 \end{bmatrix}$$

Step 7: Update the process covariance matrix (P_k)

$$P_k = (I - KC)P_{k_p}$$

$$P_k = \left\{ \begin{bmatrix} 1 & 0 \\ 0 & 1 \end{bmatrix} - \begin{bmatrix} 0.37 & 0 \\ 0 & 0.26 \end{bmatrix}\begin{bmatrix} 1 & 0 \\ 0 & 1 \end{bmatrix}\right\}\begin{bmatrix} 234 & 0 \\ 0 & 9 \end{bmatrix}$$

$$P_k = \begin{bmatrix} 147 & 0 \\ 0 & 7 \end{bmatrix}$$

Iteration 2: $(\Delta T = 2\,\text{s})$

Step 1: Initialize the states (x_{k-1}) and process covariance matrix (P_{k-1})

NOTE

$\hat{X}_k$ and P_k computed in iteration 1 will now become X_{k-1} and P_{k-1}, respectively, in iteration 2.

Thus, we have

$$X_{k-1} = \begin{bmatrix} 5336 \\ 357 \end{bmatrix} \text{ and } P_{k-1} = \begin{bmatrix} 147 & 0 \\ 0 & 7 \end{bmatrix}$$

Step 2: Predict the new state $(\hat{x}_{k_p})$

$$\hat{x}_{k_p} = A_{k-1}x_{k-1} + B_{k-1}u_{k-1} + w_{k-1}$$

$$\hat{X}_{k_p} = \begin{bmatrix} 1 & 1 \\ 0 & 1 \end{bmatrix}\begin{bmatrix} 5336 \\ 357 \end{bmatrix} + \begin{bmatrix} \frac{1}{2} \\ 1 \end{bmatrix}[6] + 0$$

$$\hat{X}_{kp} = \begin{bmatrix} 5693 \\ 357 \end{bmatrix} + \begin{bmatrix} 3 \\ 6 \end{bmatrix} = \begin{bmatrix} 5696 \\ 363 \end{bmatrix}$$

Step 3: Predict the process covariance matrix (P_{k_p})

$$P_{k_p} = A_{k-1}P_{k-1}A_{k-1}{}^T + Q_{k-1}$$

$$P_{k_p} = \begin{bmatrix} 1 & 1 \\ 0 & 1 \end{bmatrix} \begin{bmatrix} 147 & 0 \\ 0 & 7 \end{bmatrix} \begin{bmatrix} 1 & 0 \\ 1 & 1 \end{bmatrix} + 0$$

$$P_{k_p} = \begin{bmatrix} 147 & 7 \\ 0 & 7 \end{bmatrix} \begin{bmatrix} 1 & 0 \\ 1 & 1 \end{bmatrix} = \begin{bmatrix} 154 & 7 \\ 7 & 7 \end{bmatrix}$$

$$P_{k_p} \cong \begin{bmatrix} 154 & 0 \\ 0 & 7 \end{bmatrix}$$

Step 4: Compute the Kalman gain K

$$K = P_{k_p}C_k{}^T (C_k P_{k_p} C_k{}^T + R)^{-1}$$

$$K = \begin{bmatrix} 154 & 0 \\ 0 & 7 \end{bmatrix} \begin{bmatrix} 1 & 0 \\ 0 & 1 \end{bmatrix}$$

$$\times \left(\begin{bmatrix} 1 & 0 \\ 0 & 1 \end{bmatrix} \begin{bmatrix} 154 & 0 \\ 0 & 7 \end{bmatrix} \begin{bmatrix} 1 & 0 \\ 0 & 1 \end{bmatrix} + \begin{bmatrix} 400 & 0 \\ 0 & 25 \end{bmatrix} \right)^{-1}$$

$$K = \begin{bmatrix} 154 & 0 \\ 0 & 7 \end{bmatrix} \left(\begin{bmatrix} 154 & 0 \\ 0 & 7 \end{bmatrix} + \begin{bmatrix} 400 & 0 \\ 0 & 25 \end{bmatrix} \right)^{-1}$$

$$= \begin{bmatrix} 154 & 0 \\ 0 & 7 \end{bmatrix} \begin{bmatrix} 554 & 0 \\ 0 & 32 \end{bmatrix}^{-1} = \begin{bmatrix} 0.28 & 0 \\ 0 & 0.22 \end{bmatrix}$$

Step 5: Find $y_k = Cx_{k_m} + v_k$

$$y_k = \begin{bmatrix} 1 & 0 \\ 0 & 1 \end{bmatrix} \begin{bmatrix} X_{2_x} \\ \dot{X}_{2_x} \end{bmatrix} + [0]$$

$$y_k = \begin{bmatrix} 1 & 0 \\ 0 & 1 \end{bmatrix} \begin{bmatrix} 5500 \\ 375 \end{bmatrix} + [0] = \begin{bmatrix} 5500 \\ 375 \end{bmatrix}$$

Step 6: Find the current estimated state $(\hat{x}_k)$

$$\hat{x}_k = x_{k_p} + K[y_k - Cx_{k_p}]$$

$$\hat{x}_k = \begin{bmatrix} 5696 \\ 363 \end{bmatrix} + \begin{bmatrix} 0.28 & 0 \\ 0 & 0.22 \end{bmatrix} \left\{ \begin{bmatrix} 5500 \\ 375 \end{bmatrix} - \begin{bmatrix} 1 & 0 \\ 0 & 1 \end{bmatrix} \begin{bmatrix} 5696 \\ 363 \end{bmatrix} \right\}$$

$$\hat{x}_k = \begin{bmatrix} 5696 \\ 363 \end{bmatrix} + \begin{bmatrix} 0.28 & 0 \\ 0 & 0.22 \end{bmatrix} \begin{bmatrix} -196 \\ 12 \end{bmatrix} = \begin{bmatrix} 5641 \\ 366 \end{bmatrix}$$

Step 7: Update the process covariance matrix (P_k)

$$P_k = (I - KC)P_{k_p}$$

$$P_k = \left\{ \begin{bmatrix} 1 & 0 \\ 0 & 1 \end{bmatrix} - \begin{bmatrix} 0.28 & 0 \\ 0 & 0.22 \end{bmatrix} \begin{bmatrix} 1 & 0 \\ 0 & 1 \end{bmatrix} \right\} \begin{bmatrix} 154 & 0 \\ 0 & 7 \end{bmatrix} = \begin{bmatrix} 111 & 0 \\ 0 & 5 \end{bmatrix}$$

Thus, when $\Delta T = 1\,\text{s}$, $\hat{X}_k = \begin{bmatrix} 5336 \\ 357 \end{bmatrix}$ and $P_k = \begin{bmatrix} 147 & 0 \\ 0 & 7 \end{bmatrix}$

and when $\Delta T = 2\,\text{s}$, $\hat{X}_k = \begin{bmatrix} 5641 \\ 366 \end{bmatrix}$ and $P_k = \begin{bmatrix} 111 & 0 \\ 0 & 5 \end{bmatrix}$

Extended Kalman Filter (EKF)

The Kalman filter hitherto discussed estimates the state of discrete-time process represented by linear stochastic difference equations as given in equations (8.5) and (8.6) which are reproduced below.

$$x_K = A_{K-1}x_{K-1} + B_{K-1}u_{K-1} + w_{K-1} \qquad \text{(i)}$$

$$y_K = C_K x_K + v_K \qquad \text{(ii)}$$

In other words, x_k, x_{k-1}, u_{k-1} and y_k are linearly related through A_{k-1}, B_{k-1} and C_k. However, in practice, the process is nonlinear. In such a case, the process has to be represented by non-linear stochastic difference equations as given in equations (8.42) and (8.43).

$$x_K = f(x_{K-1}, u_K, w_{K-1}) \qquad \text{(8.42)}$$

$$y_K = C(x_K, v_K) \qquad \text{(8.43)}$$

In the state equation (8.42), the non-linear function 'f' consists of driving function u_{k-1}, process noise w_{k-1} and relates the state at the previous time-step $(k-1)$ to the state at the current time step (k). In the measurement equation (8.43), the non-linear function 'h' consists of measurement noise v_k and relates the state x_k to the measurement y_k. The estimation can be linearized around the current estimate by using the partial derivatives of 'f' and 'c'. In such a case, the Kalman filter is referred to as EKF. It may also be noted that the values of w_k and v_k at each time step will not be known, and hence, they can be neglected.

Steps Involved in the Design of Discrete-Time EKF

Step 1: Initialize X_{k-1} and P_{k-1}
  Step 2: Predict the new state $(\hat{x}_{k_p})$

$$\hat{x}_{k_p} = f(x_{K-1}, u_K, 0) \tag{8.44}$$

Step 3: Predict the process covariance matrix (P_{k_p})

$$P_{k_p} = A_{k-1}P_{k-1}A_{k-1}^T + w_k Q_{k-1}w_k^T \tag{8.45}$$

Step 4: Compute the Kalman gain K_k

$$K = P_{k_p}C_k^T(C_k P_{k_p}C_k^T + v_k R_k v_k^T)^{-1} \tag{8.46}$$

Step 5: Find

$$y_k = C(x_{k_m}, v_k) \tag{8.47}$$

Step 6: Find the current estimated state $(\hat{x}_k)$

$$\hat{x}_k = x_{k_p} + K[y_k - C_k(x_{k_p}, 0)] \tag{8.48}$$

Step 7: Update the process covariance matrix (P_k)

$$P_k = (I - K_k C_k)P_{k_p} \tag{8.49}$$

where

$$A_{[i,j]} = \frac{\partial f_i}{\partial x_j}(\hat{x}_{k-1}, u_k, 0) \tag{8.50}$$

$$W_{[i,j]} = \frac{\partial f_i}{\partial w_j}(\hat{x}_{k-1}, u_k, 0) \tag{8.51}$$

$$C_{[i,j]} = \frac{\partial C_i}{\partial x_j}\left(\hat{x}_{kp}, 0\right) \tag{8.52}$$

$$v_{[i,j]} = \frac{\partial C_i}{\partial v_j}\left(\hat{x}_{kp}, 0\right) \tag{8.53}$$

NOTE

To avoid complexity, the time step subscripts for A, W, C, V are not shown in equations (8.50) to (8.53).

Bibliography

B. D. O. Anderson, J. B. Moore, *Linear Optimal Control*, Prentice-Hall, Inc, 1971.

D. P. Atherton, *Nonlinear Control Engineering*, Van Nostrand Reinhold Company, 1975.

C. T. Chen, *Linear System Theory and Design*, 3rd Edition, Oxford University Press, 1998.

R. C. Dorf, R. H. Bishop, *Modern Control System*, 10th Edition, Pearson Education, 2004.

M. Gopal, *Modern Control System Theory*, 2nd Edition, New Age International, 1993.

M. Gopal, *Control Systems Principles and Design*, 3rd Edition, McGraw Hill, 2002.

M. Gopal, *Digital Control and State Variable Methods: Conventional and Neural-Fuzzy Control Systems*, 2nd Edition, Tata McGraw-Hill Publishing Company Limited, 2006.

B. S. Grewal, *Higher Engineering Mathematics*, 40th Edition, Khanna Publishers, 2014.

S. Guellal, P. Grimalt, Y. Cherruault, Numerical Study of Lorenz's Equation by the Adomain Method, *Computers & Mathematics with Applications*, Vol. 33, No. 3, pp. 25–29, 1997.

C. C. Karaaslanli, *Bifurcation Analysis and its Applications*, Intech Open Science, 2012.

H. K. Khalil, *Nonlinear Systems*, 3rd Edition, Prentice Hall, 2001.

D. E. Kirk, *Optimal Control Theory: An Introduction*, Dover Publications, Inc., 2012.

B. C. Kuo, F. Golnaraghi, *Automatic Control Systems*, 8th edition, John Wiley & Son, 2003.

F. L. Lewis, D. L. Vrabie, V. L. Syroms, *Optimal Control*, 3rd Edition, John Wiley & Sons, 2012.

I. J. Nagrath, M. Gopal, *Control Systems Engineering*, 2nd Edition, Wiley Eastern, 2006.

D. S. Naidu, *Optimal Control Systems*, CRC Press, 2003.

K. Ogata, *Modern Control Engineering*, 4th Edition, Prentice-Hall, 2001.

M. B. Rhudy, R. A. Salguero, K. Holappa, A Kalman Filtering Tutorial for Undergraduate Students, *International Journal of Computer Science & Engineering Survey (IJCSES)*, Vol. 8, No. 1, February 2017.

D. Simon, *Optimal State Estimation: Kalman, H∞ and Nonlinear Approaches*, John Wiley & Sons, 2006.

J. J. E. Slotine, W. Li, *Applied Nonlinear Control*, Prentice Hall, 1991.

C. Sparrow, An introduction to Lorentz equations, *IEEE Transactions on Circuits and Systems*, Vol. 30, No. 8, pp. 533–542, August 1983.

R. Srivastava, P. K. Srivastava, J. Chattopadhyay, Chaos in a Chemical System, *The European Physical Journal Special Topics*, Vol. 222, pp. 777–783, 2013.

F. Sziadarovszky, A. T. Bahill, *Linear System Theory*, 2nd Edition, CRC Press, 1998.

T. Y. Thanoon, S. F. AL-Azzawi. *Stability of Lorenz Differential System by parameters*, Mosul University, 2008.

G. Welch, G. Bishop, *An Introduction to the Kalman Filter*, University of North Carolina, 2001.

G. P. Williams. *Chaos Theory Tamed*, Joseph Henry Press, 1997.

Index